Metallogenic models and exploration criteria for buried carbonate-hosted ore deposits—a multidisciplinary study in eastern England

Metallogenic models and exploration criteria for buried carbonate-hosted ore deposits—a multidisciplinary study in eastern England

Edited by J. A. Plant and D. G. Jones

Contributors

T. B. Colman, J. D. Cornwell,
D. G. Jones, N. d'A. Laffoley*,
A. S. D. Walker
BGS, Keyworth

K. Smith, N. J. P. Smith
*BGS, Grange Terrace,
Edinburgh*

J. A. Plant
BGS, Gray's Inn Road, London

G. C. Brown, P. C. Webb
*Department of Earth Sciences,
Open University, Walton Hall,
Milton Keynes MK7 6AA, UK*

T. D. Ford
*Department of Geology,
University of Leicester,
University Road, Leicester
LE1 7RH, UK*

* Formerly *BGS*; now *Ashanti
Goldfields Corporation (Ghana)
Ltd, Diamond House, PO Box
2665, Accra, Ghana*

Cover illustration

Mineral veins (yellow) and faults
(black) in the South Pennine
Orefield are superimposed on
aeromagnetic data (maxima
orange, minima blue). The
outcrop of the Viséan rocks is
outlined in white.

Bibliographic reference

**Plant, J. A., and Jones, D. G.
(editors).** 1989. *Metallogenic
models and exploration criteria
for buried carbonate-hosted ore
deposits—a multidisciplinary
study in eastern England.*
(Keyworth, Nottingham: British
Geological Survey; London: The
Institution of Mining and
Metallurgy.)

London   The Institution of Mining and Metallurgy
Keyworth, Nottingham   British Geological Survey   1989

BRITISH GEOLOGICAL SURVEY

The full range of Survey publications is available through the Sales Desks at Keyworth and Murchison House, Edinburgh. Selected items can be bought at the BGS London Information Office, and orders are accepted here for all publications. The adjacent Geological Museum bookshop stocks the more popular books for sale over the counter. Most BGS books and reports are listed in HMSO's Sectional List 45, and can be bought from HMSO and through HMSO agents and retailers. Maps are listed in the BGS Map Catalogue and the Ordnance Survey's Trade Catalogue, and can be bought from Ordnance Survey agents as well as from BGS.

The British Geological Survey carries out the geological survey of Great Britain and Northern Ireland (the latter as an agency service for the government of Northern Ireland), and of the surrounding continental shelf, as well as its basic research projects. It also undertakes programmes of British technical aid in geology in developing countries as arranged by the Overseas Development Administration.

The British Geological Survey is a component body of the Natural Environment Research Council.

Keyworth, Nottingham NG12 5GG

☎ Plumtree (060 77) 6111 Telex 378173 BGSKEY G
Fax ☎ 060 77-6602

Murchison House, West Mains Road, Edinburgh EH9 3LA

☎ 031-667 1000 Telex 727343 SEISED G
Fax ☎ 031-668 2683

London Information Office at the Geological Museum, Exhibition Road, South Kensington, London SW7 2DE

☎ 01-589 4090 Fax ☎ 01-584 8270
☎ 01-938 9056/57

64 Gray's Inn Road, London WC1X 8NG

☎ 01-242 4531 Telex 262199 BGSCLR G
Fax ☎ 01-242 0835

19 Grange Terrace, Edinburgh EH9 2LF

☎ 031-667 1000 Telex 727343 SEISED G

St Just, 30 Pennsylvania Road, Exeter EX4 6BX

☎ Exeter (0392) 78312

Bryn Eithyn Hall, Llanfarian, Aberystwyth, Dyfed SY23 4BY

☎ Aberystwyth (0970) 611038 Fax ☎ 0970-624822

Windsor Court, Windsor Terrace, Newcastle upon Tyne NE2 4HB

☎ 091-281 7088 Fax ☎ 091-281 9016

Geological Survey of Northern Ireland, 20 College Gardens, Belfast BT9 6BS

☎ Belfast (0232) 666595 and 666752

Maclean Building, Crowmarsh Gifford, Wallingford, Oxfordshire OX10 8BB

☎ Wallingford (0491) 38800 Telex 849365 HYDROL G
Fax ☎ 0491-25338

Parent Body

Natural Environment Research Council

Polaris House, North Star Avenue, Swindon, Wiltshire SN2 1EU

☎ Swindon (0793) 4115001 Telex 444293 ENVRE G
Fax ☎ 0793-411501

CONTENTS

PREFACE

The report that follows presents a first statement of the results of a study sponsored by the European Economic Community and the Department of Trade and Industry as communicated to a meeting of the Institution of Mining and Metallurgy Midlands Section at which the President (Dr A. J. Robinson) and the Section Chairman (Professor N. A. Warner) were present. It was my privilege to attend and to make some concluding remarks; these form the basis of this preface.

The Geological Survey, now in its 154th year, has as one of its principal objects the provision of geological maps based on surveying in maximum detail at a scale of 1:10 000. In areas where extractive or other engineering industry is important, the surveys have been revised more than once as new subsurface data became available. The organisation is responsible for a data bank containing about a million borehole records. Since 1960 its surveying methods have been enhanced by the provision of geophysical and geochemical field methods with appropriate laboratory support. Notable examples of the results include the Gravity Map and the Aeromagnetic Map of Britain, for which Dr William Bullerwell had primary responsibility, and the Geochemical Maps, prepared under Dr S. H. U. Bowie and Dr Jane Plant. From all these sources, and from general data derived from information held as Commercial in Confidence, a huge mass of information is available upon any given area such as the East Midlands. The present study is, nevertheless, one of the few cases where a beginning has been made upon the synthesis of the data by digital computer methods with a particular strategic objective in view, in this case the possible existence of concealed metalliferous mineralisation. This move into the field of high technology should be warmly welcomed not merely for what it may eventually lead to in the East Midlands but also because the procedures could form the basis of a service industry applicable anywhere in the world.

That mineralisation may occur beneath a cover of younger rocks in the area is a fair conclusion from the existence of great numbers of fissure veins carrying galena, fluorite and baryte that have been exploited in the Southern Pennine Orefield in North Derbyshire. Lead mining there has been carried on since the Roman occupation and may have been practised earlier, and since the later years of the last century the field has been an important source of non-metallics, particularly fluorspar. It is important to be satisfied that an extension or recurrence of this Pennine-type mineralisation would be worth finding, but, unfortunately, past records of production leave much to be desired. Official statistics of lead ore output for 1845 to 1938 show only 0.69 million tonnes (Mt) galena, more than half of which came from one mine, Millclose. Good records prior to 1845 may exist in the archives of the Barmote Courts, which Dr Ford tells me have been in existence for nearly seven centuries, but the task of obtaining the information is formidable and may not be entirely practicable. If argument by analogy from the Northen Pennine Orefield is permissible, here output of 1.73 Mt for 1845–1938 may be compared with accurate records from the London Lead and W.B. companies and the Devonshire ducal estates that bring the total for 1700–1938 up to 3.7 Mt, or with the estimated all-time total of 5 Mt. The application of these ratios to Derbyshire would suggest 1.44 or 2.0 Mt, but these figures may still be too low. Official figures for fluorspar output may be interpreted as indicating 4.1 Mt for

1880–1977, which must presumably be increased to at least 5 Mt to include recent output. However, the 1880–1938 figure of 0.78 Mt is almost certainly low, owing to unreported production by small operators. It is fair to estimate the value of the past output at present prices at a figure approaching £1 × 10^9 (one billion pounds). Certainly it is legitimate to ask: Could there be another equally productive field concealed in the East Midlands? This would become more attractive if the dominant style of mineralisation were widespread limestone replacement as in the Tristate District or the New Lead Belt of Missouri.

In the Southern Pennine Orefield the strong mineralisation is virtually confined to a group of limestones of Asbian and Brigantian age with interbedded lavas ('toadstones') and volcanic clay horizons ('wayboards'). The vertical range seldom reaches as much as 200 m. It does not penetrate the overlying Namurian shale and grit sequence. At Millclose, Britain's largest individual lead mine, the series of orebodies progressed downwards beneath the Stanton Syncline, passing through 14 volcanic horizons. My impressions from visiting the mine workings before they were closed in 1938 confirmed the impression gained from the studies of Mr Traill and Dr Shirley that they could have been extended downwards but for the very heavy pumping cost, equivalent, if my memory is correct, to raising 60 t of water per tonne of ore. Here and elsewhere in the orefield, although the bottoms of some individual oreshoots have been seen, the downward limit of mineralisation has not been established. The mineralisation is strongest along the eastern flank of the limestone massif and appears not to be associated with the flat-topped culmination of the structure. The crude zonal pattern, fluorite–baryte gangue giving place westward to baryte and then to calcite, if interpreted in the light of the central position of fluorite in the Northern Pennines, argues for the solutions travelling up-dip along the eastern flank, a suggestion supported by the strong fluorite mineralisation in the periclinal line containing the Ashover and Crich inliers. These considerations are mentioned as a reminder that the problems of further exploitation in depth in the orefield itself await solution. More would have been done on the prospection of the eastern flank by at least one international mining company twenty years ago if they could have found satisfactory targets, but the abundant remains from bole-hill smelting vitiated any satisfactory exploration geochemistry. The orefield itself, nevertheless, remains 'elephant country', though it is not the primary object of the present study.

Perhaps the most pertinent question for the present exercise is: Do culminations in high Dinantian limestone occur concealed beneath the cover of younger rocks east of the orefield at a depth that might be approachable by exploration for metals? An elaborate network of seismic reflection lines (unpublished) has been used to prospect the eastern half of the East Midlands area for petroleum, and this has indeed shown up a number of culminations, some of which have become producing oilfields. The oldest of these, Eakring, is an interesting feature for two reasons: (i) several boreholes on the western flank of the structure show geothermal gradients up to 76°C/km, hailed long ago by Sir Edward Bullard as among the highest in non-volcanic terrains in the world; he concluded that water had been to greater depth before rising at 45°C. However, the deep borehole on the apex of the structure (Eakring 146) showed a gradient less than normal; (ii) this latter hole produced a good core of fluorite in limestone. Theoretically, however, structures that provide good traps for petroleum cannot be expected to provide the right conditions for ore accumulation, for in spite of the enhanced solubility probably promoted by chloride- or fluoride-complexing, enormous volumes of water must pass through the site of mineralisation. Their flow may be channelled, but it must not be prevented. Thus, the culminations that may be of present interest would be those proved not to contain oil pools.

Dr Plant introduced the study with an account of the two genetic models with which her team have mostly been concerned: the high heat production (HHP) 'granite' model and the basin-dewatering model. It appears important not to regard these models as mutually exclusive, for a combination of both is needed in Pennine metallogeny. The Alston Block in the Northern Pennines has become a classic case, perhaps the most classic, of the association of zoned mineralisation and devolatilised coals with cupolas of a HHP granite of earlier age than the mineralised host rocks. The coincidences are too perfect in detail to be accidental, and it is natural that this discovery has influenced subsequent thinking about the rest of the Pennines. Two points must be made in passing: (i) radioactive elements in the granite account for only about half its present heat flow, the rest being due to its conductivity being superior to that of the surrounding rocks; (ii) given that the granite cupolas produced hot-spots at the time of mineralisation, they could not have supplied the necessary hypersaline brines. For these, reliance must be placed on basin dewatering. In the Askrigg field, although the Wensleydale Granite may have played some part, the two belts of strongest mineralisation, the North Swaledale and Grassington/Greenhow belts, lie remote from it, respectively, to the north and south. Brines from the Stainmore Trough and Craven Basin may well have been involved, as Dr P. J. Rogers has advocated. For Derbyshire, the emphasis in the present state of knowlege must be on basinal brines, though Dr John Cornwell's indications of possible basement granites may indicate key points in the East Midlands picture.

Mr Adrian Walker provided the Section Meeting with a unique demonstration, on the screens, of the process of integration of magnetic, gravity and stratigraphic information by computer actually in progress. The pattern of strongly WNW–orientated structures to which this gave rise in the basement contrasted strongly with the dominantly E–W and ENE pattern of vein fractures in the exposed orefield, but it may be recalled that sinistral transcurrent faulting along WNW lines is capable of generating ENE Riedel fractures in superincumbent strata; a good case exists at present in Blackdene Mine, Weardale. The synthesis of deep structure and stratigraphy in this report by Messrs K. Smith and N. Smith is fundamental to the study and gives a foretaste of what will need to be done much more widely in the country. The geochemical studies of igneous rocks, chiefly from exposed basement, by Professor Brown and Dr Webb and of black shales from basement and Carboniferous by Dr Jones represent a valuable start towards identifying possible sources of the epigenetic elements. More data appear to be needed on fluorine, and the problem of the discrepancy between the Pb/Zn ratios in possible source rocks and that in the orefield (53:1 if we accept 2 Mt as total lead concentrates output). It would take over 3 Mt ZnS to restore the balance: Does this exist as deep ore, not yet discovered, or has the true source not yet been identified?

The Permian and Mesozoic rocks of the cover receive no attention in this report. The basal black shale of the Zechstein sequence in NE England is more metalliferous than any of the shales examined by Dr Jones and has more nearly the appropriate Pb/Zn ratio; but perhaps this merely confirms my contention that waste fluids from the Pennine mineralisation were discharged into the Zechstein Sea. The possibility of concentrated ore in the Magnesian Limestone cannot be ruled out, and it might be easier to reach, if it exists, than deeper culminations in the Lower Carboniferous. If it is supposed that this suggestion is not consistent with the evidence brought forward by Dr Ford, and amply confirmed by everyone's experience, of the sharp cut-off of the mineralisation in the Southern Pennines at the Namurian shale/Dinantian limestone contact, it should be recalled that in the Durham area, Pennine-type veins have been worked in Westphalian strata. Any temptation to interpret the Derbyshire field in terms of contemporary Lower Carboniferous ore genesis must be dismissed in view of the evident epigenetic style of the deposits and the absence of signs of contemporaneous erosion of the deposits at the top of the limestone. These are neither sedimentary or volcanic exhalative deposits. This is not to deny that such deposits (of Irish style) could occur in the concealed area. The first approach might well be a demonstration on a combination of stratigraphical and structural evidence that faulting contemporaneous with Lower Carboniferous sedimentation took place. It is worth recalling that in Eire the important host rocks are of Tournaisian age and these may reasonably be claimed to have had exhalative sulphides associated at at least four major producing sites. New evidence of deposition of Tournaisian on the English side of the Irish Sea and North of St. George's land suggests that massive sulphide bodies of the Irish type might be located here and it is right to continue to search; after all, it is only 25 years since Tynagh and Silvermines came into their own.

History has been made in the application by Jane Plant and her colleagues of high technology to multidisciplinary science. The hope must be expressed that this Report is only the small beginning of a sustained programme leading to the identification of targets for the future deep prospector.

Sir Kingsley Dunham, D.Sc., F.R.S., F.Eng.

Professor Emeritus, University of Durham *formerly* Director, Institute of Geological Sciences (now British Geological Survey.

FOREWORD

Since its foundation in 1835 as the Geological Survey of Great Britain the British Geological Survey (BGS) has been charged with the acquisition and curation of geological data for the British Isles. Until the 1950s these activities involved mainly geological surveying and the publication of conventional geological maps at standard scales of 1:10 560, 1:63 360 and 1:625 000. With the development of geophysical and geochemical tools of survey and many other datasets, the activities and responsibilities of the BGS have expanded greatly and the approach to regional studies is increasingly multidisciplinary. In parallel with these advances in geology there has been greater emphasis on computer manipulation of data, which are now generally captured and stored in digital form. The BGS has been at the forefront of developments in the preparation of systematic digital datasets and their manipulation by use of advanced image analysis systems as a basis for a wide range of economic and environmental applications. The present study, in which new systematic datasets covering the Derbyshire 'Dome' and the East Midlands of England have been prepared and analysed to develop models of mineral deposits and exploration criteria for buried carbonate-hosted mineral deposits, provides an excellent example of the new methods adopted by the BGS. Gravity, aeromagnetic, geological, deep geological and geochemical datasets, together with digital datasets for the mineral veins in the South Pennine Orefield, have been prepared and analysed for the region. In addition, secondary datasets derived by processing one or more primary datasets have been used: for example, the gravity datasets have been mathematically processed to remove the effects of Namurian and younger low-density sediments in order to enable anomalies in the basement to be identified. The petrogenesis of igneous rocks in the pre-Carboniferous basement, and of those contemporaneous with deposition of the Carboniferous sediments, has also been studied in collaboration with the Department of Earth Sciences of the Open University.

The study provides an important analysis not only of the geochemistry and metallogeny of the Pennine orefields but of the events in the structural and stratigraphic evolution of the region, that eventually led to the development of the mineralisation. Metallogenic models for other types of carbonate-hosted deposits are also described, including the so-called 'Irish-style' deposits of the Central Plain of Ireland. The potential for new discoveries of such base-metal associations in the region is considered in relation to prospectivity maps prepared by integrating a range of datasets by use of the BGS/Natural Environment Research Council image analysis system. The datasets prepared have many other applications. Universities, industry and others can purchase them through the BGS/Royal Aircraft Establishment, GISA, image analysis facility at Farnborough.

F. G. Larminie
Director, British Geological Survey
1 April 1989

ACKNOWLEDGEMENTS

We should like to thank Dr N. Aitkenhead, Dr D. Holliday, Dr D. Slater and Dr A. Whittaker of the BGS for advice and information and Dr P. A. Sabine, Dr R. Howarth, Mr P. J. Moore and Mr D. Ostle for support and encouragement. We are particularly grateful to Dr J. Boissonas and Dr L. Van Wambeke of the CEC for their advice and encouragement throughout the research programme. Dr Slater and Dr Van Wambeke also provided critical comments on the draft report.

XRF analyses were carried out under the direction of Dr B. P. Atkin and Dr P. K. Harvey of Nottingham University and neutron activation analyses were performed on a repayment basis at the HERALD Reactor Centre, AWRE, Aldermaston, at the London University Reactor Centre and at ICI Billingham. The rare earth elements were determined in the laboratory of the BGS Geochemistry Directorate, under the supervision of Mr B. Tait.

We are also grateful to Mr I. Stevenson for leading a field trip to the South Pennine Orefield and to Laporte Minerals and their geologists, Mr J. Hedges and Dr N. Butcher, for allowing us to visit Sallet Hole mine and opencast operations. Dr N. J. Riley carried out a palaeontological examination of samples from Bretton Clough. Samples of Carboniferous igneous rocks were kindly supplied by Dr R. MacDonald and Dr S. Kirton. We are grateful to British Petroleum, British Coal and Laporte Minerals for permission to publish information on confidential boreholes.

The final text has benefitted greatly from the efforts of M. J. Jones of the IMM. The reference list was compiled with the assistance of Rachel Williams of the BGS.

The programme was sponsored jointly by Directorate General XII of the Commission of the European Communities (Contract: MSM-90-UK) and the Minerals and Metals Division of the Department of Trade and Industry.

SUMMARY

Geochemical and isotopic data for the ore minerals and rocks of the Pennine orefields were studied in relation to new deep geological/geophysical data on the Derbyshire 'Dome' and East Midlands of England. The study was aimed at the development of metallogenic models and exploration criteria to assist in the identification of carbonate-hosted (Pennine- and Irish-style) ore deposits, including mineralisation associated with high heat production (HHP) granites, buried beneath Permian and later cover.

New systematic databases were compiled in digital form and examined by use of the BGS/NERC I^2S image analysis system (IAS), which enables statistical processing and the integration of different datasets to be carried out interactively. The primary datasets prepared and analysed included those of gravity, aeromagnetics, seismic refraction and reflection, heat-flow, structure-contour data for selected stratigraphical sequences and horizons (based on borehole data and seismic interpretation) and maps of the distribution of mineralised veins and fault systems in the exposed orefields. New lithogeochemical databases, including analyses for major and trace elements of igneous rocks and shales in the region, were also prepared with the use of outcrop and borehole samples, together with a compilation of previously published mineralogical, geochemical and isotopic data. Derived images prepared from one or more primary datasets were used extensively for the study. For example, basement lineaments reactivated during the Dinantian were identified by a combination of enhanced gravity and aeromagnetic images together with second derivative gravity plots and maps of listric faults prepared from seismic interpretations.

The preferred model developed for the Pennine mineralisation involves seismic pumping related to Lower Permian tectonism with de-watering of Viséan–Namurian shale basins and fluid reservoirs that had become overpressured as a result of Namurian–Westphalian sedimentation. Mineralised veins were formed as a result of the incursion of moderately acid, highly saline ($NaCl–CaCl_2–H_2O$) brines generated in infinite reservoir conditions and carrying hydrocarbons, Pb, Zn, Ba and F into fracture systems in limestone sequences. Sulphide precipitation probably occurred as a result of reduction of sulphate in limestone formational fluids by hydrocarbons in the brines with deposition of fluorite as a result of the release of calcium by acid neutralisation reactions associated with sulphide precipitation. Baryte was precipitated by mixing of the Ba-enriched brine with SO_4^{2-} above and around the zone of sulphide deposition.

In the North Pennines the buried Weardale and Wensleydale granites, which have calculated heat productions of 3.7 and 3.4 μWm^{-3}, respectively, locally focused fluids into hydrothermal convection cells, so there is a spatial association between mineral zonation and the subcrop of buried granites. The relationship is particularly clear in the case of the Weardale Granite, in which radioelements are relatively unfractionated and in which hydrothermal alteration/re-equilibration occurred, initially, contemporaneously with emplacement of the intrusion in the Lower Devonian. Temperature maxima coincide with areas of intense mineralisation in the fluorite zone, centred over the negative gravity anomaly associated with the granite with temperature minima in the baryte zone around the margin of the buried intrusion. Elsewhere the role of low-density chemically evolved granites in the basement may have been to increase the buoyancy of crust locally and to propagate fracture systems from basement to cover.

Such a model accounts for the common surface geological-geochemical features of the Pennine orefields, including the mineral assemblage with Pb + F > Ba + Zn and the structural control of mineral veins whereby permeability was principally related to movement of fracture systems rather than karstification or near-surface groundwater movements. It also accounts for differences between the North and South Pennine Orefields: for example, the increased temperature and K/Na ratios indicated by fluid inclusion studies in the Alston Block compared to the Askrigg and South Pennine Orefields.

Some of the most important aspects in the structural/stratigraphic evolution of the study region, which eventually led to the formation of the Pennine orefields, are thought to include: the development of basement heterogeneities (Upper Proterozoic–Silurian), particularly structural grain and major lineaments (the Pennine axis lies at the intersection of the Caledonide NE–SW-trending basement to the West and WNW–ESE-trending basement to the East); the emplacement of large low-density radiothermal granites (Lower Devonian); the development of half-graben hinged along faults following basement lineaments with thin carbonate drapes over basement highs and clastic red-bed–sandstone–thick basinal shale sequences in tectonically controlled basins (Tournaisian–Namurian); the overpressuring of the shale basins as a result of the deposition of thick sedimentary sequences and the collection of fluids in reservoirs (end Namurian–Westphalian); and, finally, the reactivation of WNW- and NW-trending basement lineaments and the propagation of predominantly E- and ENE-trending fault systems in the cover with seismic pumping of fluids into brittle fracture systems in platform limestones and sandstones (Lower Permian).

In contrast to the Pennine mineralisation, Irish-style deposits are syngenetic/syndiagenetic and were formed penecontemporaneously with the deposition of Tournaisian shelf carbonates. The Irish deposits, which generally contain Zn > Ba > Pb, are preserved in limestones that frequently overlie detrital basal sandstones in half-graben basins. The listric faults that controlled the basins comprise reactivated Caledonide (NE–SW-trending) basement lineaments. The Irish deposits are generally of higher temperature (up to 250°C) than the Pennine mineralisation and their formation from low-to moderate-salinity fluids was followed by an episode of alkaline–tholeiitic basaltic magmatism.

The geophysical and deep geological criteria derived from the models for Pennine- and Irish-style mineralisation, respectively, can be combined to prepare prospectivity maps by use of image analysis. The areas in the study region that have been identified as having potential for carbonate-hosted mineralisation at surface or at depth for Pennine-style mineralisation include the Castleton and Alport areas in Derbyshire, the Eakring Anticline (particularly adjacent to the eastwards extension of the Bakewell Fault) and the subcrop of Brigantian limestone to the NE of the Gainsborough Trough. Areas with potential for Irish-style mineralisation include the Castleton and Alport areas adjacent to the Bakewell Fault and the margins of the Gainsborough Trough. Other areas in Britain considered prospective for Irish-style mineralisation on the basis of criteria proposed for the study area include the margins of the Solway Basin, particularly near to listric basin margin faults, and the Craven Basin, where volcanism is absent but limited synsedimentary zinc and lead mineralisation has been found in Dinantian carbonates.

The Irish-style and Pennine-style deposits can be related to different phases in the tectonic evolution of the Northern Foreland of the Hercynian orogen. Hence, the Irish-style deposits were formed in geothermal systems over zones of high heat flow and tectonism in the crust. They are related to the waxing phase of a regional regime of crustal extension and basin formation associated with the rise of hot asthenosphere beneath the crust, which was characterised by alkaline basaltic magmatism, high geothermal gradients and listric faulting along reactivated basement fault systems. In contrast Pennine-style ore deposits were formed following a period characterised by declining geothermal gradients (contemporaneous magmatism becoming tholeiitic by the early Permian and, finally, ceasing) and by regional subsidence of the crust with overpressuring of basins as a result of the deposition of thick Westphalian sequences; they were deposited from fluids similar to oil-field brines expelled from overpressured basins by regional tectonism in Lower Permian times. Pennine-style ore deposits are thus considered to be a special case of Mississippi Valley type ore deposits in which tectonism rather than simple basin de-watering was involved.

1 INTRODUCTION

J. A. Plant and D. G. Jones

Metalliferous mineral exploration was, until recently, based on the application of relatively simple direct geochemical and geophysical methods. For example, geochemical exploration methods were aimed mainly at the detection of anomalously high levels of such metals as Cu, Pb, Zn, U or Mo in surface soils and stream sediments associated with exposed or near-surface ore deposits. The most common geophysical methods were also direct techniques, such as the application of induced polarisation for the location of sulphides, and the overall assessment of ore deposits was generally based on descriptive classifications, such as those of Smirnov (1976), rather than genetic systems. This general approach is useful in the detection of metalliferous minerals that are undergoing erosion and leaching, but, as was pointed out by Woodall (1984), exploration is increasingly based on the scientific understanding of the ways in which ore deposits form. This is particularly true where concealed deposits are sought, such as those of Olympic Dam in South Australia, which was located by use of predictive metallogeny beneath 300 m of cover (Roberts and Hudson, 1983). The detection of buried ore deposits requires an approach comparable with that which is used for hydrocarbon exploration whereby exploration criteria, based on sound genetic models, are applied.

It is now widely recognised as a result of stable isotope and fluid inclusion studies of rocks (e.g. Sheppard and Taylor, 1974; Roedder, 1977 and Taylor, 1977), investigations of extant geothermal fields (e.g. Fairbank and others, 1981 and Ellis and Mahon, 1964) and hydrothermal activity associated with ocean ridge systems (e.g. Mottl, 1983) that many types of metalliferous mineral deposits are formed by the interaction of large volumes of epizonal water* with rocks. Since the late 1960s fluid dynamic models have been developed with the use of computers for a considerable number of ore deposit types. These include submarine ore deposits associated with magmatism (Lister, 1972), uranium deposits formed by postmagmatic hydrothermal convection in abnormally radioactive plutons (Fehn and others, 1978), and hydrothermal ore deposits generally (Cathles, 1981). Geological models based on the integration of petrographical, field geological, geochemical and geophysical observations have also been developed in the light of experimental data (e.g. Barton, 1986; Woodall, 1984; O'Driscoll, 1986). The model for Kuroko massive sulphide deposits, associated with calc-alkaline volcanism on the sea-floor in island arc settings (Ohmoto and Rye, 1974; Sakai and Matsubaya, 1974), provides one such example. Mineral deposit models, and their geological, geochemical and geophysical characteristics, have been used as a basis for comprehensive metallogenic maps and reports prepared by the Geological Survey of Canada (e.g. Eckstrand, 1984) and similar studies are in progress in the Scandinavian countries. An expert system, entitled 'Prospector', was developed for mineral exploration, based on mineral deposit models, by Hart and others (1978) and a programme aimed at advancing knowledge and expertise in mineral deposit modelling worldwide has been initiated by the IUGS/UNESCO.

At the same time as progress has been made in formulating mineral deposit models and exploration criteria, important advances have also been made in the application of computer image analysis systems (IAS) to mineral exploration. Commercial image analysis systems, designed primarily for the study of remotely sensed data, became available in the late 1970s following their development for the analysis of data from the *Mariner* missions to Mars (Condit and Chavez, 1979). Image analysis techniques are now employed extensively in geology and, although studies continue to be based mainly on remote sensing, they have also been shown to provide a powerful method of processing a much wider range of spatially related data, including geological, geochemical and geophysical information (Guinness and others, 1983; Green, 1984). The IAS used in this study, for example, allows single images (maps) to be processed interactively, normally in less than two seconds, and the interaction of up to 12 images in a few seconds. The application of this technology to the formulation of mineral deposit models is considered more fully in Appendix 1.

The present study is designed to take advantage of recent developments in mineral deposit modelling and image analysis systems with the aim of producing genetic models and exploration criteria for mineralisation buried beneath later, barren sedimentary cover: emphasis is given in this initial study to regional investigations as a basis for the identification of areas that merit more detailed study. The development of methods of exploration for buried ore deposits is particularly important in Europe where exposed Precambrian, Caledonian and Hercynian basement and Devonian, Carboniferous and Permo-

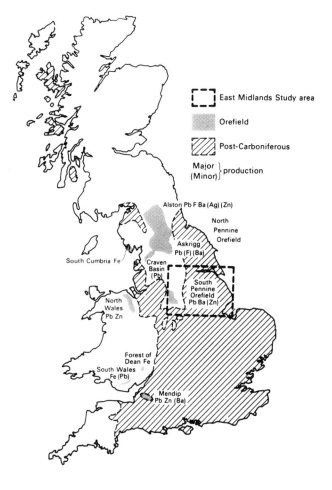

Fig. 1.1 Dinantian carbonate-hosted orefields

---

* Including metamorphic, formational and meteoric water.

Trias 'older' cover frequently contain important mineralisation, but where much of the land area, particularly in the NW, is covered by barren Mesozoic sediment. Over large areas of Europe this cover is thin and, hence, buried ore deposits are likely to occur at economically viable depths. The East Midlands of England was selected for the study because it contains the South Pennine Orefield and its probable extension beneath Mesozoic cover to the east. The metallogeny of the Pennines has been studied intensively over a long period and a considerable amount of mineralogical, isotopic and geochemical data are available. Extensive information is available (for example, seismic reflection and borehole data) on the sub-surface geology to the east of the orefield as a result of intense hydrocarbon exploration in recent years and further information is provided by regional gravity and aeromagnetic surveys. The principal study region is indicated in Fig. 1.1. The North Pennine Orefields are also indicated, since their metallogeny was considered in the formulation of a general model for Pennine-style mineral deposits.

Two types of carbonate-hosted mineralisation that are likely to occur in lower Carboniferous cover rocks in the region are considered: Pennine-style and Irish-style. The former, consisting of epigenetic Pb–Ba–F veins cutting Dinantian (Asbian and Brigantian) platform limestones and sandstones, has been classed as a fluoritic sub-type of

the Mississippi Valley genus of ore deposits (Dunham, 1983) comparable with the Illinois, Kentucky and Sweetwater fields in the USA. The latter are Zn–Ba–Pb deposits of the type that occur in the Lower Dinantian (Tournaisian and Chadian) rocks of the Midland Plain of Ireland. Their genesis is controversial, some authors (e.g. Sangster, 1986; Brown and Williams, 1985) suggesting that they represent a type of Mississippi Valley mineralisation in which at least a component of fluids was derived by basin de-watering and others favouring an origin related to deeply excavating hydrothermal convective circulation systems (e.g. Boyce and others, 1983).

Research has been concentrated on exposed or mine geology. A regional study that used digital datasets, including remotely sensed information, and covered the North and South Pennine orefields was carried out by Hunting Geology and Geophysics under a CREST programme for the CEC, but the main emphasis of this work was also the identification of mineralisation in the exposed orefields.

The role of basement characteristics, such as lineaments and fracture patterns, buried granites and variations in heat production and geochemistry is poorly understood. The identification of lineaments that represent deep faults contemporaneous with ore deposition, for example, is important in defining areas prospective for mineral deposits generally. This is partly because of their

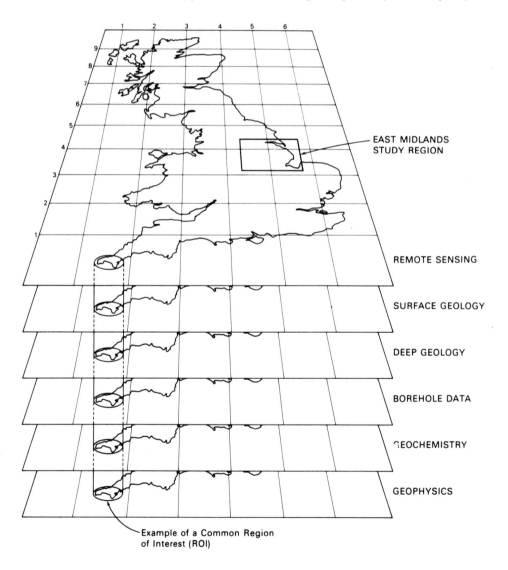

Fig. 1.2 Extraction of data from multiple sources

2

role in creating zones of high permeability with repeated seismicity continually reopening fluid pathways sealed by mineral deposition (Fehn and others, 1978). Deep lineaments are also increasingly recognised as playing a role in channelling magmas to high levels in the crust (Fettes and others, 1986) and are commonly zones of above-average heat flow (Reading and others, 1986). Lineaments may, however, be sealed and made inactive by magmatism (Reading and others, 1986).

For granites it has been suggested that high heat production (HHP) granites in basement may play a role in the development of mineral deposits in overlying unconformable cover sequences a considerable time after magmatic cooling has occurred (Simpson and others, 1979; Fehn and others, 1978). Such deposits have been related to the heat production and high thermal conductivity of the granites, which, at times of increased tectonism or heat flow from the mantle, can become centres of convective circulation (Brown and others, 1980). Geophysical anomalies consistent with the presence of large HHP granites in the basement have been reported to occur throughout Ireland (Phillips and others, 1976), England and the British North Sea (Donato and Tully, 1982).

The structural/stratigraphic evolution of the cover rocks of the region has been considered previously, but only generalised geophysical and deep geological outlines of the region have been available until recently; the role of sedimentary basins in generating ore fluids of appropriate temperature and composition has not been adequately evaluated in relation to modern basin analysis studies such as those of McKenzie (1978) and Dewey (1982), for example. The geochemical and geophysical properties of the cover sequence are also little known.

In this study an attempt has been made to develop and constrain available mineral deposit models for carbonate-hosted mineralisation in relation to the structural and stratigraphic evolution of the area by preparing and evaluating new systematic datasets for the basement and cover formations of the East Midlands. These include geophysics (aeromagnetic, gravity, heat flow and seismic refraction and reflection), geology (structure and stratigraphy from deep borehole and geophysical data, particularly seismic reflection), metallogeny (geology, geochemistry and isotopic data on the orefields) and the geochemistry of igneous rocks and shales.

An approach based on simply interacting primary raw datasets to formulate models was rejected in favour of one that involved considerable interpretation, assessment and pre-processing of information outside the IAS. For the deep geological information, for example, the structural and stratigraphic evolution of the area was studied and a tectonic model was formulated before particular features were selected for study and integration with other datasets by the IAS. Geochemical data on shales and igneous rocks in the region were too scattered spatially and temporally to enable a digital image to be prepared and these, together with isotopic and geochemical data on ore minerals in the Pennine Orefields, were also evaluated outside the IAS. The IAS was used most extensively in the processing and qualitative interpretation of regional geophysical data, particularly the gravity and magnetic results (Chapter 4), although some pre-processing was carried out on a main-frame computer before the results were transferred to the IAS (this was notably for the calculation of the reduced to the pole (magnetics) and stripped gravity surfaces (see Appendix 1)). Quantitative interpretation of the geophysical data in terms of

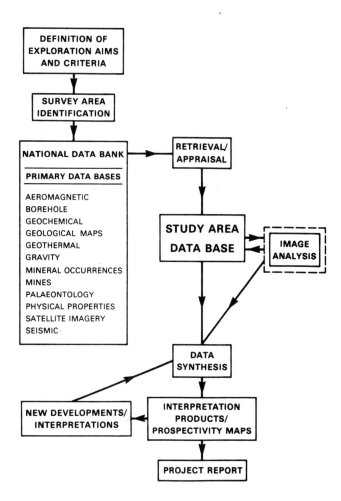

Fig. 1.3   Structure of data analysis

geological models was also carried out remote from the IAS, mainly by use of the PERQ workstation at BGS Keyworth, and measurements of rock physical properties were performed by use of standard techniques in the Engineering Geology laboratories of BGS.

In addition to the application of sophisticated methods of processing geophysical data, the main application of the IAS was in the integration of datasets to prepare mineral deposit models and apply exploration criteria to

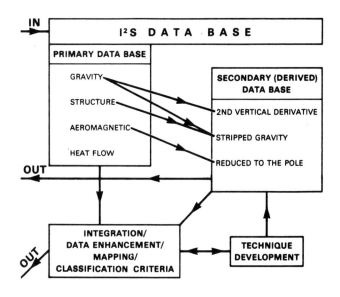

Fig. 1.4   Image analysis procedure

3

identify areas prospective for mineralisation. The data acquisition and image analysis procedures used are summarised in Figs. 1.2, 1.3 and 1.4 and examples of the images are shown in Plates 1–12.

The general approach that has been developed has relevance not only to studies of the mineralisation in rocks of the Carboniferous shelf, which extended across Europe from the Rhine to Ireland during the Lower Carboniferous, but also to studies of mineral deposits that were formed in similar environments elsewhere. The Lower Cambrian carbonate platform, which extended in two belts over Central Europe from the Pyrite Belt of Portugal to Amorica and Bohemia and from Asturia to SW France and Sardinia, provides a possible example of a similar, but older, metallogenic regime.

In the following sections previously published work on the geology and metallogeny of the study region is first reviewed. This is followed by four sections that present new geophysical, deep geological and geochemical data together with new interpretations of the structural/stratigraphical evolution of the region. In the final sections the new and previously published data are combined to formulate metallogenic models for Pennine- and Irish-style mineralisation and exploration criteria are developed suitable for the identification of areas prospective for these types of ore deposits buried beneath later cover. Finally, areas that merit more detailed study are identified by use of prospectivity maps generated by digital image analysis procedures to combine appropriate sets and sub-sets of exploration criteria for the two types of deposit.

2 GEOLOGY OF THE EAST MIDLANDS

K Smith and N J P Smith

The East Midlands area can be divided into two broad physiographic regions, which are separated by the outcrop of the basal Permian unconformity (Fig. 2.1). In the west the outcrop of Carboniferous rocks rises more than 600 m above sea-level in the Central Pennines. In the east a uniformly dipping Permian to Cretaceous sequence overlies the Carboniferous and gives rise to the subdued topography of the Eastern England Shelf.

In the following section some of the previous geological and geophysical investigations of the East Midlands are outlined and the stratigraphy of the area is briefly described. Particular emphasis is placed on studies of the Carboniferous, which is mineralised at outcrop and forms the main target for buried mineralisation in the area.

2.1 Lower Palaeozoic and Precambrian

Lower Palaeozoic and Precambrian strata do not outcrop within the East Midlands study area, but an indication of the possible composition of the concealed pre-Carboniferous basement is provided by the outcrop of Precambrian rocks in the adjacent area of Charnwood Forest (see Fig. 7.1). At Charnwood the Precambrian sequence is largely composed of volcaniclastic sediments with interbedded andesitic and dacitic lavas, agglomerates and breccias. A series of greywackes and mudstones with minor quartzite forms the youngest part of the sequence.

The main Charnian outcrop is intruded by a varied group of diorites of probable late Precambrian or Cambrian age (Sylvester-Bradley and Ford, 1968; Le Bas, 1982; Institute of Geological Sciences, 1982).

South-west of Charnwood Lower Palaeozoic strata outcrop in a small inlier at Nuneaton (see Fig. 6.11), where a thin basal sequence of quartzite and limestone rests un-

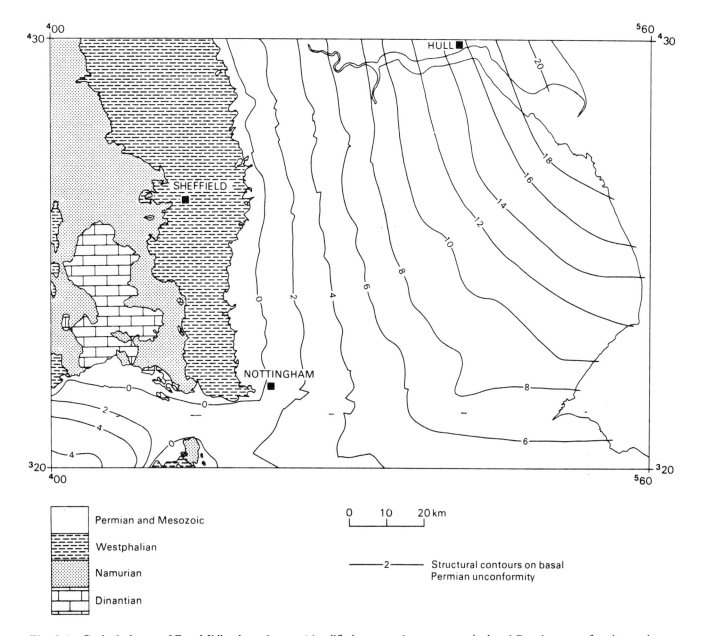

Permian and Mesozoic

Westphalian

Namurian

Dinantian

0   10   20 km

———2——— Structural contours on basal Permian unconformity

Fig. 2.1 Geological map of East Midlands study area (simplified structural contours on the basal Permian unconformity are in hundreds of metres below sea-level)

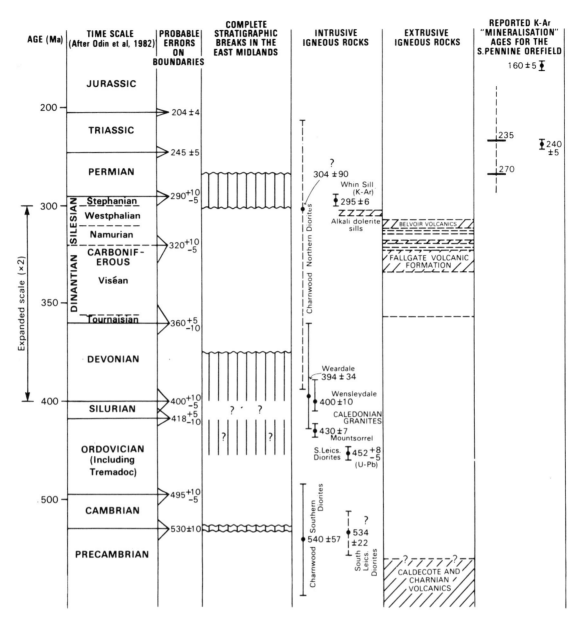

Fig. 2.2 Radiometric age dates obtained from eastern England in relation to geological time scale of Odin and others (1982) and to periods of extrusive igneous activity. Data compiled from Aitkenhead, 1977; Aitkenhead and others, 1985; Burgess, 1982; Cribb, 1975; Dunham, 1974; Fitch and Miller, 1967; Hampton and Taylor, 1983; Ineson and Mitchell, 1972; Kirton, 1984; Pidgeon and Aftalion, 1978; Thorpe and others, 1984; Trewin, 1968; Trewin and Holdsworth, 1972; Walters, 1981

conformably on the local equivalent of the Charnian Supergroup (Caldecote Formation), and is overlain by thick argillaceous sediments of Cambrian-Tremadocian age (Allen, 1968; Taylor and Rushton, 1971). Similar Lower Palaeozoic sediments exist at shallow depth in the Charnwood area (Butterley and Mitchell, 1945).

Isotopic age dates obtained from an outcrop of granodiorite at Mountsorrel at the north-eastern margin of Charnwood Forest and also from a group of diorites and tonalites that are exposed in South Leicestershire suggest that these igneous bodies form part of a widespread suite of Caledonian plutonic intrusions in eastern England (Le Bas, 1968, 1982; Pidgeon and Aftalion, 1978; Hampton and Taylor, 1983; Thorpe and others, 1984; and see Fig. 2.2).

Geophysical studies of the concealed pre-Carboniferous rocks of the East Midlands began in 1936 when gravity and ground magnetometer surveys were carried out in the vicinity of Kelham Borehole in an attempt to define areas of shallow basement (Lees and Cox, 1937). More exten-

sive gravity surveys subsequently located a large positive gravity feature at Nocton, near Lincoln (Lees and Taitt, 1946). In 1943 Nocton No. 1 Borehole intersected phyllitic strata below a thin Carboniferous sequence. Shallow basement was also proved by drilling geophysical anomalies at Foston, Stixwould and Sproxton (Kent, 1967; see Fig. 5.3). These boreholes indicated that a high velocity layer mapped by seismic refraction, in the south-eastern part of the East Midlands, corresponded to pre-Carboniferous basement (Kent, 1966). Other scattered boreholes in the area (see Fig. 6.11; 7.8) have since proved deformed Lower Palaeozoic rocks (e.g. Eyam; Dunham, 1973) or undated strata that were lithologically compared to the Charnian Supergroup (e.g. Great Osgrove Wood and Galley Hill boreholes; Burgess, 1982).

Seismic refraction data obtained in a major north–south traverse of northern Britain suggest that much of the Central Pennine area was underlain by a thick high-velocity refractor (5.8 – 6.0 km/s), which was identified as Lower Palaeozoic basement (Bamford and others, 1976).

A similar refractor in southern Derbyshire was compared to Charnian basement by Whitcombe and Maguire (1981a). Rogers (1983) proposed that the Charnian refractor was downfaulted by the Cronkston–Bonsall faults within the Derbyshire Massif (see Fig. 5.4) and was concealed by Lower Palaeozoic basement further north.

Regional aeromagnetic and gravity anomaly maps (Institute of Geological Sciences, 1977a, 1977b) provide recent compilations of other geophysical data, which have been used to model basement structure in the East Midlands area (Maroof 1976; Evans and Maroof, 1976; Cornwell in Aitkenhead and others, 1985).

Deep seismic reflection profiles are a major source of basement information. The use of the seismic reflection method was abandoned in the East Midlands after an initial trial in the late 1930s, but subsequent improvements in data acquisition have made it the most valuable tool of hydrocarbon exploration in the area (Kent, 1966). Coherent reflections obtained from the top of the Lower Palaeozoic basement in the deeper parts of the East Midlands Carboniferous basin have contributed greatly to the model of basement structure that is presented in this report.

2.2 Carboniferous

This section includes strata of probable Devonian age that are preserved locally at the base of the Dinantian sequence (George and others, 1976).

2.2.1 DINANTIAN

The Dinantian strata of the East Midlands comprise a varied sequence of mostly carbonate rocks that is diversified by minor sandstones and calcareous mudstones and by widespread basic volcanic rocks and intrusives. The Dinantian sequence is considerably mineralised at outcrop and forms the main target for metalliferous mineral exploration in the area (see Chapter 3).

Dinantian rocks outcrop extensively in the southern Pennines, where a major inlier (the Derbyshire 'Dome' or Massif) forms the rugged scenery of the Peak District (Stevenson and Gaunt 1971: Aitkenhead and others, 1985). Other outcrops near Derby (Frost and Smart 1979), Crich and Ashover (Smith and others, 1967), and Mixon (Aitkenhead and others, 1985) coincide with minor anticlines flanking the main uplift. Small Dinan-

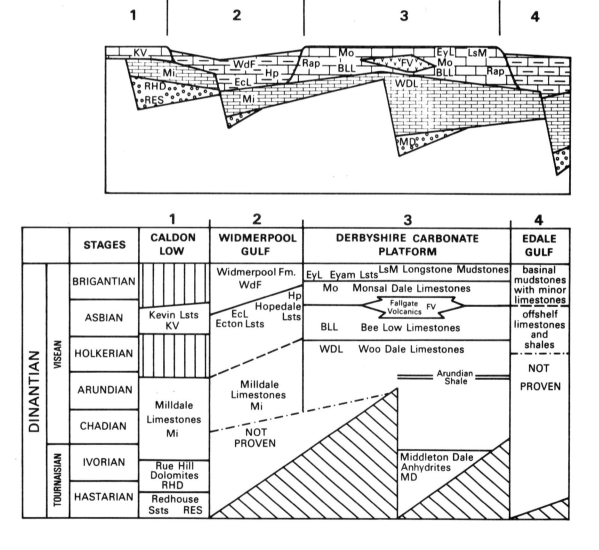

Fig. 2.3 Schematic geological section and stratigraphic table of Dinantian rocks in the East Midlands. The latter, modified from Aitkenhead and Chisholm (1982), includes Middleton Dale Anhydrite Series of Strank (1985). Geological section, showing inferred reationships of formations in Derbyshire at end of Dinantian, modified after Smith and others (1985)

tian inliers also occur to the east of the Leicestershire Coalfield (Mitchell and Stubblefield, 1941). Coral, brachiopod and goniatite zones form the basis of Dinantian stratigraphic correlation, but schemes based on other zonal fossils, such as conodonts, foraminifera and miospores, have recently found increasing application (George and others, 1976; Aitkenhead and Chisholm, 1982; Fig. 2.3).

Hydrocarbon exploration in the East Midlands was originally stimulated by the recognition that Dinantian carbonates capped by Namurian shales potentially provide a good hydrocarbon trap.

This expectation was partly fulfilled by the discovery of oil at the top of the Dinantian in Hardstoft No. 1 Borehole (Giffard, 1923), but subsequent investigations have established that Dinantian hydrocarbon reservoirs are poorly developed and are largely controlled by the presence of fracture porosity (Lees and Taitt, 1946).

During the second major campaign of oil exploration on the Eastern England Shelf seismic refraction profiles were used to map a high-velocity refractor that corresponded to the top of the Dinantian massive limestones. A structural contour or isochron map of the Dinantian refractor identified a number of structural highs, which were subsequently tested by drilling (Lees and Taitt, 1946). By 1945 numerous short intersections of Dinantian strata had been obtained by deep hydrocarbon boreholes (Ineson and Ford, 1982). The complete Dinantian sequence was intersected by Foston, Nocton and Stixwould No. 1 boreholes, which together demonstrated that a broad thinning of the Dinantian succession took place from the outcrop south-easterly across the East Midlands Shelf. Eakring No. 146 Borehole intersected a thicker complete Dinantian sequence that comprised 500 m of carbonate rocks resting upon about 900 m of varied conglomerates and sandstones with some interbedded limestones (Falcon and Kent, 1960). Similar coarse, pebbly sandstones were intersected at the base of the Dinantian in Caldon Low Borehole at the south-western margin of the Derbyshire Massif (Aitkenhead and Chisholm, 1982). In both cases the unfossiliferous red beds may be partly of Devonian age.

Facies analysis of the exposed Dinantian suggested that most of the Derbyshire Massif formed a carbonate shelf, which was separated from surrounding Dinantian basins by a series of patchily developed fringing reefs (Hudson, 1931). Boreholes at Alport and Edale subsequently proved late Dinantian off-shelf facies to the north of the Castleton reef belt (Hudson and Cotton, 1945a, 1945b; Stevenson and Gaunt, 1971). Woo Dale Borehole, which was drilled in the Derbyshire Massif, proved pre-Carboniferous basement below a notably thin Dinantian sequence (Cope, 1973, 1979; Strank, 1985), which seemed to confirm that the Derbyshire platform carbonates had developed above an area of shallow basement analogous to that which formed the pre-Carboniferous floor of the Askrigg and Alston blocks. Kent (1966) used a variety of geological and geophysical criteria to delineate the boundary between Dinantian block and basin facies in the subsurface of the East Midlands. The Eyam Borehole (1970–72), which was drilled in the Derbyshire Massif to provide further stratigraphical information about the early Dinantian, intersected more than 1800 m of mostly carbonate rocks before proving pre-Carboniferous basement (Dunham, 1973; Strank, 1985). This succession, which greatly contrasted with that known from the Woo Dale area, suggested that the block and basin model did not adequately explain Dinantian facies variation in

Derbyshire. It became apparent that a uniform late Dinantian carbonate shelf masked considerable thickness variations in the early Dinantian sequence. Miller and Grayson (1982) proposed that the difference between the depths of the pre-Carboniferous basement in Woo Dale and Eyam boreholes indicated that most of the Derbyshire Massif was underlain by a basement block that was tilted broadly north-eastwards. Early Dinantian sediments formed a north-easterly thickening wedge on the basement ramp and were overlain by a late Dinantian carbonate shelf that had prograded towards the deeper part of the basin in the Edale Gulf. A second basement block, also dipping north-eastwards, was inferred to have controlled the development of early Dinantian Waulsortian facies in south-west Derbyshire. A new tectonic model of Dinantian sedimentation is presented in this report (Chapter 5).

2.2.2 Namurian

The Namurian of the East Midlands largely comprises interbedded shales, siltstones and sandstones laid down by a series of overlapping deltas, which advanced into the area from the north and east. Minor rock types include coal, with associated seatearths and ganisters, limestone (especially towards the base of the sequence) and mudstone marine bands that provide a reliable basis for regional correlation (Ramsbottom and others, 1978; Fig. 2.4). In places thin layers of high potassium bentonite, derived from volcanic ash, form a minor component of the early Namurian sequence (Trewin, 1968; Trewin and Holdsworth 1972, 1973; Aitkenhead, 1977). Sedimentological and petrological studies indicate that most Namurian sandstones were probably derived from the Caledonian basement, of granitic and granitoid-gneissic affinity, which is presently preserved at depth in the area of the central North Sea (Gilligan, 1920; Stevenson and Gaunt, 1971; Frost and others, 1981).

Namurian sediments outcrop extensively in the high dissected plateau of the Central Pennines. At the base of the sequence in Derbyshire the Edale Shales were deposited in a deep basin that was starved of coarse detritus (Stevenson and Gaunt, 1971). The overlying formations comprise a coarsening-upwards deltaic cycle of Kinderscoutian age. The Mam Tor Beds are distal turbidites that mark the onset of deltaic deposition in late Kinderscoutian times (Allen, 1960). Proximal turbidites and delta slope deposits make up the Shale Grit and Grindslow Shales (Walker, 1966), and the Kinderscout Grit was deposited by major distributary channels (McCabe, 1978). Delta-top coals are sparsely developed at the top of the sequence. The Marsdenian deltas, which overlapped the deposits of the Kinderscout Delta, show a similar pattern of progradational facies variation (Collinson and others, 1977; Chisholm, 1977; Jones, 1980; Okolo, 1983). At outcrop in Derbyshire, and in the subsurface to the east, thin, condensed sequences of early Namurian mudstones rest unconformably on a late Dinantian carbonate platform (Smith and others, 1967). Kelham Coal Borehole (1910) first indicated a general south-eastward thinning of the Namurian sequence against a northern extension of the emergent Midland Massif (Lees and Cox, 1937). Subsequently, geophysical well log correlation of oil exploration boreholes beneath the Eastern England Shelf confirmed the easterly thinning sequence and suggested that late Namurian sediments overlap Dinantian carbonates towards the margin of the basin (Falcon and Kent 1960; Downing and Howitt, 1969).

The development of a deep Namurian basin to the north of the Derbyshire carbonate platform was first demonstrated by the presence of thick Edale Shales in the Alport and Edale oil exploration boreholes (Hudson 1931; Hudson and Cotton, 1945a, 1945b). Similarly expanded early Namurian sequences were also intersected in the subsurface south of Nottingham by Long Clawson and Widmerpool No. 1 boreholes, and in the Gainsborough area by Gainsborough and Walkeringham No. 1 boreholes (Lees and Taitt, 1946; Falcon and Kent, 1960). Kent (1966) proposed that these expanded sequences were deposited in three linear basins, the Gainsborough

Trough, the Edale Gulf and the Widmerpool Gulf, which extended east–south-eastwards across the East Midlands Shelf from the Central Pennine Basin (see Figs. 5.4, 5.8). These areas of thick Namurian deposition coincided with Dinantian basins in which the Brigantian and Asbian strata characteristically comprised mudstones interbedded with turbiditic limestones. Elsewhere, thin Namurian sequences rested, often unconformably, on late Dinantian platform carbonates of 'block' facies. The model of Carboniferous basin formation that has been devised during this project considerably refines this interpretation of the relationship between Namurian and Dinantian facies (Chapter 5).

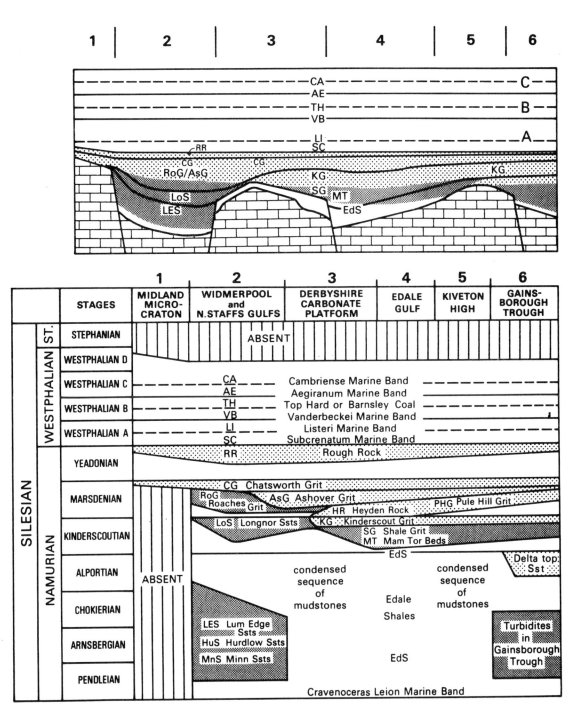

Fig. 2.4 Schematic geological section (*X-X'* in Fig. 5.7) and simplified stratigraphic table of Silesian rocks in East Midlands. Namurian formations composed predominantly of sandstones and turbidites are indicated by coarse and fine stippling, respectively. Stratigraphic position of Top Hard coal (main structural datum in East Midlands area) shown together with more important Westphalian marine bands

2.2.3 WESTPHALIAN

The Westphalian sediments of the East Midlands are predominantly siltstones and mudstones with numerous interbedded coals and laterally impersistent sandstones. They form a repetitive sequence of minor fluvio-deltaic cycles, or cyclothems, which increase in thickness and frequency towards the centre of the Westphalian basin. Regional stratigraphic correlation is based on the presence of marine bands, which are thin, laterally extensive mudstones, each with a characteristic goniatite-bivalve fauna, which were deposited on the Westphalian deltaic plain during periodic marine transgressions (Ramsbottom and others, 1978).

Westphalian rocks outcrop extensively in the Yorkshire and East Midlands Coalfield and are also preserved in synclinal areas in the Western Pennines. In the south-western part of the area, the exposed coalfields of Cheadle, South Derbyshire and Leicestershire are separated by the Permo-Triassic deposits of the Needwood Basin (Fig. 2.5). A vast amount of geological data has been acquired as a result of coal exploration, which has extended progressively eastwards from the Westphalian outcrop into the concealed part of the East Midlands Coalfield. Numerous cored boreholes and shafts have provided good stratigraphic control of the coal-bearing sequence. Large-scale structural contour maps of the major economic coal seams are available for all the areas that are currently being exploited. The widely developed, Top Hard, or Barnsley, coal seam of Westphalian B age is commonly used as a regional structural datum (Wilson, 1926; Edwards, 1951). Variations in coal geochemistry have also been assessed on a regional basis (Suggate, 1976), and some progress has been made towards a comprehensive sedimentological analysis of Westphalian facies variation across the area (Elliott, 1968, 1969; Guion, 1978).

The widespread occurrence of oil seepages within the Westphalian led to the investigation of the surface anticlines of the East Midlands Coalfield, during the first campaign of oil exploration on the British mainland between 1918 and 1922. The first deep borehole, which was drilled to test the Hardstoft Anticline, found oil in 1919. Other anticlines, including those at Ironville, Heath, Brimington and Norton-Ridgeway (see Fig. 5.12) were tested without further success (Giffard, 1923). Oil exploration was resumed in 1935, when the basis for the renewed interest was an interpretation of the East Midlands Coalfield as a broad synclinal structure separating the anticlines flanking the Pennines (which had been tested during the earlier campaign) from an area of similar anticlines concealed beneath the Eastern England Shelf (Lees and Cox, 1937). The presence of a buried anticlinal axis in the Eakring area, which was originally suggested by analysis of coal exploration data, was confirmed in the first successful application of deep geophysical techniques to oil exploration in the East Midlands. Two seismic refraction lines and a seismic reflection traverse across the area located a closure near Eakring village. In 1939 Eakring No. 1 Borehole, which was drilled to test the closure, encountered oil near the base of the Westphalian (Lees and Taitt, 1946). This discovery stimulated a search for oil in the East Midlands that has continued to the present day.

Deep borehole data acquired during the second oil exploration campaign (Lees and Taitt, 1946) and in subsequent exploration (Falcon and Kent, 1960) has confirmed the extension of the East Midlands Coalfield as far as the east coast.

The introduction of geophysical well logging to the East Midlands by British Petroleum in 1949 led to the use of gamma and sonic logs in regional correlation of the Westphalian (Howitt and Brunstrom, 1966; Downing and Howitt, 1969). The increased radioactivity of the marine bands (Knowles, 1964), and in particular of the Listeri marine band near the base of Westphalian A, facilitated their widespread recognition by gamma logging in uncored boreholes. Coal seams, which are often difficult to recover by coring, are readily recognised in uncored sequences by their sonic response. A synthesis of coal exploration, hydrocarbon borehole and seismic information in the form of a structural contour map of the Top Hard coal seam, extending over much of the East Midlands area, was compiled by Falcon and Kent (1960). The eastward thinning of the various subdivisions of the Westphalian across the East Midlands (Howitt and Brunstrom, 1966) forms part of a larger-scale variation about the main Westphalian depocentre in the Western Pennines (Calver, 1969; Leeder, 1982) (see Fig. 5.11).

An important geological result of the oil exploration campaigns was the discovery of a major basic igneous province in the concealed late Namurian and early Westphalian A sequence east of Nottingham (Falcon and Kent, 1960). Recent coal exploration in the Vale of Belvoir (Fig.2.5) has helped to define the extent of early Westphalian basic lavas, tuffs and sills (Burgess, 1982; Kirton, 1984).

2.2.4 PERMIAN AND MESOZOIC

The Permian and Mesozoic formations of the Eastern England Shelf (Fig.2.5) were deposited at the western

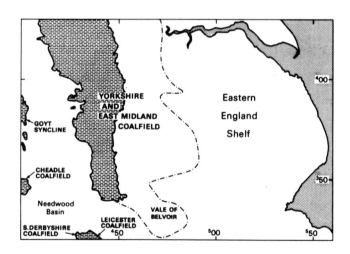

Fig. 2.5 Major coalfields of East Midlands study area. Dashed line encloses approximate area where intensive exploration has provided detailed information on concealed Westphalian strata

margin of a major basin that underlies the southern North Sea. Isopachyte and structural contour maps of the post-Carboniferous sequence are incorporated in a recent atlas of onshore sedimentary basins (Whittaker, 1985). Local stratigraphic details and additional references are provided by Sylvester-Bradley and Ford (1968) and Kent (1980). The stratigraphy and structure of the onshore sequence are well constrained by deep borehole and seismic reflection data, which indicate that the Permian to Cretaceous succession is tilted broadly east–north-eastwards, but is otherwise little deformed. At the east coast the pre-Permian surface lies more than 2 km below sea-level (Fig.2.1). In the south-west part of the East Midlands more than 600 m of Permo-Triassic rocks accumulated in the Needwood Basin, which formed a minor separate area of subsidence (Fig.2.5).

3 METALLOGENY OF PENNINE OREFIELDS

T. B. Colman, T. D. Ford and N. d'A. Laffoley

3.1 Introduction

The North and South Pennine orefields have a long history of research, and the literature on their geology, mineralogy, mining history, ore parageneses and genesis is extensive. Comprehensive data for the North Pennine orefields were given by Dunham (1948) and Dunham and Wilson (1985) for the Alston and Askrigg areas, respectively. No comparable compilation is available for the South Pennine Orefield, but numerous theses—for example, Butcher (1976), Worley (1978), Atkinson (1983) and Mostaghel (1984)—and many other papers, some of which are referred to below, describe most of the parameters that have a bearing on the formation of the mineral deposits. Ineson and Ford (1982) summarised the previous work on the South Pennine Orefield and discussed possible genetic models, and Dunham (1983) provided a comprehensive overview of the three Pennine orefields, which he characterised as a 'fluoritic subtype' of the Mississippi Valley genus of Pb/Zn ore deposits.

The mineral deposits of the three Pennine orefields, those of Alston, Askrigg and the South Pennines, although broadly similar in style of mineralisation, display a number of differences that have a bearing on genetic models for the Pennine mineralisation generally. They are all composed of large numbers of long (up to several km), narrow (<20 m), steeply dipping sulphide/fluorite/carbonate/sulphate vein oreshoots, of limited vertical extent, confined to a small number of massive, competent limestone or sandstone horizons of Asbian, Brigantian and Pendleian age. Semi-concordant flats are also developed in the Alston and South Pennine orefields, iron carbonates predominating in the former orefield above the main sulphide-bearing horizons. The main ore minerals are fluorite and galena, with subsidiary, although locally important, baryte, calcite, sphalerite, witherite, chalcopyrite and quartz. The orefields are presently restricted to relatively small areas of the total Viséan carbonate outcrop, although they may have been more extensive before uplift and erosion took place. The mineralisation has been suggested, on the basis of isotopic dating, to have taken place at intervals between 290 and 170 Ma (Westphalian to Jurassic) in the Alston field and somewhat later in the South Pennines (Permian to Jurassic), where some of the mineralisation postdates the dolomitisation of limestone, which is believed to have taken place in Zechstein times (Dunham, 1952). Structural and other geological evidence, however, suggests that the main mineralising event was during the early Permian (Dunham, 1983).

This chapter provides a summary of the main characteristics of the mineralisation in the three orefields and examines the major differences between them. Genetic models are reviewed with particular reference to sources of fluids, heat, metals and ligands and the transport mechanism involved in the formation of Pennine mineralisation; comparisons are also made with Irish-style deposits. A preferred genetic model is described in Chapter 8 and the potential for Pennine- and Irish-style mineralisation to the east of the South Pennine Orefield is considered in Chapter 9.

3.2 The Pennine Orefields—a summary

The main mineralogical, geochemical and isotopic data from the three orefields are summarised in Table 3.1, which is based on Dunham (1983), Dunham and Wilson (1985), Atkinson (1983) and Mostaghel (1984). ·

The main differences between the three areas are noted below.

3.2.1 FLUORITE

Only limited amounts of fluorite occur in the Askrigg Orefield. It is present in many of the veins (Dunham and Wilson, 1985), but has only rarely been worked on a large scale, except at Greenhow, adjacent to the North Craven Fault. The total output of 0.14 Mt can be compared with that of the Alston region at 2.2 Mt and the South Pennines at 4.3 Mt; the galena outputs are 1 Mt, 4 Mt and 2.5 Mt, respectively. Fluorite is also absent from the Irish carbonate-hosted synsedimentary Pb/Zn sulphide deposits and many other Pb/Zn 'Mississippi Valley type' (MVT) orefields.

The disparity in production of galena per unit area for the Askrigg field may be attributed in part to the main productive horizon, the Main Limestone, having been eroded over much of the area.

Fluorite is not found in quantity in any other British Pb/Zn mining areas, including those of Leadhills, North Wales, Shropshire and the Mendips (see Fig. 1.1).

3.2.2 BARYTE

As in the case of fluorite, only small amounts of baryte have been recovered from the Askrigg Orefield. The total output of 0.038 Mt can be compared with the Alston region at 1.03 Mt (and with 0.88 Mt witherite) and the South Pennines at 0.88 Mt. It is unlikely that this disparity is entirely historical, although the Askrigg Orefield had largely been abandoned by the end of the nineteenth century when demand for fluorite and baryte began to rise. If any substantial deposits had been available, they would have been exploited.

The relative scarcity of fluorite and baryte in the Askrigg field may be due to three factors: (1) the composition and temperature of the mineralising fluids, and thus their capacity to dissolve and transport the Ba^{2+} and F^- ions, (2) the composition of the rocks through which the fluids passed and (3) the nature and composition of the source rocks.

Available fluid inclusion data indicate that the composition of the mineralising fluids in the three areas was similar, so the disparity may be related to differences in the composition of the source rocks.

The other major areas of baryte production in the British Isles have been SW Scotland (Muirshiel and Gasswater) in structures into which Tertiary dolerite dykes were later emplaced, and Shropshire, where baryte veins cut Precambrian Longmyndian sediments. The Irish carbonate-hosted stratiform Ballynoe deposit is also a major producer of baryte (Taylor and Andrew, 1978). In the Grampian Highlands of Scotland the Aberfeldy deposit in the Middle Dalradian Ben Eagach Schist, which was discovered by the British Geological Survey (Coats and others, 1980) is a major stratiform baryte deposit with resources that exceed the total amount of baryte produced in Britain to date.

Table 3.1

Orefield	Alston	Askrigg	South Pennine
Recorded production Mt			
Galena (Estimated total)	4.0	1.0	2.5
Galena (Recorded total)	3.01	0.56	0.78
Fluorite	2.3	0.14	4.3
Baryte	1.03	0.038	0.88
Witherite	0.87	—	—
Sphalerite	0.28	—	0.06
Calcite	—	0.001	0.75
Area sq. km.	1400	950	400
Area km²/Galena Mt	350	950	160
Area km²/Fluorite Mt	636	6785	93
Area km²/Baryte Mt	736	25000	454
Galena/Fluorite	1.8	7.1	0.58
Galena/Baryte	2.1	26.3	2.84
Galena/Sphalerite	14.2	—	41.7
Number of vein oreshoots	685	615	>3200 veins
Host rocks	Asbian to Westphalian	Asbian to Kinderscoutian	Asbian and Brigantian
Main mineralised horizon	Great Lst. (Base of Namurian)	Main Lst. = Great Lst. (Base of Namurian) Grassington Grit (where unconformable on Brigantian)	Monsal Dale Lst. (Brigantian)
% of oreshoots in main horizon	30%	24% (Main Lst.) 31% (inc. Grassington Grit)	High >50%
Sandstone host to mineralisation	Yes	Yes – in south of area	No (Shale cover intact at time of mineralisation)
Igneous rocks	Whin Sill dolerite	No	Extensive basic lavas, tuff and sills contemporaneous with limestone deposition
Age	296 Ma	—	Holkerian to Westphalian 325 – 290 Ma
Alteration of igneous rocks	Removal of Fe, Mg, Na, Ca Addition of K, CO₂, H₂O (Ineson 1968)	—	Removal of Ca Addition of Fe, K, Al, H₂O (Walters 1981)
Granites proved beneath orefield	Yes	Yes	No
Age	394 ± 34 Ma	400 ± 10 Ma	—
Basement	Ordovician slates, Caledonian granite	Ordovician – Silurian slates, Caledonian granite	Ordovician slates and lavas of uncertain age
Structure			
Major structures bounding orefield	Yes N, W, S	Yes N, W, S	No Not in same sense
Major directions	E – W, NW – SE	E – W, NE – SW	NW – SE faults within & on W side of orefield
Vein directions	ENE, E – W	WNW, E – W	ENE, E – W, WNW, NW
Maximum vein length	21 km Slitt vein	29 km North Swaledale mineral belt – linked system of veins	8 km Long Rake
Vein density/sq.km.	2	<1	5 – 10
Stratiform deposits (flats)	Yes especially Great Lst. and including iron ore	Few	Yes
Pipe veins	Few	Few	Yes. South of Long Rake and Castleton area
Mineralisation (MAIN) (subsiduary)	QUARTZ FLUORITE GALENA BARYTE WITHERITE ANKERITE Sphalerite	CALCITE GALENA Fluorite Baryte Quartz	CALCITE FLUORITE GALENA BARYTE Sphalerite Quartz

3.2.3 Contemporaneous Igneous Rocks

The South Pennine area has approximately 30 contemporaneous alkali to tholeiitic basaltic lavas, and associated tuffs, dolerite sills and vents, intercalated at intervals through the Viséan limestone succession in the northern and eastern parts of the orefield (Walters, 1981). Ten of the lavas have an areal extent greater than 10 km², and are up to 100 m thick with volumes in excess of 4 km³.

Four main centres have been identified by Aitkenhead and others (1985) (Map 1): (1) Matlock (Bonsall Moor), (2) Alport (including Ashover), (3) Longstone Edge and (4) Tunstead/Calton Hill.

Two other centres have been tentatively located by

Orefield	Alston	Askrigg	South Pennine
Pattern of mineral zones	Elliptical Fluorite (inner) Baryte (outer) Sharp contacts	Irregular Fluorite (inner) Baryte Calcite (outer) Contacts diffuse and overlapping	North–south Fluorite (east) Baryte (central) Calcite (west) Contacts diffuse and overlapping

Geochemistry
1) Isotope and fluid inclusion geochemistry

Galena PbS			
Sulphide $\delta^{34}S$ ‰	-5.9 to $+15$	-3.8 to -0.6	-23.2 to $+7$
$^{206}Pb/^{204}Pb$	18.15 to 18.33	18.44 to 18.67	18.36 to 18.63
$^{207}Pb/^{204}Pb$	15.48 to 15.61	15.56 to 15.81	15.55 to 15.69
$^{208}Pb/^{204}Pb$	37.96 to 38.28 (Moorbath, 1962) as recalculated by Mitchell and Krouse (1971)	38.25 to 39.03 (Mitchell and Krouse, 1971)	37.88 to 38.55 (Coomer and Ford, 1975)
Model ages Ma from Pb isotopes	62 to 127 (Moorbath, 1962) as recalculated by Mitchell and Krouse (1971)	-89 to 164 (Mitchell and Krouse, 1971)	-72 to 17 (Moorbath, 1962) as recalculated by Mitchell and Krouse (1971)
Baryte BaSO$_4$			
Sulphate $\delta^{34}S$ ‰	$+16.5$ to $+22.1$	n.d	$+4.4$ to $+22.9$
Oxygen $\delta^{18}O$ ‰	$+10$ to $+15.3$	n.d	$+11.9$ to $+25.5$
Fluorite CaF$_2$			
$^{87}Sr/^{86}Sr$	0.7113 – 0.7093 (Shepherd and others, 1982)		0.7101 – 0.7082
Fluid inclusion homogenisation °C	210 to 70	164 to 88	170 to 74
Salinity wt. % NaCl Na/Cl eq.	20 to 27	15 to 25	18 to 25
K/Na	0.145 to 0.08	0.18 to 0.018	0.034 to 0.008
Calcite CaCO$_3$			
Carbon $\delta^{13}C$ ‰			-2.4 to $+3.9$
Oxygen $\delta^{18}O$ ‰			$+19.0$ to $+27.7$

2) Mineral chemistry

Galena PbS			
Ag (mean) ppm	150	40	9
Ag (range) ppm		<20–2000 (Small, 1978)	2–18
Bi ppm			140
Sb (mean) ppm			670
Sb (range) ppm		<300–2800 (Small, 1978)	12–4562
Sphalerite ZnS			
Ag ppm			2–9
Cd ppm	250–400	3800–14 000	500–970
Fe %	0.05–9.85	0.03–0.26	0.04–0.74
Mn ppm			300–400
Fluorite CaF$_2$			
Y ppm	100–1200	12–290	4–38
Total REE ppm	20–800		10–50

3) Age of mineralisation from K and Ar isotopes

	Alston	Askrigg	South Pennine
	$^{39}Ar/^{40}Ar$ 284 ± 40 Ma 230 Ma 170 Ma (Dunham and others, 1968)		K–Ar 270 Ma 235 Ma 180 Ma (Ineson and Mitchell, 1972) 240 ± 5 Ma 160 ± 5 Ma (Walters, 1981)

4) Age of mineralisation from Rb-Sr isochron
206 ± 9 Ma
(Shepherd and others, 1982)

Walters (1981) near Tideswell Moor and on Bradwell Moor.

Two of the centres—Alport and Tunstead—lie on the trace of the 'Bakewell Fault' proposed by Smith and others (1985) and in Chapter 5, whereas the Matlock centre is cut by the Bonsall Fault. Most of the centres occur in the most intensely mineralised and productive areas of the orefield.

However, the extensive lavas and sills around Buxton occur in almost unmineralised Asbian/ Holkerian limestones below the main ore-bearing Brigantian Monsal Dale limestones.

The lavas and tuffs currently act as aquicludes with springs along their upper surfaces; it has been suggested (Dunham, 1983) that they played a similar role during the mineralising events, channelling and ponding mineralising fluids before their release upwards or onwards, when increasing hydrostatic pressure opened faults and joints.

The tuffs are commonly altered to a stiff green clay or 'toadstone'. Millclose Mine 129 Fathom orebody was developed beneath a 'toadstone', although mineralisation occasionally passed up via fractures into the overlying limestone (Traill, 1939).

Unlike the South Pennines, the Alston Orefield has no lavas but was intruded, prior to mineralisation, by a major quartz-dolerite sill, the Whin Sill, and associated dykes. The sill has partly controlled the mineralisation and has been extensively metasomatised, and in places demagnetised, by the mineralising fluids (Bateson and others, 1984), losing large amounts of Fe, Mg and Ca. These elements were redeposited above the main sulphide horizons as extensive ankerite and limonite flats, which have been worked for iron ore. The Whin Sill cuts strata that range from the basal Brigantian to the Westphalian. It is centred on Teesdale and extends over more than 5000 km^2 with a maximum thickness of 75 m and has been dated at 296 Ma (Stephanian according to the Odin (1982) timescale). The Askrigg Orefield has no contemporaneous igneous rocks.

The spatial association of fluorite and Carboniferous volcanic rocks in the South Pennine Orefield may be significant. Although the Alston Block has produced considerable quantities of fluorite, it has no known Carboniferous volcanism. The igneous rocks could have simply influenced fluid flow or they may have been sources of such elements as F, and perhaps Ba, in the fluids. They may also have been local centres of increased heat flow and permeability along the lineaments into which they were emplaced.

Other areas of Britain, such as Northumberland and the south of Scotland, which have extensive Carboniferous volcanism, have been reported to be almost unmineralised (Dunham, 1983). The North Wales Viséan limestones are well mineralised with Pb/Zn veins, but have no associated volcanics and contain insignificant quantities of fluorite and baryte. The source of the F has been ascribed to muscovite in the granite (Alston) or Namurian sandstone (South Pennines) with Pb and Ba also being derived from feldspars. Russell and Smith (1979) argued for a contemporaneous juvenile magmatic source to supply F to deep formation waters. Fluorite and volcanics rarely occur in Mississippi-type deposits, but they appear to show a spatial association in some areas, e.g. the Hicks Dome in the Illinois–Kentucky orefield in the USA, where the most intense fluorite mineralisation (which is rich in rare earth elements (Hall and Heyl, 1968)) with Ag-rich galena, is centred over the cryptovolcanic dome with abundant fluorite, to depths in excess of 600 m, in breccia dykes and pipes (Heyl, 1973; Trace, 1973). The Lexington Dome in the Central Kentucky orefield has a central fluorite zone over the dome, surrounded by baryte- and galena-dominant areas (Heyl, 1973).

3.2.4 GRANITES

Major Caledonian granite intrusions form the cores of the Alston and Askrigg Blocks beneath the North Pennine orefields. The emplacement of the intrusions, at 394 ± 34 Ma in the case of the Alston granite and 400 ± 10 Ma for the Wensleydale granite (Dunham, 1974), predated deposition of the Viséan limestones by 60 Ma and the mineralisation by at least 100 Ma. The granites have, nevertheless, been assigned a localising role in the formation of some ore deposits (Brown and others, 1980; Dunham and Wilson, 1985). It is considered unlikely that granite occurs beneath the South Pennine Orefield, although granites are postulated on the basis of interpretations of gravity data (Chapter 4) to underlie the Market Weighton, Newark and The Wash areas. The nearest of these to the South Pennine Orefield is Newark, 50 km east of Matlock.

The two northern granites are both relatively large volume, high heat production granites (Chapter 7), but the geochemical evidence suggests that the Alston Granite was fractured and hydrothermally altered during emplacement, whereas the Wensleydale Granite remained largely unaltered (Chapter 7). Brown and others (1987) and Dunham (1983) argued that radiothermal granites provided local heat sources to drive hydrothermal systems in the regime of generally increased heat flow and tectonism of the Lower Permian. As outlined above, the Weardale Granite has been suggested to have been the source of F in the Alston Block and the absence of large quantities of fluorite in the Askrigg field may reflect the lack of fractures and alteration in the underlying Wensleydale Granite (see Chapter 8). Hence extraction of substantial amounts of F from the granite has not occurred. The existing fluorite deposits occur mainly around Greenhow, adjacent to the Craven faults, and around Askrigg, in Swaledale, associated with major E–W structures parallel to the northern margin of the granite. An alternative source of F must be sought, at least for the South Pennine Orefield, and the possibility of derivation from shales is discussed below.

3.2.5 MINERAL ZONATION

Alston

A clear concentric mineral zonation occurs in the Alston field, a central fluorite zone, which contains no baryte, being surrounded by a baryte zone that lacks fluorite (Dunham, 1948). The zones are centred above the Alston granite (Fig. 3.1). Within the central fluorite zone the Great Sulphur Vein appears to have been a major locus of mineralising fluids, the highest fluid inclusion temperatures being in the quartz–iron sulphide–chalcopyrite asemblage (210°C). A second locus occurs in Weardale (Solomon and others, 1971). These loci correspond with cupolas in the buried Weardale Granite and even minor outlying cupolas underlie mineralised areas. The baryte (and witherite) mineralisation extends eastwards into Westphalian rocks of the Durham coalfield.

Askrigg

Zonation of the mineralisation and its spatial relationship with the gravity anomaly, representing the buried granite, are less clear than in the case of the Alston Block (Dunham and Wilson, 1985). Fluorite, generally associated with baryte, occurs in several distinct areas above and marginal to the granite (Fig. 3.1) surrounded by baryte and then calcite zones. The mineralisation occurs in two main centres—the Greenhow mining district to the south and Swaledale to the north. The former is more likely to associated with the Craven Fault zone, although the elongated Swaledale zone lies over the northern margin of the granite. The overall distribution is sporadic and uneven.

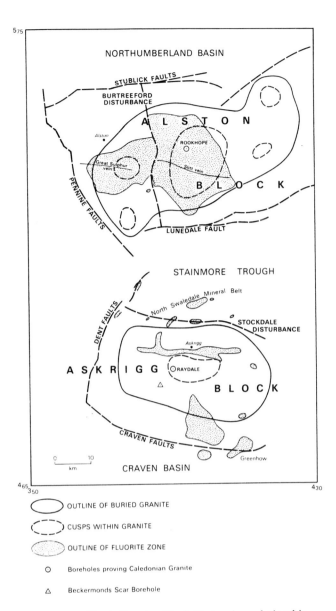

○ OUTLINE OF BURIED GRANITE

⊂⊃ CUSPS WITHIN GRANITE

▱ OUTLINE OF FLUORITE ZONE

○ Boreholes proving Caledonian Granite

△ Beckermonds Scar Borehole

Fig. 3.1 North Pennine Orefield showing relationship between buried granites and mineralisation

South Pennine

The mineral zonation of the South Pennine Orefield is quite different from that of the northern fields.

The mineral zones are approximately N–S with the fluorite zone in the east, baryte in the centre and calcite in the west of the orefield. The distinctions between the zones are poorly defined and gradational, showing considerable overlap in many places. Since the recognition of the zonation by Dunham (1952) and Mueller (1954) mining of lower-grade fluorite deposits (down to 10–15% CaF_2) has moved the western boundary of the fluorite 'zone' further west, and an early, almost ubiquitous, calcite phase also complicates the interpretation of the zonation. The broad outlines of the N–S zonation are consistent over the whole orefield, however, although not for individual areas or veins. The zonation does not appear to reflect temperature gradients, declining regionally from the east to west during mineralisation. Atkinson (1983) and Rogers (1977) reported a homogenisation temperature mean of 92°C for fluorite in Viséan limestone in boreholes in the Eakring area, 30 km east of the orefield. This is at the lower end of the South Pennine homogenisation temperatures (Atkinson, 1983).

3.2.6 Caprocks

During the mineralising events the Askrigg and South Pennine Orefields had a thick cover of Namurian shales, which appears to have precluded mineralisation of the overlying thick competent Namurian sandstones. This cover has been removed in the south part of the Askrigg Orefield by intra-Pendleian erosion that has cut out successively lower beds to the Main Limestone at the base of the Namurian, allowing ore shoots to develop in the Grassington Grit above the unconformity (Dunham and Wilson, 1985).

In the Alston Orefield the shales between the ore-bearing limestones and sandstones were thin enough to allow ore fluids to pass through and no caprocks equivalent to those of the Askrigg and South Pennine fields are present. The intensity of the mineralisation declines rapidly in horizons above the Great Limestone, although it persists in one mine through the full sequence of 'Millstone Grit'. Baryte and witherite veins have been worked in the Westphalian near Durham (Dunham, 1948).

3.2.7 Structure

The mineralisation is strongly controlled structurally, the dominant form of orebody being semi-vertical veins or 'rakes'. These are up to 20 m wide and extend over several km in length in a small number of preferred orientations—mainly E–W, NW–SE and ENE–WSW (Map 1.) Dunham (1983) observed that the northern orefields represent gentle domes if the effects of the Mesozoic bounding faults are removed, and that the doming was pre-Permian, as the Permian rests with unconformity on the underlying Westphalian. The South Pennine Orefield was also affected by Variscan compression, which produced a series of mainly east–west folds over most of the orefield. The vein density of the South Pennine Orefield is much higher than in the North Pennines (Table 3.1) and Dunham (1983) suggested that this may be due to the increased folding. Large numbers of 'scrins' or minor veins occur, generally infilling joints, especially around Wirksworth in the south east and Flagg in the central part of the orefield. Weaver (1974) described the joint pattern for the south part of the area. He suggested that the mineralisation exploited the dominant joint sets, which formed, in response to the main Variscan E–W compression, as a set of conjugate shears. Preliminary fieldwork in the north part of the orefield indicates that some of the pipe deposits, such as Hubbadale (Map 1), have a parallel orientation to the major NW joint direction, but few of the major veins have similar orientation to the joints. The deep structure of the orefield is described in Chapter 5. All of the mapped veins and faults of the South Pennine orefield have been digitised from published 1:10 560 and 1:10 000 scale geological maps, together with additional information from published plans of old mining areas, to prepare a database of more than 4300 fractures. This database has been studied by use of the I^2S image analysis system (Appendix 1) in relation to the deep structure of the orefield.

3.2.8 Age of Mineralisation

The geological evidence for the timing of mineralisation is sparse. The Whin Sill (295 Ma) is cut and altered by the mineral veins (Dunham, 1948; Bateson and others, 1984), providing evidence that the Alston orefield was mineralised after the intrusion of the sill. Dunham and others (1968) have dated the clay minerals, produced during alteration of the Whin Sill by mineralising fluids, by the ^{40}Ar/^{39}Ar method. Three periods of hydrothermal activity were found (Table 3.1). Mitchell and Krouse (1971) criticised the use of very fine-grained clays and suggested that the 'ages' may represent differential Ar loss. The general difficulties associated with the use of isotopic age dating on low-temperature deposits, in which isotopic equilibrium may have been only poorly established, has been discussed by Sangster (1983). Mitchell and Krouse (1971) considered an age of about 280 Ma (Lower Permian) to be the most likely time of mineralisation and this age was also favoured by Dunham and others (1968). There is no geological evidence for mineralisation after the deposition of the Permian Marl Slate. A late Triassic age of 206 ± 9 Ma has been suggested by Shepherd and others (1982) on the basis of a fluid inclusion Rb–Sr isochron from the Great Sulphur vein in the west of the orefield (Fig. 3.1). This is a much later age than any previously described and places the mineralisation well after the Variscan orogeny.

The mineralisation of the South Pennine Orefield occurred after the dolomitisation of Asbian and Brigantian limestones in the south of the area, believed to have taken place in Zechstein times (Dunham, 1952). Mitchell and Krouse (1971) recalculated the Pb isotope data of Moorbath (1962), producing a set of anomalous J-type 'ages', including negative ages. This was later criticised by Coomer and Ford (1975) on the basis of the simple assumptions used and that the assumed homogeneous mantle source of Pb isotopes is not a reliable indicator of a closed system of radiogenic Pb production. Ineson and Mitchell (1972) used K/Ar dating on hydrothermally altered lavas in the area. Their data show several peaks (Table 3.1), which extend from late Carboniferous to Jurassic times. Similar work has been carried out by Walters (1981), who reported two main mineralising events at 240 and 160 Ma. These 'dates' may suffer from the problems mentioned above by Mitchell and Krouse (1971) and cannot be regarded as definitive. New evidence on the timing of mineralisation is presented in Chapter 8.

The timing of the mineralisation of the Askrigg orefield is the most poorly constrained on the basis of either geological or isotopic evidence. Dunham and Wilson (1985) suggested the Lower Permian as the most likely date for the main mineralisation on the assumption that the Wensleydale Granite was reheated just prior to the Whin Sill intrusion and provided the heat engine to drive the hydrothermal circulation. Mitchell and Krouse (1971) calculated model lead ages of 164 to -89 Ma, implying a significant contribution of radiogenic Pb.

3.2.9 GENETIC MODELS

A number of genetic models have been proposed for the Pennine orefields and were reviewed by Dunham (1983). Any genetic model requires the following information: (a) fluid source, (b) a source of metals and ligands such as fluoride, (c) an energy source to drive the hydrothermal system, (d) adequately high permeability to permit fluid flow and (e) a suitable depositional environment.

Fluid sources

The ore-forming fluids can be derived from a variety of sources: (a) juvenile magmatic fluids or fluids that have equilibrated with magmas, (b) metamorphic water, (c) connate formation water and (d) meteoric water.

Magmatic or juvenile fluids rising from buried Hercynian granites comparable with those of SW England were invoked originally for the North Pennine Orefield, but the discovery of the Caledonian age of the underlying granites made this unlikely. Russell and Smythe (1978) have argued in favour of juvenile fluids derived from alkaline magmas emplaced at depth beneath the Pennines contemporaneously with such volcanism in the Oslo Graben. The generally positive δ^{34}S (sulphide) values, which show a wide range (Solomon and others, 1971), are, however, inconsistent with magmatic fluids.

Metamorphic fluids have not been considered as a source as the Lower Palaeozoic rocks that underlie most of the area were folded and cleaved during the Caledonian orogeny, which predates the Pennine mineralisation by at least 100 Ma. Samson and Russell (1983) dismissed a metamorphic source for the Irish Silvermines deposit on similar grounds; maximum fluid production from the underlying Lower Palaeozoic rocks would have been 40 Ma earlier than the mineralisation. However, Phillips (1986) has argued for the derivation of fluids from lithified Lower Palaeozoic shales of only 5% porosity to form the Pb/Zn veins of the Central Wales mining field.

Formation fluids from a variety of sources have also been suggested as potential mineralisers. Descending evaporite brines from the thick halite and anhydrite sequences in the overlying Permian and Triassic formations were proposed by Davidson (1966), but in the North Pennines it is unlikely that they would have been able to move down through the intervening Westphalian and Namurian formations to the Viséan limestones. In the South Pennine area the limestone is believed to have been partly exposed and dolomitised during the Permian, so this source may have been available in the south of the area, although the mineral veins postdate the dolomitisation and, hence, a multi-stage fluid movement would be necessary. Such a fluid could not be involved in the formation of the mineral deposits in the north, which were formed under Namurian shale cover. In the South Pennine Orefield Worley and Ford (1977) argued for mineralisation during the Permo-Triassic by expulsion of metal-rich Viséan carbonate formation waters from the North Sea Basin, which moved eastwards and upwards to meet sulphate brines in the uplifted, near-surface South Pennine Viséan limestones; this model is shown by Fig. 3.2. In the North Pennine Orefield Solomon and others (1971) attributed the Alston mineralisation to the mixing of cold intra-Carboniferous connate brines with hot chloride brines of connate origin, but which have been differentiated by filtration during movement through the underlying Lower Palaeozoic sediments and heated by the Devonian granite. Fig. 3.3 shows their model for the genesis of the Alston mineralisation.

The presence of thick—up to 3.5 km—sequences of Viséan, Namurian and Westphalian shales in basins adjacent to the positive blocks with Viséan platform carbonate cover has analogies with other carbonate-hosted Pb/Zn provinces such as Pine Point and the Mississippi Valley deposit type area. In these provinces it has been suggested that mineralisation was produced by the expulsion of formation waters during early compaction and

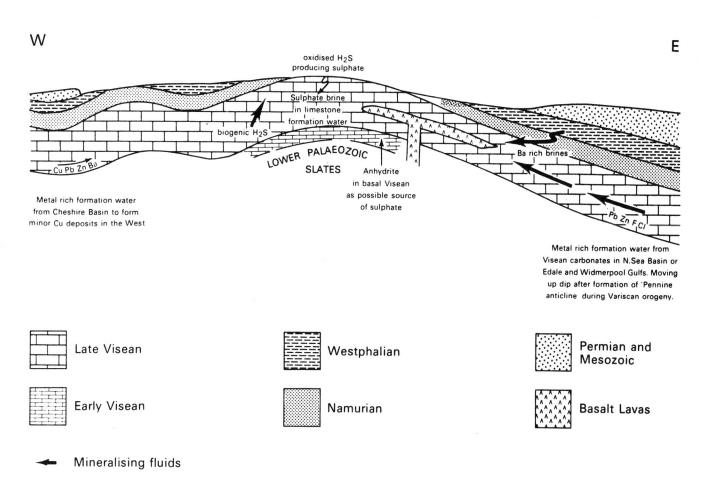

Fig. 3.2 South Pennine Orefield: model as proposed by Worley and Ford (1977)

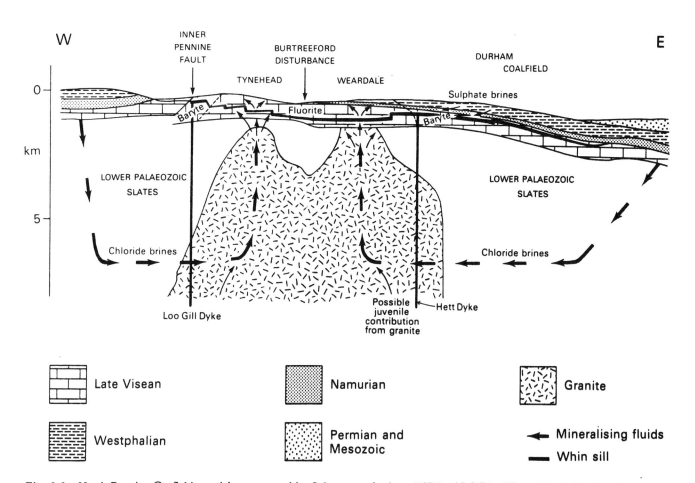

Fig. 3.3 North Pennine Orefield: model as proposed by Solomon and others (1971) with Whin Sill modified after Francis (1982) in the manner of Dunham (1987)

diagenesis of shales in adjacent basins (Beales and Jackson, 1966). Such a model is difficult to apply directly if the proposed timing of the episodic Pennine mineralisation, from late Carboniferous to early Mesozoic times, is accepted. In the case of the Irish-style deposits Russell (1978) argued for deep hydrothermal circulation using high salinity Lower Carboniferous sea water to leach metals from the underlying 15 km thick Lower Palaeozoic sequence. Such a model may explain the mineralisation at the base of the Tournaisian in the Irish deposits when only thin Carboniferous sediments had been deposited. Samson and Russell (1983) subsequently discussed the possibility that a shallow magma source could have provided heat to expel formation waters from the thin Carboniferous sequence, but considered it unlikely. Brown and Williams (1985) quoted Jacob and others (1985), who considered the depth to the Precambrian basement to be only 4 km, the Lower Palaeozoic crust consisting mainly of massive volcanic rocks, the permeability of which was probably too low in Viséan times to permit large-scale fluid movement. Brown and Williams considered that Carboniferous basin dewatering, possibly assisted by a porous basal sandstone layer, was the source of the mineralising fluids. Irish-style mineralisation is considered further in Chapter 8.

Evidence that the vitrinite reflectance values of coal seams were affected by the intrusion of the Whin Sill (Creaney, 1980) suggests that the Carboniferous sediments of the Pennine area had been compacted by the time the area was affected by earth movements associated with the Variscan orogeny. Dewatering must therefore have been advanced by the end of the Carboniferous. The proposed growth faults of Smith and others (1985) in the South Pennines would provide a mechanism for the transfer of large amounts of fluids from the Edale and Widmerpool Gulfs and the Gainsborough Trough and it may be that some mineralisation was taking place at an early stage in the Carboniferous. The implications of this in relation to Irish-style mineralisation are discussed further in Chapter 8. Mineralisation in the Lower Permian or in the Mesozoic, as suggested by Ineson and Mitchell (1972), requires that the expelled diagenetic waters remained in the basin sediments until their movement up dip to the orefield. Movement of fluids in the Mesozoic could have been triggered by the partial removal of the impermeable shale caprock—this would also explain the pre-mineralisation dolomitisation of the area south of Long Rake in the centre of the South Pennine Orefield. This would imply expulsion of the ore fluids at the surface.

Mixtures of the various different fluid sources have also been considered as the potential mineralising agents. In the Durham coalfield baryte and witherite veins occur in the Westphalian Coal Measures. Dunham (1983) discussed the possibility of interaction between rising juvenile metal chloride and sulphate brines with barium chloride brines such as those which still exist in the coalfield. Schofield (1982) suggested that decomposition of the underlying volcanics provided Fe and Mg for the dolomitisation of the Woo Dale Limestones. The amounts of fluids are unlikely to have been adequate for mineralisation, however.

Descending meteoric water is unlikely to have played an important role (Dunham, 1983), mineralisation occurring under impermeable caprocks in the northern part of the South Pennine Orefield and in the Askrigg Block. However, Edmunds (1971) and Sheppard and Langley (1984) have discussed the origin of modern Ba-bearing brines in groundwater in NE England; this is developed further in Chapter 8.

Metal sources

The metals and anions could be derived from a variety of sources. The dominance of fluorite and galena and relative absence of sphalerite have been explained either by the presence of alkaline volcanic rocks (Russell and Smythe, 1978) or ascribed to the conditions of deposition in the host rocks (Atkinson, 1983). Dunham (1983) listed the contents of the principal elements associated with mineralisation in a variety of likely source rocks in the Northern Pennines. Additional data for a range of potential source rocks, including igneous rocks and shales, are given in Chapters 6 and 7. Pb isotope ratios have been used to assign a source to the lead in metal deposits (Gulson, 1986). Coomer and Ford (1975) have shown that the South Pennine Pb isotopes are extremely homogeneous with no significant local or orefield variations. They suggested that the Pb was well mixed before precipitation. Vaasjoki and Gulson (1986), in discussing Pb isotopes from some Australian carbonate-hosted Pb/Zn deposits, noted that the deeply circulating fluid model of Russell (1978) would, with progressive increase in depth of circulation, produce varying Pb isotopic compositions as the fluids sampled different lithologies. They argued that deeply circulating brines can be considered only for carbonate-hosted deposits with heterogeneous Pb isotope ratios and, thus, this type of model would not appear to be applicable to the South Pennines.

Energy source

Heat to raise the fluids to the required temperatures of 100 to 200°C could be derived at depth in shale basins, even under a normal geothermal gradient. This could be enhanced locally by contemporaneous volcanism or by radiothermal granites. Additional heat sources would be particularly important if the mineralisation was early rather than late because of the absence of a thick pile of sediments to provide a fluid-driving mechanism through dewatering. To achieve temperatures of 200°C the fluids would have to have travelled to 6 km depth in an average heat flow area. However, this could be attained at much shallower levels under a higher than normal thermal regime, and for the South Pennine area the highest fluid inclusion homogenisation temperatures are only 170°C. The existence of the easterly migrating basic volcanic province in the South Pennine area from Viséan to Westphalian times may also be significant, not from the point of direct heat input from magma but by providing sources of high heat flow adjacent to Carboniferous basins. Other sources of energy include gravity (Garven and Freeze, 1985; Bethke, 1986) and seismicity. Seismic pumping of fluids is important in the transport of fluids in the formation of several types of vein deposits (Phillips, 1972).

Transport

In the case of the Pennines various transport mechanisms have been envisaged, including compaction under increasing sediment load with fluid movement upwards and outwards to the basin edges assisted by faulting (Worley and Ford, 1977). Convective flow over zones of high heat flow has been proposed, in the case of the Irish deposits , which formed early in the Tournaisian. The requirement for deep leaching of metals from Lower Palaeozoic rocks has been related to zones of high permeablity in the extensional tectonic regime prevailing at the time (Boyce and others, 1983), the return flow being via rapidly subsiding growth faults with exhalation into highly saline brine pools on the Tournaisian sea-floor. However, Brown and Williams (1985) favoured a wholly Dinantian source for the Irish mineralisation, with dewatering of the Dinantian sedimentary pile, which may be substantially thicker than was envisaged by earlier workers (Chapter 8). The conditions for either model could be met in the vicinity of the South Pennine Orefield at the margins of sedimentary basins such as those of the Edale or Widmerpool Gulfs.

Depositional environment

The depositional environment of the Pennine mineralisation was mainly competent limestones and sandstones and this has been attributed to their ability to maintain open fracture systems, coupled with the reactive nature of the limestone. Sporadic mineralisation occurs in rocks of up to Triassic age (Ineson and Ford, 1982), but generally has no economic significance. Deposition may be triggered by reduction in pressure or temperature, mixing of fluids of different composition, the presence of reducing biogenic sulphur or change in pH or Eh. The common association of liquid or solid hydrocarbons in many of the deposits may indicate a close connection between oil formation and metal ores.

In the South Pennine area the presence of deposits in the uppermost limestone suggests that one of the controls of mineralisation was ponding beneath an impermeable shale cap.

The Pennine deposits appear to be more dispersed in space and time than the Irish ones, which are restricted to a particular time period and structural setting.

Chemistry

In the North and South Pennine Orefields rare earth element (REE) chemistry, fluid inclusion and stable isotope studies have been used to argue in favour of formation waters, derived from shale-dominated sedimentary basins, as the most likely source of the mineralising fluids (Shepherd and others, 1982; Atkinson, 1983). The total REE content of the South Pennine fluorites (Atkinson, 1983) is low compared to those of the North Pennines (Shepherd and others, 1982, and Fig. 3.4. It has been suggested by Atkinson (1983) that the ore fluids in the South Pennines may have originally had the same REE content as those in the north, but that prolonged contact with ion-exchange media, such as clays, has led to adsorption of most of the REE. The REE patterns of the two orefields are substantially different, however, with a positive Eu anomaly in the North Pennine and slight negative Eu anomaly (Eu/Eu* = 0.70) in the South Pennines. This disparity could be due to the derivation of REE from a feldspar (?granite) source in the North Pennines and a shale source in the South Pennines. The REE patterns of feldspar and whole rock samples of the Weardale granite (Shepherd and others, 1982) and Fig. 3.4 demonstrate similar patterns to those observed in the fluorites, whereas the South Pennine fluorite REE pattern is similar to that of shales (see Chapter 6).

The S and O isotope ranges for the three orefields are listed in Table 3.1, together with the homogenisation temperatures for fluid inclusions. The Alston field shows two distinct areas of higher temperatures centred over cusps in the granite. The South Pennine field shows little systematic variation in a generally lower-temperature

regime and Atkinson (1983) envisaged a quiet saturation of the entire orefield. The Irish deposits generally tend to have higher homogenisation temperatures of up to 250°C (Samson and Russell, 1983).

The Pb isotope ratios of the Pennine orefields show some variation; those of the South Pennines and Askrigg give some negative 'ages' indicating J-type, radiogenically enriched Pb. As a whole the ratios are less variable than the range for Mississippi Valley deposits in the type region. The Irish deposits give older 'ages' (Boast and others, 1981). At Tynagh the age of 348 ± 22 Ma is identical with that of the enclosing sediments, but Silvermines and Navan show increasing amounts of radiogenic Pb, leading Boast and others (1981) to invoke selective leaching of radiogenic Pb from the source rocks for these deposits.

The trace-element content of sulphide minerals of the Pennine orefields has been described by Ixer and Townley (1979) and Vaughan and Ixer (1980). The Askrigg and South Pennine Orefields have a simple and uniform sulphide mineralogy, but the Alston Orefield is considered to have a more diverse fluid composition with systematic changes in fluid composition and Vaughan

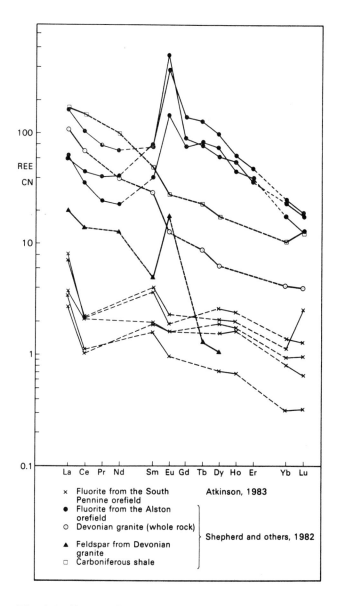

Fig. 3.4 Rare earth element content of fluorites from North and South Pennine Orefields and of selected potential sources of F

and Ixer (1980) considered that it has at least a partly magmatic origin, with higher formational temperatures of up to 250°C. This was later criticised by Dunham (1983), who suggested that the sphalerite – pyrrhotite assemblage in baryte, used for the temperature determination, did not represent an equilibrium assemblage or that it might be related to the Whin Sill dolerite wallrock. The widespread occurrence of Ni-bearing sulphides, including bravoite, was noted in both the North and South Pennine Orefields and it was suggested that this could be a common feature of several MVT orefields.

The sulphide $\delta^{34}S$ ratios for the Alston field are generally positive. The values are inconsistent with a local biogenic S source, resulting from reduction of connate sulphate, since a greater degree of local variation would be expected. Moreover, a Carboniferous pyrite source would give negative sulphide $\delta^{34}S$ values (Solomon and others, 1971). In the case of the Irish deposits the sulphide $\delta^{34}S$ ratios for the Navan deposit show a range from -22 to -10 in 'early' sulphides to $+8$ to $+13$ in 'later' formed minerals (Andrew and Ashton, 1985). The early sulphides are thought to have formed from biogenically reduced sea-water sulphate, whereas the values in the later sulphides are consistent with derivation from the local Lower Palaeozoic basement.

The sulphate $\delta^{34}S$ and $\delta^{18}O$ ratios from baryte in the Alston and South Pennine Orefields are generally uniform, being comparable with the values obtained from the Irish deposits, and the sulphate source is attributed to Carboniferous connate water that remained within the host rocks. This sulphate source is required in both the Solomon and others (1971) and Worley and Ford (1977) models for the two orefields.

The wider variations in sulphide and sulphate isotopic ratios in the South Pennine Orefield led Robinson and Ineson (1979) to argue for a mixture of sources, the sulphide S being provided by biogenic and connate sources and the sulphate S being derived from the mixing of fresh water and connate sea-water sources. Their proposed model requires Carboniferous formation water, from the Edale and Widmerpool Gulfs, to mix with sour petroleum fluids in the limestones and Ba-rich brines descending from the Westphalian, as shown in the model of Worley and Ford (1977) (Fig. 3.2). This conflicts with the model of Atkinson (1983), who argued against fluid mixing as a depositional mechanism since galena is the dominant sulphide mineral and he suggested that the ore fluid probably had a Zn/Pb ratio >1. In order to explain the dominance of galena Atkinson argued that the fluids were slightly acid (pH <6) and deposition was due to changes in pH during interaction with the limestone, galena having a lower solubility than sphalerite in such conditions. This explanation ignores the problem of the disposal of a Zn-rich fluid that has left no trace of substantial Zn deposits elsewhere in the orefield, including those areas which were capped by Namurian shale. This is discussed further in Chapter 8.

3.3 Evidence of other carbonate-hosted base-metal mineralisation in or near the study region

3.3.1 Cu Mineralisation

Cu – Pb – Zn – calcite – baryte – fluorite mineralisation occurs in the Arundian-Asbian, Ecton Limestone, at Ecton in the Manifold valley in the south-west of the South Pennine Orefield (Map 1). The deposits have mainly formed in near-vertical pipe-like bodies in contrast to the sub-horizontal 'manto' or pipe deposits in the east of the orefield. The mineralisation, mainly calcite and chalcopyrite with chalcocite and malachite, has surrounded partly dissolved blocks of limestone; formation of cavities may have occurred before or during mineralisation by rising hydrothermal fluids channelled into a tight anticlinal axis (Critchley, 1979). Similar mineralisation occurs 5 km to the west at Mixon (Aitkenhead and others, 1985). Several other Cu deposits occur in Viséan limestone, to the south and south-west of the South Pennine Orefield, especially at the unconformity with the overlying Permo-Trias (King, 1968). At Snelston, where the limestone is unconformably overlain by Triassic Mercia Mudstone Group (Dewey and Eastwood, 1925), malachite and minor chalcopyrite, with oxidised galena, occurred in sandstones within the Trias, and galena in thin veins and clay-filled cavities in the limestone. The deposits may have been formed after deposition of the Triassic sandstone, but they may also represent weathering and erosion of pre-existing Viséan hosted deposits during the Triassic. The origin of the Cu is unknown. Mineralisation in rocks of Triassic age is widespread in the English Midlands (King, 1968). It consists mainly of baryte-cemented sandstone with local, often fault-related, concentrations of Cu, Pb, Co and V, such as those of Alderley Edge (Carlon, 1979). The mineralisation usually occurs in the upper part of the Sherwood Sandstone Group and Holmes and others (1983) have suggested it formed by diagenetic processes in these continental red beds. The Caldon Low borehole (Chapter 5) intersected thin veinlets of chalcopyrite in grey beds in the ?Devonian Redhouse Sandstones Formation (Chisholm and others, in press). It is possible that the source of the Cu in Ecton and other Viséan limestone-hosted Cu deposits was the the Carboniferous basin to the south-west of the South Pennine Orefield, which could explain the restriction of Cu mineralisation to that area.

3.3.2 Pb–Zn–Fluorite Mineralisation

Boreholes in the Eakring area, 50 km east of the South Pennine Orefield, intersected mineralised Brigantian limestone close to the contact with the overlying Namurian. The mineralisation was described as intense by Lees and Taitt (1946) with fissures and cavities lined with crystals of fluorite, baryte and calcite. The discovery of this mineralisation has been important in encouraging the proposition that buried Pennine-style mineralisation may exist to the east of the exposed orefield. Ineson and Ford (1982) have described the occurrence of traces of galena, fluorite or pyrite in numerous boreholes in the East Midlands, though the number of mineralised Dinantian boreholes is small.

Mineralisation occurs in Norfolk, on the north side of St Georges Land, where boreholes have intersected base-metal (marcasite – galena – sphalerite – calcite) mineralisation infilling karst-like vugs, and forming thin veins, in Viséan limestone, immediately beneath an unconformity with overlying Westphalian to Permian rocks. The general geology of the area has been described by Allsop (1984) and preliminary investigations of the mineralisation in the Somerton borehole (see Fig. 6.11) indicate that it is low Ag galena with pyrite – marcasite in a calcite/dolomite matrix. There has been no post-Variscan emergence to form secondary minerals, as in the formerly important orefields of the Belgian Brabant Massif, where

Pb and Zn have been worked in veins and strata-bound deposits in Devonian (Frasnian) and Upper Dinantian (Asbian and Brigantian) carbonates underlying a major pre-Namurian unconformity. The metals have been suggested by Balcon (1981) to have been derived from continental leaching and deposited in evaporite basins in the Holkerian to give stratiform metal enrichment. Anhydrite and sedimentary breccias from solution of preexisting evaporites are known from this horizon. The metals were subsequently remobilised and emplaced in fault-controlled veins, replacements and karst-related cave systems during the Variscan orogeny by the movement of connate brines. A subsequent Tertiary weathering period caused oxidation of the sulphides to form important smithsonite ($ZnCO_3$) deposits. Pb isotopes show similar ratios to the Pennine mineralisation with little variation between the synsedimentary Devonian and vein-style Devonian and Viséan mineralisation (Cauet and others, 1982) and the δS^{34} sulphate sulphur ratios are consistent with a sea-water sulphate source (Dejonghe and others, 1982). A common origin was suggested for both types of deposit (Cauet and others, 1982).

3.3.3 IRISH-STYLE MINERALISATION

Irish-style mineral deposits have not been reported in the study region, or in England generally. In Ireland such deposits formed early in the Carboniferous, before the establishment of deep basins with a thick infill. They are largely stratiform, though with a significant epigenetic component from feeder veins (Andrew and Ashton, 1985). The orebodies are localised by major growth faults, which underwent periodic and rapid movement, and occur in Waulsortian-type carbonate mudbanks straddling faults. They have thick underlying Lower Palaeozoic sequences of 4 to 15 km (depending on whether the thicknesses proposed by Russell (1978) or Jacob and others (1985) are accepted) and were formed in a higher than normal heat flow regime, associated with penecontemporaneous volcanism. The Irish deposits are also Zn-dominant with galena, some baryte and minor Cu, contain no significant fluorite and formed early in the development of contemporaneous basins. In contrast, the Pennine deposits were formed much later and, although they are structurally controlled, are associated with wrench rather than normal faulting. They are predominently vein deposits with dominant galena and/or fluorite and baryte; sphalerite is a minor component. The presence of thick (3.5 km) basinal deposits in the Gainsborough Trough and Edale and Widmerpool Gulfs associated with proposed early Viséan growth faults (Chapter 5) may prove to be the locus of as yet unknown mineralisation analogous to that in Ireland. This is discussed further in Chapter 8.

4 REGIONAL GEOPHYSICS

J. D. Cornwell and A. S. D. Walker

4.1 Introduction

In the East Midlands study region a large proportion of the area is concealed by barren Mesozoic cover (Chapter 2) and an understanding of the regional geology must rely on the interpretation of geophysical data to supplement evidence from scattered deep boreholes.

Regional geophysical data for the region comprise Bouguer anomaly and aeromagnetic maps that form part of the published 1:250 000 scale series for the UK (Fig. 4.1). These data provide information on the distribution of major bodies of anomalous magnetisation or density and can be used to define broad geological structures, often below the depths effectively investigated by seismic reflection surveys.

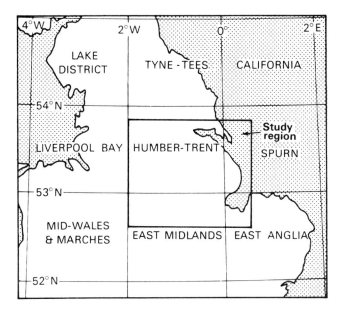

Fig. 4.1 Key to BGS 1:250 000 scale geological and geophysical (Bouguer anomaly and aeromagnetic) maps

The regional geophysical data were converted into digital format in order to facilitate interpretation and to combine them with other data sets, particularly by the use of the image analysing system (IAS). The IAS was used extensively in the processing and qualitative interpretation of the regional geophysical data (mainly gravity and magnetic) (Appendix 1). Some processing was also carried out on a main frame computer and the results were transferred to the IAS, notably the calculation of the reduced to pole magnetics and stripped gravity.

Quantitative interpretation of the geophysical data in terms of geological models was carried out mainly using a PERQ workstation (ICL) at BGS, Keyworth.

Measurements of rock physical properties were performed using standard techniques in the laboratories of the Engineering Geology Research Group, Keyworth.

Interpretation of the Bouguer anomaly and aeromagnetic data has been carried out previously in parts of the study region, but there has been no systematic study of the entire data sets. One of the most important features of the gravity data for the purpose of the present study is the significance of the larger Bouguer anomaly lows. A particularly well-defined low at the northern margin of the area has been interpreted by Bott and others (1978) as a concealed granite (probably of Caledonian age) on the basis of its form and its association with a group of magnetic anomalies that appear to be due to magnetic rocks draped over the concealed intrusion. The postulated granite lies near, but not directly beneath, the Market Weighton structure—an axis of greatly reduced Jurassic sedimentation. Rollin (1978, 1982) produced a similar interpretation for this gravity feature and also suggested that a pronounced gravity low in the Newark area (see Plate 10) could be due either to another concealed granite or to a basin of low-density sediments of probable Devonian or lower Carboniferous age. To the south-west an anomaly coincident with the Widmerpool Gulf can be explained by the known thickening of low-density Namurian rocks, together with a contribution from the argillaceous Dinantian basin sediments (Cornwell in Frost and Smart, 1979; Rollin, 1980).

In the south-east of the region a group of pronounced Bouguer anomaly lows occurs in an area of relatively shallow (*ca* 1.0 km) basement rocks beneath The Wash. The anomalies have been examined by Allsop (1983), but it was not possible to differentiate between a granite or a basin origin, although the presence of magnetic rocks partially surrounding the anomalies suggests that they represent intrusions. In the south-west of the study region Maroof (1976) interpreted the gravity high in Derbyshire as being due to a concealed ridge of high-density (2.80 Mg/m^3) rocks, probably of Precambrian age, lying at depths of between 0.1 and 1.0 km.

Bullerwell (in discussion of Bott, 1961) first drew attention to the elongated zones of anomalies that characterise the aeromagnetic map of the East Midlands, pointing out that one zone could be traced from The Wash to the Lake District, a distance of more than 300 km. It has been suggested that the zones are related to belts of igneous or metamorphic rock; Wills (1978), for example, interpreted the magnetic zones as Precambrian basement ridges.

In the south-west of the region a zone of pronounced magnetic anomalies has been interpreted as being due to Caledonian granodiorite intrusions (Evans and Maroof, 1976) similar to the intrusion exposed at Mountsorrel in Leicestershire. A map showing interpreted depths to magnetic basement has been prepared by Decca Resources Ltd. (in Kent, 1974) and it covers part of the study area.

4.2 Geophysical characteristics of mineralised areas

4.2.1 REGIONAL GEOPHYSICAL SURVEYS

Past applications of regional geophysical surveys to the exploration for economic base-metal mineralisation has often been based on the correlation between granitic bodies indicated by Bouguer gravity anomaly lows and observed epigenetic mineral zonation. A classic example of this is the North Pennine Orefield, where the existence

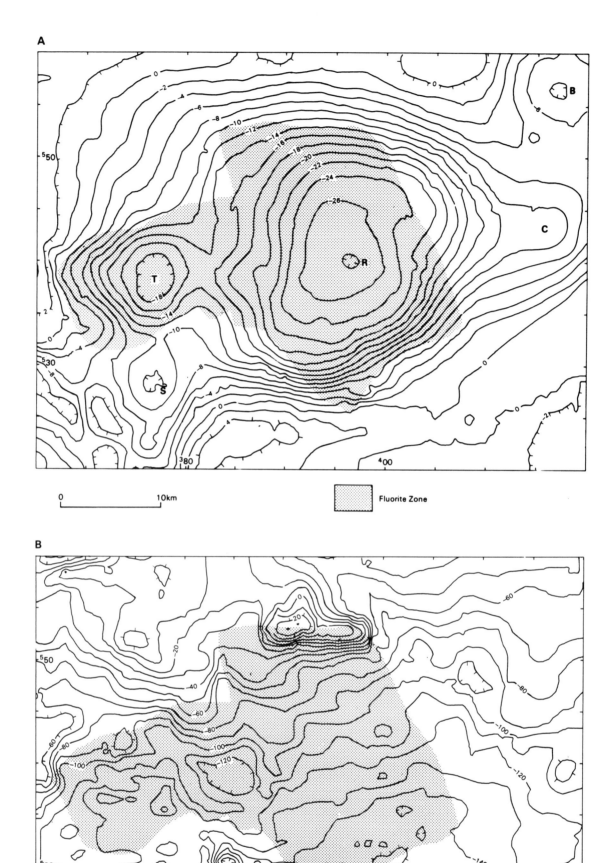

Fig. 4.2 Bouguer gravity (*A*) and aeromagnetic (*B*) anomaly maps of Alston Block based on BGS data. Contours at 1 mGal and 10 nT intervals, respectively. Key to abbreviations: B—Burnopfield, C—Cornsay, R—Rookhope, S—Scordale and T—Tynehead

26

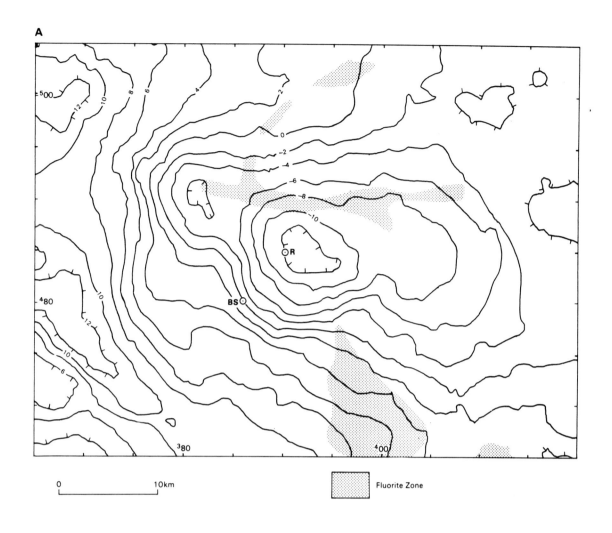

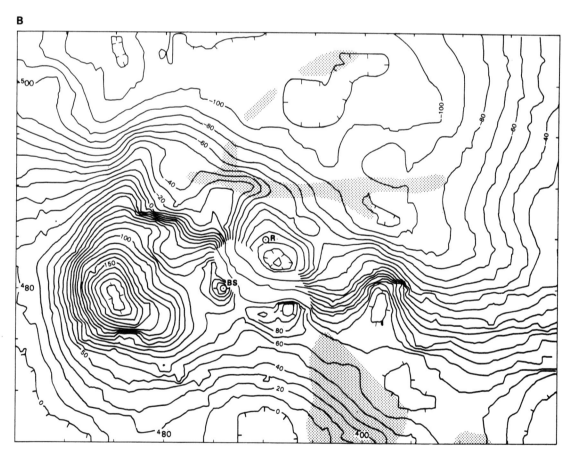

Fig. 4.3 Bouguer (*A*) and aeromagnetic (*B*) anomaly maps of Askrigg Block. Contours at 2 mGal and 10 nT intervals respectively. Key to abbreviations for boreholes: BS—Beckermonds Scar, R—Raydale

of a concealed granite was first postulated by Dunham (1934) on the evidence of the type and distribution of the mineral zones (Chapter 3). The Bouguer gravity field of the area was examined by several workers, but the most comprehensive survey and interpretation were published by Bott and Masson-Smith (1957a). The pronounced Bouguer anomaly low discovered corresponds remarkably well with the distribution of the main mineral veins and particularly with the fluorite zone. Superimposed on the main low, two gravity minima (Rookhope and Tynehead), interpreted as the highest parts of the granite body, coincide precisely with two main centres of mineralisation described by Dunham (1934). Three other gravity minima occur, one of which coincides with an isolated group of veins at Scordale (Fig. 4.2), whereas the other two (Cornsay and Burnopfield) occur where Dinantian limestones are covered by younger Carboniferous rocks.

The low-density Weardale Granite also appears to give rise to an area of low magnetic values and Bott and Masson-Smith (1957b) suggested that these are due to a reversed remanent magnetisation of the granite. However, no evidence has subsequently been presented to indicate that the granite is magnetic. An alternative explanation for the magnetic low, which is seen as a broad regional pattern of contours on the aeromagnetic map (Fig. 4.2), is that it represents a non-magnetic body, perhaps the deeper part of the Weardale Granite, set in a magnetic basement. Most of the shorter wavelength magnetic anomalies in Fig. 4.2 are due to the Whin Sill, an outcropping or near-surface dolerite intrusion in Carboniferous sediments.

The fault-bounded Askrigg Block (Fig. 4.3) is similar to the Alston Block in many geological aspects and in its coincidence with a pronounced Bouguer anomaly low, which has also been interpreted as a concealed granite (Bott, 1961). The anomaly is smaller in amplitude and area than that on the Alston Block and suggests the existence of a central cupola with two flanking 'shoulders' of granite, giving the intrusion an overall E–W elongation. This granite (the Wensleydale Granite) was, like the Weardale Granite, shown by drilling to have a weathered surface beneath the Carboniferous unconformity and to be of Caledonian age (Dunham, 1974). The Wensleydale Granite is characterised on the aeromagnetic map (Fig. 4.3) by a central non-magnetic zone over the main granite cupola flanked by a semi-circle of marginal anomalies. The combination of Bouguer anomaly and magnetic anomalies suggests that the granite was intruded into a continuous zone of magnetic rocks, which can be traced with a SE trend into the study region. The clear spatial coincidence of the mineral zones and the position of the concealed granite that characterises the Alston Block is absent on the Askrigg Block. The mineralisation (Dunham and Wilson, 1985) occurs in two main isolated centres, the Greenhow mining district to the south and Swaledale to the north. The former is more likely to be associated with the Craven Faults zone, whereas the elongated zone to the north lies over the northern margin of the granite and of the zone of magnetic basement rocks.

The Dinantian limestones of the Derbyshire 'Dome', which, as in the north Pennines, hosted the mineral veins, tend to coincide with high gravity values or gradient zones (see Plate 11a). The high values are partly due to the effect of the higher density Dinantian limestone at the surface, but it has also been suggested that the Derbyshire 'Dome' is underlain by a concealed ridge of dense basement rocks (Maroof, 1976). Although the Bouguer anomaly data do not indicate a low in Derbyshire comparable with those seen in the North Pennines, a poorly defined low occurs at the eastern side of the dome, near Bakewell (Cornwell in Aitkenhead and others, 1985). The anomaly lies adjacent to the main zone of mineralisation in the Dinantian limestones, but there is insufficient evidence to decide if it is due to a small acidic intrusion or a basin of (?) early Carboniferous or older sediments.

Evans and Maroof (1976) have argued that magnetic anomalies in the area (Fig. 4.16) are due to concealed granodiorite intrusions, similar to those in Leicestershire, which appear to have little density contrast with their host rocks. These authors suggested that concealed intrusions of this type, although Caledonian in age, could have influenced the distribution of mineralisation in younger rocks by providing channelways for the mineralising solutions and pointed to the coincidence of the Derbyshire fluorite zone with the postulated magnetic granodiorites. It seems more probable from an examination of the depths and locations of the magnetic bodies in Derbyshire (Cornwell in Aitkenhead and others, 1985) that they represent intrusions of Carboniferous rather than Caledonian age.

The geological environment of the Pennine mineralisation is similar in several respects to that of the Irish Midlands. Brown and Williams (1985) have recently examined regional gravity and magnetic data for Ireland that they interpreted to provide a model for the distribution of the Irish-style mineralisation. They related elongated low-amplitude (ca 5 mGal) Bouguer anomalies to the presence of basins of low-density Carboniferous sediments and ridges with cores of higher density (and magnetic susceptibility) Ordovician volcanic rocks. The main ore bodies occur in the regions interpreted from the gravity evidence as raised block areas, particularly near the block/trough margins.

4.2.2 Detailed Geophysical Surveys

Mineral deposits that are or have been economically viable in the Pennines are largely confined to the Dinantian limestones or the immediately overlying Namurian strata. The deposits dominantly comprise narrow, steeply inclined veins or, less commonly, 'flats' parallel to the bedding (Chapter 3). These contain a combination of electrically conductive minerals (commonly galena), which are detectable by use of induced polarisation (IP) and electromagnetic (EM) geophysical techniques, and those which are electrically resistive (sphalerite and gangue minerals) and are therefore not distinguishable from their host rocks.

Exploration for vein-type deposits has been carried out in several areas of the Pennines by use of a range of geophysical techniques, including EM, VLF-EM, IP, SP and resistivity (e.g. Wadge and others, 1983; Evans and others, 1983). These surveys showed that although it is difficult to locate mineralisation directly because of the lack of a distinct response from conductive ore mineral assemblages, it is frequently possible to locate the faults that can contain the mineralisation. The response, which is usually detected most effectively by use of EM methods, is believed to be due to the presence of more conductive material (clay, water in fractured rock) in the fault zones and, more commonly, to contrasts in the rock types across the structure. Wadge and others (1983), for example, described how a 4 km long extension of the Bycliffe vein structure near the southern margin of the Askrigg Block was detected primarily by VLF-EM, but produced no response on detailed IP traverses. A systematic examination of the application of various geophysical and geochemical methods (Brown and Ogilvy, 1982) in detecting a fluorspar vein in Derbyshire confirms that VLF-EM is the most suitable geophysical technique, but also revealed the presence of a small localised Bouguer anomaly low over the veins. Significant magnetic anomalies also occur due to the displacement at depth of a magnetic lava along the mineralised fault. Magnetic surveys have been used elsewhere to detect fault structures where the Dinantian sequences contain igneous rocks. This approach has been particularly successful at the northern margin of the Alston Block where structures giving rise to displacement of the Whin Sill have subsequently provided channels and sites for the concentration of mineralising fluids, sometimes within the dolerite intrusion itself (Evans and Cornwell, 1981; Bateson and others, 1984).

Experiences with geophysical methods in the Irish Republic have been broadly similar to those in the Pennines but, in addition, surveys there have been carried out over known economic PbZn mineralisation in Lower Carboniferous limestones. IP, combined with resistivity surveys, has proved the most useful tool for direct and indirect exploration (Williams, 1982). Sink-holes formed by karstification of the limestones and filled with clays produce anomalies similar to those of known ore bodies and gravity methods are used for screening IP responses. The presence of conductive shale horizons limits the value of EM surveys, although the VLF method has been used widely, as in the Pennines, to aid structural interpretations in areas covered by extensive drift. None of the Irish ore deposits contains pyrrhotite (the magnetic sulphide mineral), but magnetic surveys have been used to aid mapping of Ordovician volcanic rocks.

4.2.3 Detection of Mineralisation at Depth by Geophysical Methods

In the East Midlands the Lower Carboniferous rocks that host the mineralisation extend from outcrop to depths of several kilometres. As the depth increases, the geophysical methods used change from those designed to detect the mineralisation directly to those more suited for indirect exploration. In the Pennines there are few examples, however, even for near-surface deposits, where the mineralisation can be detected directly and emphasis in the application of geophysical methods is therefore placed on the recognition of suitable structures or environments for ore deposition.

In exploration for near-surface mineralisation the limiting factor in the use of frequency domain EM methods is the effective depths of penetration with the frequencies available with commercially developed equipment. This is a severe limitation for higher-frequency equipment, such as VLF-EM, especially in the type of geological environment commonly encountered in the East Midlands, where the Lower Carboniferous limestones are often overlain by conductive shale sequences and, higher in the geological column, by clay/mudstone sequences. Experiments with time-domain EM equipment still need to be carried out to enable the value of this technique in the area to be assessed.

Another distinct group of mineral deposits occupies what appear to have been cavities in the Dinantian limestones and includes the copper deposits once worked at Ecton, on the west side of the Derbyshire 'Dome'. No unmined deposit remains where geophysical methods can be applied, but the suggestion that mineralisation of this type was formed during Permo-Triassic times restricts their location to limestone surfaces exposed at that time. Geophysical techniques could be used to locate this surface in geologically suitable environments, but such an investigation was beyond the scope of the present study.

The use of seismic reflection methods has been considered in recent years in exploration for deeply buried mineral deposits. Disappointing results have been reported from trial high-resolution surveys in Ireland (by Williams, 1982), mainly because of the difficulties of obtaining good data and of interpreting the results obtained. A reflection survey across the Craven Fault, bounding the Askrigg Block in England (Wadge and others, 1983), produced data of variable quality but, in places, several reflecting horizons in Lower Carboniferous rocks could be traced down to depths of 1.3 km. Fold structures known from surface mapping were recognised on the sections, but there was no obvious response to the major fault (the Middle Craven Fault) in the area.

Exploration for more deep-seated mineralisation will depend upon the interpretation of regional geophysical data, supplemented by additional observations in critical areas. Concealed features of potential interest recognisable from the data are discussed in detail in this chapter (cf. Chapter 8), including (a) acid intrusions, (b) shale basins, (c) deep lineaments and faults affecting basement and/or cover rocks, (d) uplifted blocks and (e) basic intrusions and volcanic rocks (particularly where they may have sealed faults).

4.3 Regional geophysical data bases

Gravity and aeromagnetic data provided the basis for most of the interpretation described in this chapter, although account was also taken of heat flow and seismic refraction data. The nature and results of seismic reflection surveys are more appropriately dealt with in Chapter 5.

4.3.1 Gravity

Much of the East Midlands study region has been covered previously by BGS gravity surveys with an observation density of 1 station per 1.5–2 km². A few isolated areas have been covered by surveys carried out by oil companies. The data, which are available as Bouguer anomaly maps at a scale of 1:250 000 (Fig. 4.1) (e.g. Institute of Geological Sciences, 1977a), were re-processed for the present study.

4.3.2 Magnetics

The East Midlands region was included in the systematic aeromagnetic surveys of the UK carried out in 1955, 1956 and 1958 for the Geological Survey (now BGS). The data were mostly collected at an altitude of 305 m above the ground surface along flight lines 2 km apart.

Both the gravity and the magnetic data were re-processed for the present study and examined by use of the IAS (Appendix 1).

4.3.3 Seismic Refraction Data

The East Midlands region has been explored by seismic methods since the 1930s, when the application of refraction techniques led to the discovery of oil fields at Kelham Hills and Caunton and provided detailed information on the Eakring structure (Kent, 1985). The results of these surveys have been presented as isochron maps (Kent, 1966; Bullerwell in Stevenson and Mitchell, 1955) that show variations in the topography of a concealed high-velocity refractor, usually interpreted as Lower Carboniferous limestones.

The deep N–S refraction profile of the UK (LISPB) (Bamford and others, 1977) indicated the existence of a refractor with a velocity of about 5.8 km/s extending down from a depth of 1–2 km to 8–10 km. The base of this refractor, which is interpreted as Lower Palaeozoic sediments (overlying pre-Caledonian basement with a velocity of 6.25 km/s), shallows southwards. The base of the crust is poorly defined in the north of England between Buxton and and the Southern Uplands, but appears to be at a shallower depth than on sections of the LISPB profile to the north and south. More detailed refraction profiles, with the use of quarry blasts as an energy source, have produced sections of the upper crust in the south-west corner of the study region (Whitcombe and Maguire, 1980; 1981a; 1981b). One of these profiles indicated the presence of a 5.64 km/s refractor beneath a variable thickness (up to 2.1 km) of Palaeozoic and Triassic rocks between Charnwood Forest and the Derbyshire 'Dome'. Whitcombe and Maguire (1981a) interpreted the feature as being mainly Precambrian (Charnian) in age, but the nature of the high-velocity basement is problematical, especially in areas where Dinantian limestones occur.

Rogers (1983) had one possible solution to the problem of interpreting the nature of this high-velocity refractor by suggesting that the Precambrian rocks are downfaulted by the Cronkston–Bonsall faults within the Derbyshire 'Dome' and are covered to the north by Palaeozoic rocks.

4.3.4 Heat Flow and Heat Production

Above average temperature gradients (in the UK) are found in areas of low thermal conductivity sediments or in areas of above average heat flow. High heat flow often reflects an upper crust enriched in heat-producing radioelements (uranium, thorium and potassium) that tend to occur in significant quantities in evolved high heat production (HHP) granites or in black shales. The heat-producing characteristics of British granites, notably those from SW England, N. England and the East Highlands of Scotland, have been reviewed by Lee (1986). There are few published data for other rock types, but new results have been obtained from the chemical analysis of shales and a variety of igneous rocks in this report (Chapters 6 and 7).

In one of the first attempts to compile UK heat flow data, Richardson and Oxburgh (1978) indicated two belts of higher values (>60 mW m^{-2}). One of these extended from the Alston Block south-eastwards across the East Midlands to The Wash (cf. Brown and others, 1980). Subsequent heat flow measurements (Downing and Gray, 1986) replaced the 'belts' with more localised areas of high values—for example, over the Alston Block and the Lake District, where the extensive granite intrusions are concealed at depth (Lee, 1986). The Askrigg Block is also

underlain by a Caledonian granite, but with a slightly lower heat production than that beneath the Alston Block; the lower heat flow could reflect either fractionation or the smaller size of the Wensleydale Granite.

Within the East Midlands heat flow values reach a maximum in a north–south zone centred on the Eakring Anticline (see Plate 9c). Bullard and Niblett (1951) suggested that this anomaly was due to the eastward and upward migration from the Pennines of heated water within Carboniferous limestones. The eastward movement of water may also be responsible for the generally higher than average heat flow values in east Nottinghamshire and Lincolnshire, but Richardson and Oxburgh (1978) suggested that there is probably also a higher heat flux from the basement in the area. Lee (1986) pointed out that the temperature gradient for the upper 2–3 km of the crust in this area is comparable with that of some radiothermal granites. Lower than average heat flows were reported (Richardson and Oxburgh, 1978) for a borehole in the granodiorite at Croft (one of the south Leicestershire diorites (Appendix 2)) and in the Eyam Borehole, where downward movement of cooler water is required to explain the decrease of temperature with depth below 470 m.

Heat flow measurements have been made in about 26 boreholes in the study region.

4.3.5 Physical Properties of Rocks

The most important physical properties (velocity, density and magnetisation) of rocks for the interpretation of geophysical data for the study area can be derived from geophysical borehole logs and laboratory measurement of outcrop and core samples. The borehole logs provide adequate information for the Mesozoic and most of the Carboniferous rocks, but data for pre-Carboniferous basement rocks are sparse. A special study was carried out using samples from various sources, the results of which are listed in Appendix 2.

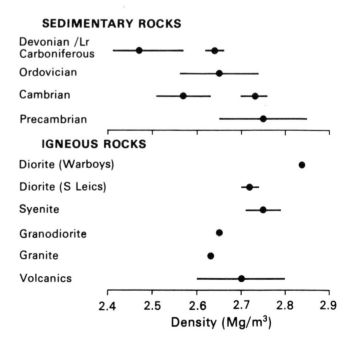

Fig. 4.4 Mean densities and standard deviations of sedimentary and igneous rocks

Table 4.1 Densities and velocities of Carboniferous and younger rocks in the East Midlands

		Density Mg/m^3	Seismic velocity, km/s	Seismic reflector
CRETACEOUS	UPPER	2.02	2.9	Base chalk
	LOWER	2.28	2.7	
JURASSIC	UPPER	2.26	2.2	
	MIDDLE	2.26	2.5	Top Lincolnshire Limestone
	LOWER	2.38	2.5	Top Mercia Mudstone
TRIASSIC	MERCIA MUDSTONE	2.38	3.0	Top Sherwood Sandstone
	SHERWOOD SANDSTONE	2.49	3.2	
PERMIAN		2.55	3.8	Top Upper Magnesian Limestone / Base Permian
CARBONIFEROUS	WESTPHALIAN	2.51	3.5	Top Hard Coal / Top Rough Rock
	NAMURIAN	2.51	4.0	Top Carboniferous Limestone
	DINANTIAN	2.66	5.0	Arundian Shale Marker / Base Carboniferous

Mesozoic and Carboniferous rocks

Oil exploration boreholes in the East Midlands have been drilled through the entire Mesozoic sequence and in many places through parts of the Westphalian and Namurian. The velocities, in particular, have been examined in detail, following the work of Wyrobek (1959), to provide the necessary control on the seismic interpretation. The velocities and densities used in the interpretation of the geophysical data are summarised in Table 4.1.

Pre-Carboniferous basement rocks

Basement has been intersected in comparatively few deep boreholes in the study area and several of these were drilled at a time when the full range of geophysical logs was not available. New data were obtained as part of the present study on samples from boreholes and also from outcrops in the Charnwood Forest and Nuneaton areas and these are given, together with values reported elsewhere (notably by Maroof, 1973) in Appendix 2. Many samples were found to be suitable only for density determinations owing to their limited size or friable nature.

Data were obtained for sedimentary and igneous rocks ranging in age from Precambrian (Charnian) to Devonian. The results from different sites are difficult to group even when samples of uncertain age or affinity are omitted. The difficulty arises from the wide range of values observed, but several general conclusions can be drawn.

The data for the Palaeozoic rocks summarised in Table 4.2 have been divided, largely on the basis of the densities, into two groups. One group (*A*) comprises higher-density, lower-porosity samples that were probably subjected to greater pressures. Group (*B*) samples have higher porosities and are less indurated and tectonised. They occur only in the south-west of the region, but are probably widespread in southern England. This general division of Lower Palaeozoic sedimentary rocks is also reflected by the geochemical data (Chapter 6). Density data for Devonian rocks are also variable, but the lower-density samples occur mainly south of the study area. The density values of group *A* samples overlap many of those of the Charnian rocks and it is therefore difficult to predict the existence of Precambrian rocks of this type solely on the basis of gravity evidence.

The densities of the igneous rocks also overlap those of Lower Palaeozoic and Precambrian sediments, but granites and some diorites and other basic rocks are potential sources of significant gravity anomalies.

Although fewer velocity data are available, the values for the sediments tend to increase systematically with the densities (Fig. 4.5). This relationship is less clear in case of the igneous rocks and is probably related to their mechanical alteration by brecciation and weathering.

Table 4.2 Average densities and velocity ranges of pre-Carboniferous basement rocks in the East Midlands (n is number of sites sampled)

		Saturated density, Mg/m³	n	Velocity km/s	n
Sedimentary rocks					
Devonian (to early Carboniferous)	A	2.64 ± 0.02	5		
	B	2.47 ± 0.10	1		
Ordovician	(?A + B)	2.65 ± 0.09	3	3.4 – 5.7	5
Cambrian (including Tremadoc)	A	2.73 ± 0.03	5		
	B	2.57 ± 0.06	8		
Precambrian (?)					
Phyllites and slates		2.75 ± 0.08	6	5.4	1
Igneous rocks					
Granite		2.63	1		
Granodiorite (Mountsorrel)		2.65	1		
Syenite (?Precambrian)		2.75 ± 0.04	4		
Diorite (S. Leicestershire)		2.72 ± 0.02	3	5.60	1
Diorite (Warboys Borehole)		2.84	1	6.10	1
Volcanics and pyroclastics		2.70 ± 0.10	6	4.6 – 5.6	2

Some of the igneous rocks sampled have magnetic susceptibilities high enough to give rise to anomalies (i.e. >0.01 SI units), and in several cases they can be shown to be associated with features on the aeromagnetic maps. The Mountsorrel granodiorite and the Warboys diorite (Appendix 2) are particularly important in this respect, but some of the other diorites and the syenites also have significant magnetisations. Measurements on samples indicate that the remanent magnetisation does not appear to contribute significantly to the total magnetisation of rocks older than Carboniferous. Ground magnetic surveys over several of the igneous rocks in the Derbyshire 'Dome' in-dicate that these too can be satisfactorily interpreted, assuming the dominance of the induced magnetisation components. Surveys over the Whin Sill in the north Pennines, however, indicate that the total magnetisation of this extensive late Carboniferous/early Permian intrusion is dominated by the remanence component (Evans and Cornwell, 1981).

Detailed data are not available for the concealed Namurian/Westphalian igneous rocks in the East Midlands region, but there are few obvious indications of anomalies due to remanent magnetisation (e.g. a negative 'reversed' anomaly).

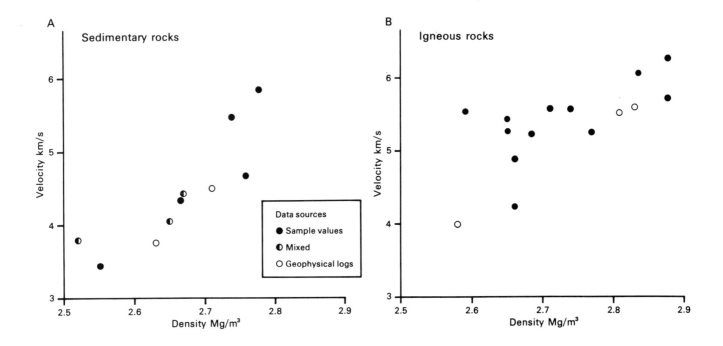

Fig. 4.5 Density/velocity cross-plots for rocks from East Midlands study region

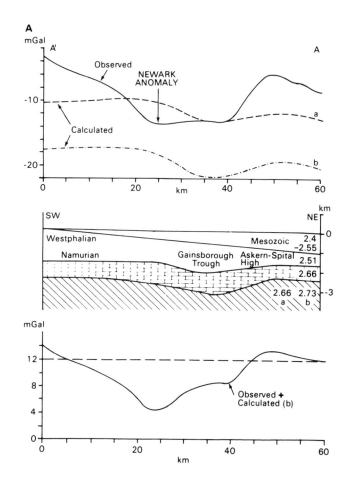

4.4 General interpretation of regional geophysical data

4.4.1 INTRODUCTION

Systematic interpretation of the gravity and magnetic anomalies in the study area was made by use of the following data:

Original Bouguer and aeromagnetic anomaly maps (shown as images, Plates 4a and 6a);
Processed versions of these: ('stripped') gravity (Map 2 and Plate 5b), reduced to pole magnetics (Map 3 and Plate 6b), second vertical derivative (Plate 4b);
Enhanced images on the IAS (pseudotopography, etc.);
Combined images on the IAS (e.g. Plates 7 and 12).

(Procedures used in deriving the images by the I²S image analysing system are described in Appendix 1).

The quantitative interpretation of the geophysical data involved the use of standard techniques—for example, in estimating depths to source rocks and the angles of slopes. The interpretation of individual profiles was based on a programme for modelling the gravity and magnetic effects of bodies with limited strike extent (2.5 dimensional) (Busby, 1985). The accuracy of the interpretation by use of this procedure can be maintained for three-dimensional bodies by subdividing the body into polygons of differing strike extent.

The lack of geological control at depth in the East Midlands means that the interpretation of the regional gravity and aeromagnetic anomalies must be tentative in many cases. Examples of alternative sources of anomalies are discussed below.

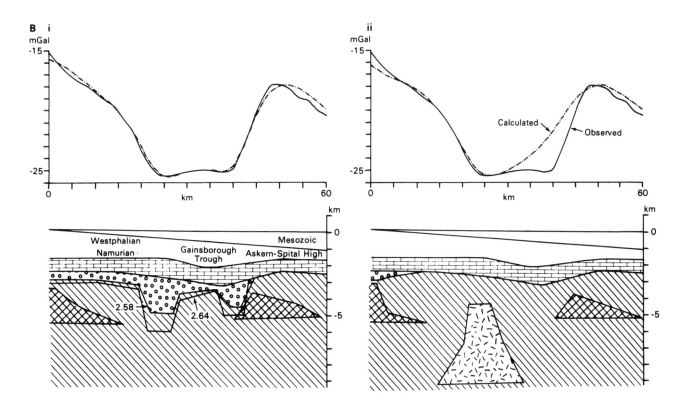

Fig. 4.6 Gravity stripping and anomaly interpretation, – profile *A-A'* (see Fig. 4.7 for location and Fig. 4.10 for key to symbols).

A Observed profile and calculated gravity effect of Carboniferous and younger rocks with densities indicated for model

B Observed Bouguer anomaly profile (after removal of background field of 12 mGal) and alternative interpretations of main gravity low (Newark anomaly). In (i) this is assumed to be due to pre-Dinantian limestone basin of sediments with density of 2.64 Mg/m³ (Table 4.2) or 2.58 Mg/m³ (value at Eakring (Appendix 2)) and in (ii) to granite intrusion

Gravity data

The observed Bouguer anomaly data (see Plate 4a) for the study region indicate variations from + 20 mGal to − 30 mGal. Part of this is due to the gravity effect of variations in the thickness of comparatively low-density Upper Carboniferous and Mesozoic sediments. The process of gravity stripping (Appendix 1) removes these effects but leaves many anomalies to be explained in terms of pre-Namurian geology (Map 2). The process of gravity stripping and the nature of the anomalies remaining is illustrated in profile form in Fig. 4.6.

The densities assigned to Namurian and younger rocks for the gravity stripping are listed in Table 4.1. There is more uncertainty in deciding the sources of the main gravity anomalies for older rocks, but the density values in Table 4.3 were adopted for interpretation purposes. The nature of the 'high-density' basement is unknown and its existence is postulated in order to explain the

Table 4.3 Densities adopted for interpretation

Rock unit	Density (Mg/m^3)
Dinantian limestone sequences	2.66
Dinantian and older sandstones/ conglomerates	2.64/2.58
Lower Palaeozoic sediments	2.73
'High-density' basement rocks	2.80
Granite	2.63
Diorites	2.73
Basic intrusives	2.90

steepness of some observed gravity gradients in places (Fig. 4.24) and the existence of some Bouguer anomaly highs without any corresponding magnetic signature. The interpreted extent and thickness of this basement are, however, largely dependent upon the background value chosen for the study region. The high-density material

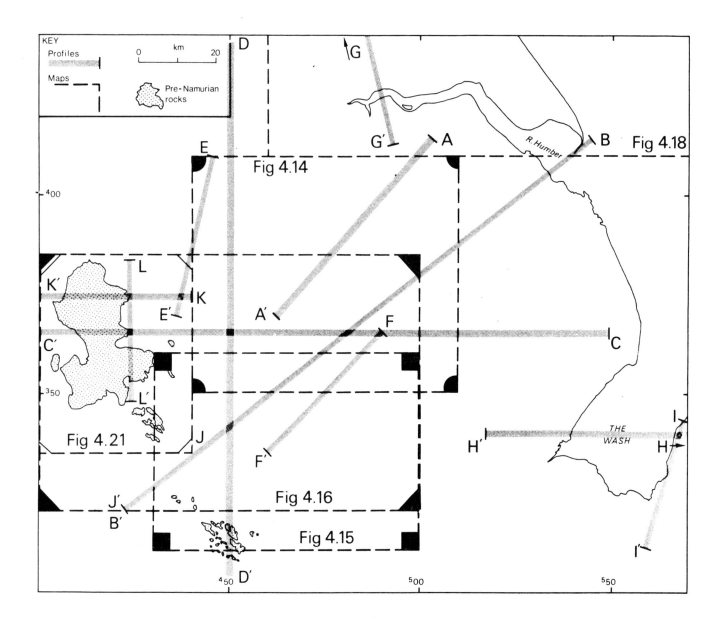

Fig. 4.7 Map of East Midlands study region with locations of profiles and areas referred to in interpretation of regional geophysical data

could be highly indurated Lower Palaeozoic sediments, although Maroof (1973) considered it to represent Precambrian Charnian basement. Elsewhere it has been assumed that the basement comprises rocks with an average density of 2.73 Mg/m³ and, mainly, of probable Lower Palaeozoic age.

Several of the problems of interpreting anomalies remaining on the stripped gravity map are illustrated on the profile shown in Fig. 4.6A and reappear on several other profiles. In this the low-density rocks responsible for the pronounced gravity low at about 25 km (Fig. 4.6B) (the Newark anomaly) could be interpreted as a concealed granite of probable Caledonian age, or a basin of early Carboniferous/Devonian sediments. For reasons given later in this chapter the granite model is preferred, but the gravity profile can be explained by many alternative models. Some of these are shown in Fig. 4.6B, but it is noted that certain models, such as the basin filled with sediments with a density of 2.64 Mg/m³, produce calculated profiles that are more difficult to reconcile with the observed data. A greater consistency can be achieved by assuming the sediments to have a lower density and the value of 2.58 Mg/m³, reported by Maroof (1973) for the lower Carboniferous sediments at Eakring, has been adopted.

In the geological model for Lower Carboniferous sedimentation the existence of Dinantian basins of lower-density sandstones/conglomerates is postulated on the downthrow sides of Dinantian growth faults (Chapter 5). If the alternative interpretation of the gravity lows (pre-Carboniferous granitic intrusions) is valid, the features are more likely to indicate the central part of a tilt block.

A second gravity low appears in Fig. 4.6B at about 40 km and coincident with the thick Namurian sediments in the Gainsborough Trough, one of the major sedimentary basins in the area (Kent, 1966). This coincidence suggests that in this case the anomaly is more likely to be due to density or thickness changes within the Carboniferous sequence, perhaps even the use of an incorrect density for the Namurian rocks. A comparatively shallow origin for the anomaly is also indicated by the form of the anomaly, notably the steepness of the gradient abutting the Askern–Spital high.

In much of the East Midlands region little is known about the thicknesses of the Dinantian rocks, although models have been prepared from indirect evidence (Chapter 5). As the dominant Dinantian rocks are limestones, with a small density contrast with the underlying basement, accurate interpretation of their thicknesses from the gravity data is not possible and the models for the Dinantian (e.g. Fig. 6.4) are very generalised.

The regional Bouguer anomaly field of the western part of northern England is characterised by a general increase towards the west; within the East Midlands region, however, an examination of long profiles of 'stripped' gravity values suggests that the regional field is either level or decreases northwards (by about 0.6 mGal/10 km). In the present study a level background field of 12 mGal was assumed for most interpretations.

Magnetic data

The most important potential sources of magnetic anomalies in the region comprise basic intrusive and extrusive rocks of Carboniferous age, Caledonian granodiorites, diorites, such as those of south Leicestershire and the Warboys Borehole (Appendix 2), and a deep basement rock of unknown nature. For interpretation purposes only induced magnetisation with typical suscep-

tibilities of 0.01–0.05 SI units is assumed to exist.

Combination of regional geophysical data

The Bouguer gravity and magnetic maps indicate that the Palaeozoic and older rocks in the East Midlands contain complex and variable structures that occur over a wide range of depths. Attempts to examine the relationship between the gravity and aeromagnetic anomalies (see Plate 7) show that the strongest correlation occurs between the lineaments indicated by the two data sets rather than between the anomaly highs or lows (see Plate 10). This suggests that many of the magnetic anomalies are due to relatively minor igneous bodies and/or that they arise from separate sources, perhaps at different levels in the crust but showing the same structural trends. It is possible therefore to indicate only in very general terms potential sources for gravity and magnetic anomalies (Table 4.4). The main gravity and magnetic anomalies referred to in the text are indicated in Plate 10b and their general interpretation is shown in Fig. 4.9.

Table 4.4 Possible sources of gravity and magnetic anomalies

Gravity	
Highs	Lows
Precambrian basement of unknown composition	Precambrian acid volcanic/granitic basement
Lower Palaeozoic mudstones	Devonian/L. Carboniferous arenaceous sediments
	L. Carboniferous shales
Magnetics	
Highs	Lows
Igneous material of Precambrian/Caledonian/Carboniferous age	Sediments of all lithologies and ages
Metamorphic rocks of unknown type	Granites

4.4.3 GEOPHYSICAL LINEAMENTS

One of the more significant uses of the regional geophysical data is in providing evidence for lineaments that could represent major faults or shear zones affecting Carboniferous or basement rocks. Such features are of considerable importance for regional structural analysis, but can be difficult to recognise from surface mapping or seismic reflection surveys. Lineaments identified in this study from geophysical maps and IAS images are shown in Fig. 4.8, where they are also named and indications are given of the apparent displacement of the more magnetic or higher-density material. The dominant trend of the lineaments is SE to SSE, with minor groups aligned to the S or SW.

Lineaments of NW–SE trend

The central part of the study region is dominated by lineaments trending NW–SE that tend to swing round to a more E–W direction in the NW. Several of these lineaments are closely associated with gravity variations

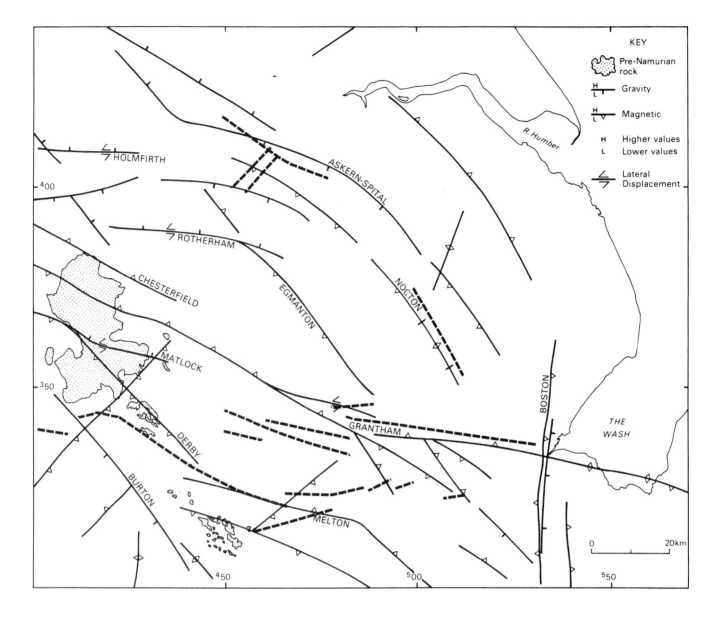

Fig. 4.8 Map of East Midlands study region showing interpreted locations of main geophysical lineaments in Carboniferous and older rocks. Key indicates apparent displacement of bodies defined by geophysical evidence. Lineaments, full lines; faults based on geological evidence, broken lines

that cannot be completely explained by the known variations in the thicknesses of the Carboniferous rocks. More significantly, several magnetic anomalies have estimated source depths greater than that of the base of the Carboniferous and occur in zones with a NW–SE trend. This trend is therefore considered to represent a dominant grain of the Caledonian basement rocks imposed by the closing of the Tornquist Sea. It is probable that the lineaments represent major lithological boundaries or strike-slip faults that in some cases were subsequently reactivated during Carboniferous times.

Evidence in the NW of the region and in adjacent areas indicates that NW-trending anomalies continue for some distance—the E–W trends also observed there may therefore have been superimposed on the older basement, perhaps during Carboniferous times.

Lineaments of SW–NE trend

Lineaments with SW to NE trends are characteristic of the Caledonian basement of much of Scotland, Wales and the Welsh Borders. A few lineaments with the same trend occur within the study region, notably in the SW, where

they probably represent the extremities of well-defined structures in western central England and the Welsh Borders, such as the Church Stretton and Red Rock faults. SW–NE trends reappear in the NE of the study region and are particularly obvious in the elongation of the Market Weighton and Hornsea gravity lows (Fig. 4.18A). There is no strong geophysical evidence for the continuation of SW–NE structures in the intervening area, but a pronounced group of faults and folds (including the Don Monocline) with this direction occurs in the Westphalian rocks of the Yorkshire Coalfield, north of grid line 390N. SW–NE-trending faults in the basement could be postulated to explain changes in the depth to the magnetic rocks along the strike of the NW–SE magnetic zones. The magnetic high at Rotherham (c.f. Fig. 4.13), for example, occurs where the Don Valley Monocline intersects the Rotherham lineament.

Lineaments of N–S trend

Evidence for N–S lineaments is largely confined to the SE of the study region, where they appear to coincide with some of the margins of postulated granites around The

36

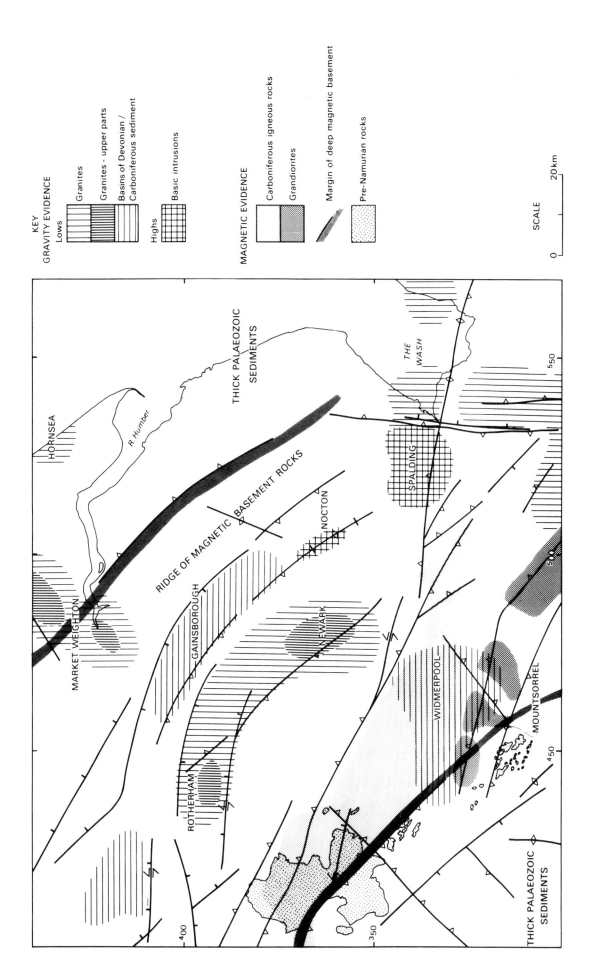

KEY

GRAVITY EVIDENCE

Lows

Granites

Granites - upper parts

Basins of Devonian / Carboniferous sediment

Highs

Basic intrusions

MAGNETIC EVIDENCE

Carboniferous igneous rocks

Grandiorites

Margin of deep magnetic basement

Pre-Namurian rocks

SCALE

0   20km

Fig. 4.9 Map summarising interpretation of main Bouguer gravity and magnetic anomalies. Geophysical lineaments as in Fig. 4.8

37

Wash. Gravity data indicate possible N–S lineaments (Fig. 4.21) at the margins of the Derbyshire 'Dome'.

Lineaments of WNW–ESE trend

Lineaments with ESE trends are most apparent in the SW and S of the area and include the Grantham lineament, which extends across the study region from the Derbyshire 'Dome' to The Wash (cf. Plate 8c). This structure appears to mark a general change in the character of the magnetic map, dividing an area to the south of complex and high-intensity anomalies from one to the north with longer-wavelength anomalies due to more deep-seated magnetic bodies. To the west the Grantham lineament seems to have been important in determining the distribution of igneous activity during the Carboniferous (Fig. 4.16) and it is possible that associated faults were sealed, preventing any subsequent movements. In the east of the region the Grantham lineament adopts a more E–W trend and intersects an area of shallower crystalline basement where no Carboniferous rocks have been preserved. To the east and SE of the study region the ESE trend dominates the structure of the basement of East Anglia and also appears to have been an important direction of faulting in Mesozoic rocks of the southern North Sea basin. Other ESE lineaments occur in the Derbyshire 'Dome' and, in the south of the area, one appears to have controlled the distribution of the granodiorite intrusions of 'Mountsorrel' type. As the Mountsorrel Granodiorite itself has been dated as Caledonian, it is probable that the ESE lineaments existed before Carboniferous times.

General comments on lineaments

The locations of faults in the study region derived independently of the regional gravity and aeromagnetic evidence from the results of boreholes and seismic reflection surveys are shown in Fig. 4.8 and in Plate 8. In several places these coincide with geophysical lineaments, notably the Askern–Spital structure, and the Barkston Fault (an E–W fault recognised from deep geological evidence and forming part of the Grantham lineament).

A similar comparison has also been made of the relationship between the main structural elements recognised in the East Midlands by Kent (1966) and geophysical anomalies (Table 4.5). This evidence suggests that many of the lineaments are related to structures affecting Carboniferous rocks. However, the form of several of the geophysical anomalies associated with the lineaments suggest that contributions from more deep-seated rocks exist.

The dominant trend of the major Carboniferous structures recognised by Kent (1966) and in this report is ESE–WNW and it is where the geophysical lineaments show the same trend that the strongest spatial correlation between the two occurs. It seems probable that the stress field responsible for the formation of the Carboniferous structures selectively reactivated older basement faults with this trend, perhaps leaving other lineaments relatively undisturbed.

4.5 Systematic regional interpretation

4.5.1 Mesozoic Rocks

The Mesozoic rocks that cover about two-thirds of the study area have been considered mainly in relation to the effect that they have on the general appearance of the Bouguer anomaly map. The densities of these rocks are known (Table 4.1) as a result of numerous geophysical borehole logs in the region and their thicknesses can be determined by use of borehole and seismic data, so the stripped gravity map could be prepared with some confidence. Although the Mesozoic rocks generally dip evenly and gently to the east, faulting and minor folding occur in places that might give rise to Bouguer anomaly variations. These are likely to be local in character and would require detailed surveys for their delineation, particularly to remove the effects of variable drift cover and of the older underlying rocks. In the east, where the basement is deep, some short-wavelength anomalies probably indicate structures within Mesozoic rocks.

The Mesozoic sequence contains no magnetic rocks and therefore does not produce any anomalies on the

Table 4.5 Response of main structural elements recognised by Kent (1966)

Structure	Magnetic	Gravity
Askern–Spital high	Mainly only its westward extension	Pronounced
Tickhill–Gainsborough fold	No	In places, but with oblique trend
Gainsborough Trough	?Yes	Yes; but part of more extensive anomaly
Kiveton Anticline	Continuous magnetic zone	Pronounced zone in west
Egmanton Anticline		No
Eakring–Foston Anticline	No	In places
Welbeck Basin	No	No
Hardstoft Anticline	No	Yes
Nocton high	Pronounced	Pronounced
Grantham high	Yes	?Yes—by low
Widmerpool Gulf	Yes; flanked by magnetic anomalies	Yes
Sproxton high	Yes	?Yes
Thistleton high	No	Yes

aeromagnetic map. The sequence, however, includes a high proportion of argillaceous sediments (Mercia Mudstones, Lias and Oxford Clay), with low resistivities (<10 ohm-metres), which would inhibit deep exploration by use of electrical methods.

4.5.2 CARBONIFEROUS ROCKS

Carboniferous rocks and structures are of primary importance to this study, but the amount of information available decreases with depth, particularly in the east, where Mesozoic cover is thickest. The Bouguer gravity and aeromagnetic anomaly data tend to provide information on different aspects of Carboniferous geology: the former largely reflect the present thicknesses of sediment, whereas some of the magnetic anomalies appear to be due to Carboniferous igneous rocks.

Carboniferous sedimentary sequences

In assessing the densities of the main geological units, Namurian and Westphalian rocks have been assigned similar values, with a more significant density interface at the base of the Namurian. The distribution and thickness variations of the Carboniferous sediments above this interface have a significant effect on the Bouguer anomaly map, but that, as in the case with the Mesozoic rocks, can be compensated on a regional scale by gravity 'stripping' (Appendix 1). The stripped map (Map 2) therefore indicates variations in the gravity field that largely reflect density variations in rocks of Dinantian or greater age. Errors in the map may occur because of lack of adequate information on the thicknesses of post-Dinantian rocks due to poor geological control or to local changes that are inadequately represented by the gridding interval used to process the data. Reference was therefore made to the original gravity data, as well as to the stripped gravity, during interpretation.

Westphalian rocks

Rocks of Westphalian age dip gently to the east and the general increase in thickness is reflected by a broad change in the observed Bouguer gravity anomaly values. Most of the major structures that affect Westphalian rocks, such as the Don Monocline and the Norton–Ridgeway Anticline near Sheffield, are not associated with distinct anomalies, although the trends of the structures generally appear to coincide with those of the Bouguer anomaly contours. The absence of anomalies reflects the lack of major density contrasts within the Westphalian sequence, although detailed gravity surveys might reveal anomalies that arose from the juxtaposition of differing lithologies.

Namurian rocks

The Namurian sequence, like that of the Westphalian, contains no marked density discontinuities, although lateral variation could occur due to facies or porosity changes. The significant changes in thickness of these rocks (Chapter 2), together with their density contrast with the underlying Dinantian limestones, means that there is a relationship in places between Bouguer gravity anomalies and major Namurian structures. However, this is also apparent on the stripped gravity map (Map 2), which indicates that these structures probably extend down into the Dinantian and in some cases into the pre-Carboniferous basement.

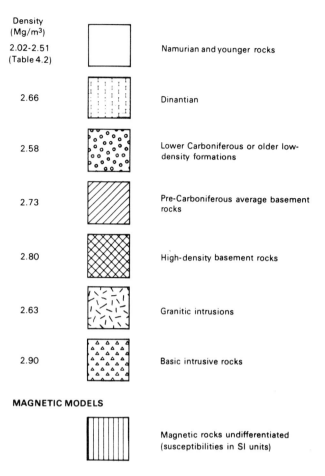

A

GRAVITY MODELS

Density (Mg/m³)		
2.02–2.51 (Table 4.2)		Namurian and younger rocks
2.66		Dinantian
2.58		Lower Carboniferous or older low-density formations
2.73		Pre-Carboniferous average basement rocks
2.80		High-density basement rocks
2.63		Granitic intrusions
2.90		Basic intrusive rocks

MAGNETIC MODELS

Magnetic rocks undifferentiated (susceptibilities in SI units)

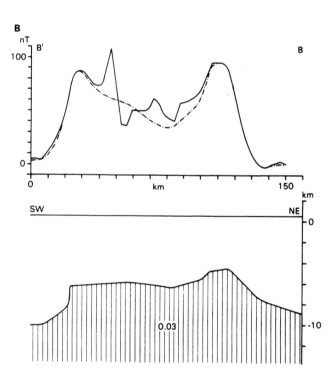

Fig. 4.10 Geophysical profiles across East Midlands

A Key to symbols and abbreviations used for interpretation profiles

B Aeromagnetic profile *B-B'* (see Fig. 4.7 for location) and general intepretation as a deep basement block. Linear background field increasing from − 100 nT at south end of profile to + 20 nT at north end removed from original data

39

The three main basins relevant to this study are Gainsborough, Edale and Widmerpool (see Fig. 5.7). These are separated by the Eakring–Foston and the Erewash Valley highs.

Gainsborough Trough

The Gainsborough Trough was a major depositional basin in which Namurian sediments are two to three times as thick as on the flanks (Kent, 1966). The trough coincides with a Bouguer anomaly low that is particularly well defined by a zone of steep gradients to the north (the Askern lineament, Fig. 4.6B and Plate 8). Neither the steepness of the gradient nor the amplitude of the anomaly is explained by the known change in the Namurian thicknesses. Alternative interpretations of this residual low include an unrecorded thickening of the basinal shales, the presence of a basal Carboniferous sandstone sequence and a pre-Carboniferous basement boundary. The main interpreted source of the anomaly must lie within about 2 km of the surface, but further seismic reflection profiling or drilling is required to resolve the uncertainty. The Askern lineament is closely paralleled by a magnetic lineament to the south (Fig. 4.8, ?a continuation of the Nocton lineament), which suggests the existence of igneous rocks at depth beneath the Gainsborough Trough.

The regional geophysical data suggest thát the Gainsborough Trough is restricted to the area of ESE-trending basement structures; to the east it terminates where the structural grain swings around towards the south to form the pre-Carboniferous basement high at Nocton (Fig. 4.9) and to the west it decreases in amplitude as the strike becomes more E–W. The Gainsborough Trough is flanked to the south by a pronounced Bouguer anomaly low that extends from near

Newark to the Rotherham area. This feature has been interpreted as either a concealed granite or a basin of low-density sediments of lower Carboniferous or greater age (Rollin, 1978, 1982 and Fig. 4.11) (cf. section on granite in this chapter). In the western part of the Gainsborough Trough the absence of a residual gravity low (at about [460 405] in Map 2) indicates that the accepted thicknesses for the Namurian are probably realistic.

In the extreme north-west it is suggested (Chapter 5) that the deep structure is similar to that in the Gainsborough Trough area, although evidence is sparse. The Bouguer anomaly low at Huddersfield is comparable with that at Rotherham, with a particularly well-defined southern margin (the Holmfirth lineament, Figs. 4.8 and 4.9). It is separated, however, from the Rotherham anomaly by a north-easterly trending gravity high (Map 2), perhaps due to thinning of the Namurian or Dinantian sequence and rise in the basement.

Edale Trough

There is no distinct Bouguer anomaly feature associated with the more restricted basin of Namurian sediments on the NE side of the Derbyshire Dome. Bouguer anomaly variations in the area are, however, dominated by the westward increase in values towards the dome (reflecting uplift along the Pennine axis, including possibly that of the high-density basement). The southern margin of the trough coincides with the postulated Carboniferous igneous intrusions along the Chesterfield magnetic lineament (Fig. 4.21) as well as the NNW-trending Mansfield Anticline.

Widmerpool Gulf

The most southerly of the Namurian basins, the Widmerpool Gulf, is associated with a pronounced Bouguer

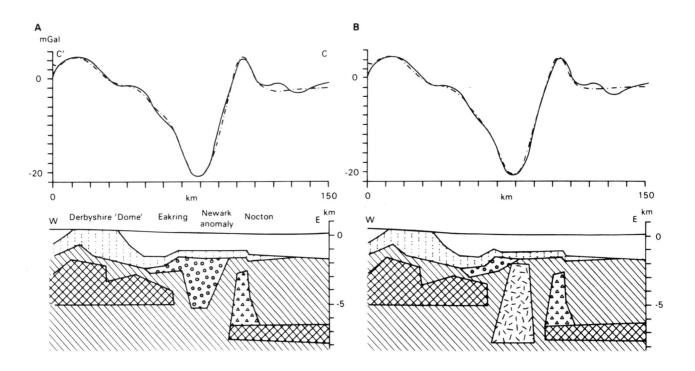

Fig. 4.11 Geophysical profiles across East Midlands

A Bouguer anomaly profile *C-C′* (Fig. 4.7) and general model in which lows are interpreted as basins of low-density Lower Carboniferous, or earlier, sediments. In this and most of the subsequent profiles gravity data have been compensated for effect of Namurian and younger rocks (i.e. as in Map 2) and regional field removed. Key to symbols in Fig. 4.10*A*

B Profile as for Fig. 4.11*A* but with low interpreted as an acid intrusion

anomaly low. The amplitude of the anomaly is significantly greater than can be explained by the known thicknesses of Namurian strata, and to the west the low is terminated by the N–S Bouguer anomaly ridge related to the presence of high-density Lower Carboniferous and pre-Carboniferous rocks at shallow depth.

One explanation of the amplitude of the anomaly is that it includes a contribution from a predominantly argillaceous sequence of rocks of Dinantian age (the Widmerpool Formation). Modelling of the gravity data (Rollin, 1980) indicates that a thickness of at least 1 km of these Dinantian rocks is required in addition to the 0.8 km of Namurian sediments (Kent, 1966) (Fig. 4.15). It is possible, however, that the gravity data are distorted in the area by other density variations, perhaps an extension of the high-density basement beneath the Derbyshire 'Dome' or, alternatively, variations associated with the presence of igneous intrusions. This second alternative is suggested by the presence of magnetic anomalies, particularly along the southern margin of the basin (Fig. 4.16), which are thought to be due to Caledonian granodiorites.

Dinantian rocks

In the study region, rocks of Dinantian age are predominantly platform limestones with shales or shale/limestone sequences developed in basinal settings. The variation from carbonate to argillaceous rocks is reflected by changes in density; samples of pure limestone almost invariably give values of about 2.70 Mg/m³, but this is reduced to 2.60–2.68 Mg/m³ by the effects of jointing, cavities, etc., whereas the shales probably have densities similar to those of Namurian rocks (about 2.51 Mg/m³). Dolomitisation might be expected to increase densities (the density of pure dolomite is 2.80 Mg/m³), but there is little evidence of this on a significant scale—a detailed gravity profile across a dolomite/limestone boundary in Derbyshire, if anything, indicates the reverse. The Dinantian limestones are underlain at a few localities (boreholes at Caldon Low and Eakring) by sandstones and conglomerates of basal Carboniferous or Devonian age with densities of about 2.64 Mg/m³ (Table 4.2). There are thus several possible sources of Bouguer gravity anomalies in the Dinantian sequence.

The Dinantian limestones of the Derbyshire 'Dome' coincide approximately with high Bouguer anomaly values, but evidence suggests that the latter are due largely to high-density pre-Carboniferous basement rocks; for example, the gravity high persists on the stripped gravity map, whereas the margins of the high correspond only generally with those of the limestone (Fig. 4.21). Moreover, the high can be traced southwards into the Leicestershire Coalfield, where Dinantian limestones are largely absent from the Carboniferous sequence.

The relationship between geophysical anomalies and the structure and mineralisation of the Derbyshire 'Dome' is discussed in greater detail later in this chapter.

Carboniferous igneous activity

Intrusive and extrusive igneous rocks ranging in age from late Dinantian to Westphalian occur at several horizons in the Carboniferous sequence. They are known over an extensive area eastwards from the Derbyshire Dome, including a large part of the concealed coalfield. In many places they have been proved only in scattered boreholes and their subsurface extent is uncertain. For the purpose of the present study it is relevant to understand the distribution of the igneous rocks and their relationship to the major structures of the region.

Many of the basic rocks are magnetic, but it is not always possible to decide if they are responsible for significant anomalies without detailed ground magnetic surveys. Where igneous rocks occur at the surface in the Derbyshire 'Dome' they are indicated on the aeromagnetic map by relatively minor anomalies, which are

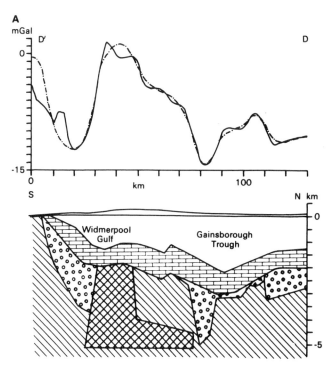

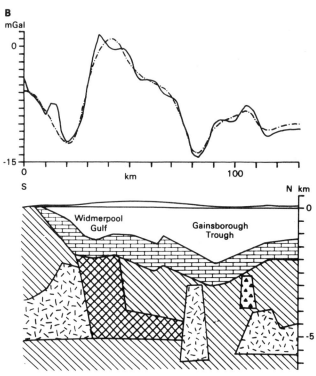

Fig. 4.12 Geophysical profiles across East Midlands

A Bouguer anomaly profile *D–D'* (Fig. 4.7) and general interpretation in which lows are interpreted as basins of low-density sediments

B As for *A* but with lows interpreted as acid intrusions

41

largely confined to an area underlain by a more deep-seated magnetic body (Fig. 4.16). Ground magnetic surveys reveal pronounced anomalies over the exposed lavas and sills (but not the tuffs) and the susceptibilities of these rocks are high (*ca*0.04 SI units); the conclusion is therefore that the bodies are not large enough to produce anomalies at the altitude of the aeromagnetic survey (305 m).

Drilling has proved greater thicknesses of volcanic rocks in the subsurface, which, in places, are at shallow depth (Aitkenhead and others, 1985) and it is probable that these thick accumulations are responsible for some of the magnetic anomalies to the east of the Derbyshire 'Dome'. The volcanic centres recognised by Aitkenhead and others (1985) do not coincide precisely with the magnetic anomalies (Fig. 4.16), but the geological evidence for the distribution of the thickest accumulations is incomplete. It is likely (Cornwell in Aitkenhead and others, 1985) that the elongated magnetic anomaly $A-A'$ (Fig. 4.16) represents a linear accumulation of thick volcanic rocks, with a high proportion of basalt, extending to the WNW from near Fallgate, where more than 290 m of these rocks was proved. This interpretation is supported by depth estimates (Fig. 4.16) on magnetic profiles across some of the anomalies which place the source rocks within the Carboniferous sequence. The anomaly $A-A'$ (Fig. 4.16) forms part of an extensive structure (the Grantham lineament) crossing the East Midlands area (Fig. 4.8), suggesting that this could have been an active fault and source of igneous material.

To the east additional evidence of extensive Carboniferous volcanic activity comes from drilling in the Vale of Belvoir, where suites of both extrusive and intrusive rocks have been proved in the Namurian and Westphalian sequences (Edwards, 1951; Burgess, 1982). Two pronounced magnetic anomalies (B1 and B2 in Fig.

4.16), however, are of unknown origin, but indicate the existence of strongly magnetised localised bodies, perhaps plug-like in form, within the Carboniferous. Anomaly C is more extensive and follows the margin of the Widmerpool Gulf (Fig. 4.15), but interpretation suggests that the source may not be in Carboniferous rocks. An 18 km long NNW-trending anomaly ($D-D'$ in Fig. 4.16) coincides in part with the thick sequence of volcanic rocks and dolerite sills beneath the Vale of Belvoir (Burgess, 1982). These may be responsible for some of the shorter-wavelength anomalies in Fig. 4.17 (possibly including the effect of a strong remanence component), but a more deep-seated source is also suspected.

To the east and south-east a group of local small-amplitude anomalies probably represents less extensive ?Carboniferous igneous centres. Anomaly E (Fig. 4.16) shows a particularly pronounced elongation along a NW–SE trending fault, one of a group of such structures in this area. This SE group of anomalies indicates relatively small magnetic bodies, perhaps at the eastern limit of the igneous province, but there is a general tendency otherwise for the amplitude and size of the magnetic anomalies to increase to the SE in a belt extending from the Derbyshire 'Dome' to anomaly $D-D'$ (Fig. 4.16), suggesting that the magnetite-bearing component, at least, of the igneous activity increased with decreasing age during the Carboniferous.

North of the SSE-trending belt of magnetic anomalies attributed to Carboniferous igneous activity, low-amplitude anomalies extend in a north-westwards direction through Bothamsall and Egmanton. Although they are attributed in this report, because of their coincidence with Bouguer anomaly lows (Fig. 4.14), to pre-Carboniferous intrusions, borehole evidence (Edwards, 1967) indicates volcanic activity in the area in Namurian times. The form of the magnetic anomaly can be mod-

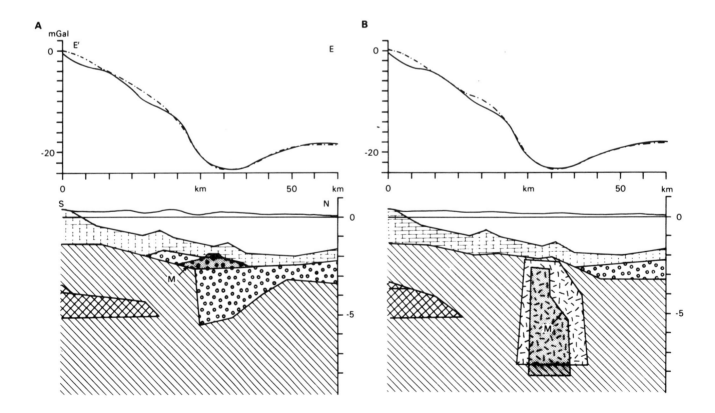

Fig. 4.13   Observed Bouguer anomaly profile $E-E'$ (Fig. 4.7) and alternative models. The magnetic bodies (M), derived from interpretations of an aeromagnetic profile, could represent volcanic rocks in A or part of low-density igneous intrusion in B

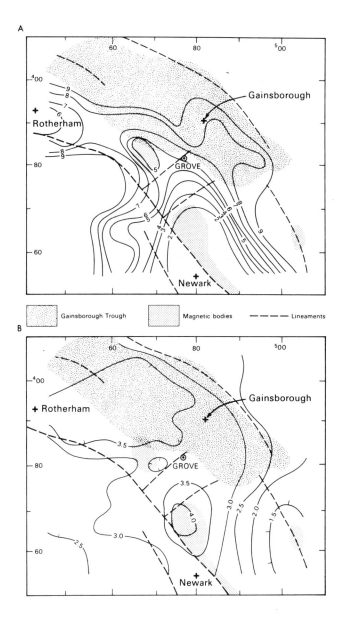

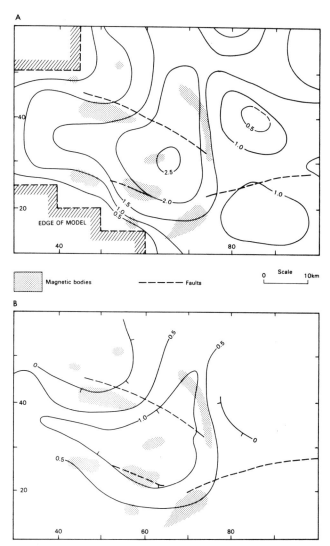

Fig. 4.14 Models (with surfaces contoured at 1 km and 0.5 km intervals below OD) derived from interpretation of Bouguer anomaly data for Newark–Rotherham area (from Rollin, 1978), magnetic anomalies and main geophysical lineaments

A Model for granite body with density contrast of −0.10 Mg/m³

B Model for sedimentary basin with density contrast of −0.35 Mg/m³ (compared with −0.15 Mg/m³ in Fig. 4.11)

Fig. 4.15 A, model derived from interpretation of Bouguer anomaly data for Widmerpool Gulf with depth contours (at 0.5 km intervals) (Rollin, 1980) on base of low-density sediments (density contrast −0.27 Mg/m³) and B, interpreted thicknesses of pre-Namurian low-density sediments derived from the model in (A). Faults based on geological evidence

elled either by a tabular body of magnetic material in the lower Carboniferous or a more deep-seated body (?the Newark granite).

4.5.3 Pre-Carboniferous Basement Rocks

General

The sparse geological information provided by the scattered boreholes that have penetrated pre-Carboniferous rocks indicates a basement of Lower Palaeozoic shales and mudstones, quartzites and volcanic rocks, some with Charnian (late Precambrian) affinities. Wills (1978) proposed the existence of two ridges of Precambrian rocks, trending approximately NW–SE, beneath a cover of Cambrian sedimentary rocks. Away from these highs,

Ordovician and Silurian sediments are preserved in synclinal structures, the most easterly of which extends beneath the North Sea.

The aeromagnetic data produce the most convincing evidence for the deeper basement rocks in the form of long-wavelength anomalies and smooth widely spaced contours, particularly in the north-east of the area. The sources of the anomalies are suggested to be at depths of 3–4 km on the basis of interpretation of discrete anomalies in the east. It is probable that the magnetic basement drops away towards the River Humber to depths in excess of 10 km across a boundary in the NE of the area. The south-western boundary is well defined in places by the Derby lineament (Fig. 4.22). The 100 km wide block thus defined (Fig. 4.10B) appears to become more complex to the south as it approaches the surface, with a central non-magnetic zone. Immediately to the south of the study area the magnetic basement rocks are truncated by a major east–west lineament and it is unlikely that this type of basement continues beneath the London–Brabant Platform. It is probable that the non-magnetic rocks overlying the magnetic basement are

43

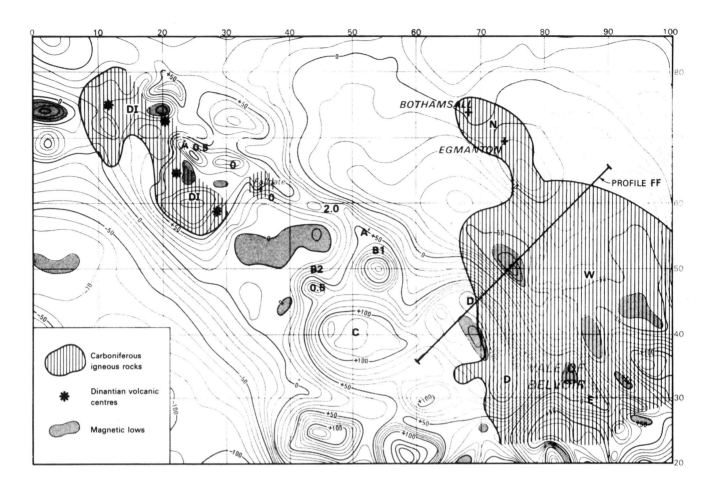

Fig. 4.16 Aeromagnetic map of Derbyshire 'Dome'–Vale of Belvoir area and the extent of Carboniferous igneous rocks (DI—Dinantian, N—Namurian and W—Westphalian). Lettered anomalies referred to in text. Interpreted depths rounded off to 0.5 km intervals

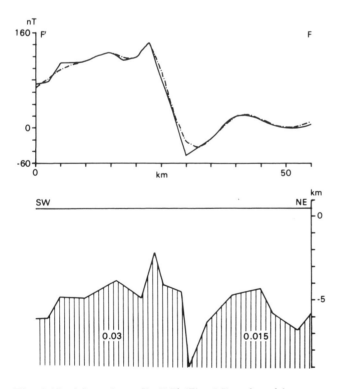

Fig. 4.17 Magnetic profile *F-F'* (Fig. 4.7) and model comprising magnetic basement rocks with two susceptibilities. The interpreted ridge at about 25 km probably represents igneous rocks within the Carboniferous sequence.

Lower Palaeozoic sediments, and thus reach their maximum thickness in the study region in the extreme NE and SW.

The magnetic basement is most likely to be Precambrian in age and probably represents a continuation at increasing depths of the crystalline basement rocks that occur around The Wash (Fig. 4.20). This basement differs from the Charnian, which has no distinct magnetic signature, but a comparison can be made with the Uriconian rocks or to the Malvernian of the Welsh Borders, both of which are associated with pronounced magnetic anomalies.

The Bouguer anomaly data are more difficult to interpret in terms of deep structure, partly because of the existence on the stripped gravity map (Map 2) of anomalies that could be due to thickness and lithological variations of Dinantian rocks, to basement structures or to a combination of both.

Another difficulty in interpreting the gravity data is the lack of evidence from the density measurements (Table 4.2) for well-defined differences between the main basement rock types. The general 'background' Bouguer anomaly level, which is reasonably uniform over the area, is thought to reflect the widespread occurrence of Lower Palaeozoic sediments with an average density of $2.73\,\mathrm{Mg/m^3}$. A high-density basement is required, particularly in the west of the study region (e.g. Fig. 4.11), to satisfy the gravity data, but there is little evidence from the density measurements for its nature. The interpreted extent of this basement is dependent upon the background value chosen for modelling the data; in the profiles shown (e.g. Fig. 4.11) the high-density basement is comparatively restricted in extent. The clearest evidence for the nature of some of the basement rocks is provided by gravity anomalies that are interpreted as being due to igneous intrusions.

Igneous intrusions

There is evidence from the regional gravity and aeromagnetic data for the existence of intrusions of acid, intermediate and basic composition in the pre-Carboniferous basement of the East Midlands. The age of the intrusions is conjectural, but Caledonian seems most likely. A BGS borehole at Warboys (Institute of Geological Sciences, 1966, p.49), immediately to the south of the study area and outside the area where Car-

boniferous sediments have been preserved, proved the source of distinct gravity and magnetic highs to be a diorite intrusion. Dating of samples gave a Carboniferous age, although Le Bas (1972) suggested that the rock represents a hydrothermally altered intrusion similar to the Leicester Caledonian suite.

Granites

Granites frequently occur as large homogeneous bodies with densities of about 2.63 Mg/m³, which are lower than most Palaeozoic and Precambrian rocks. The Bouguer anomaly lows are usually ovoid in plan and indicate a considerable depth extent. They are frequently difficult to distinguish from lows due to sedimentary basins, although they can sometimes be recognised by use of derivatives of the gravity field (e.g. Bott, 1962). The ma-

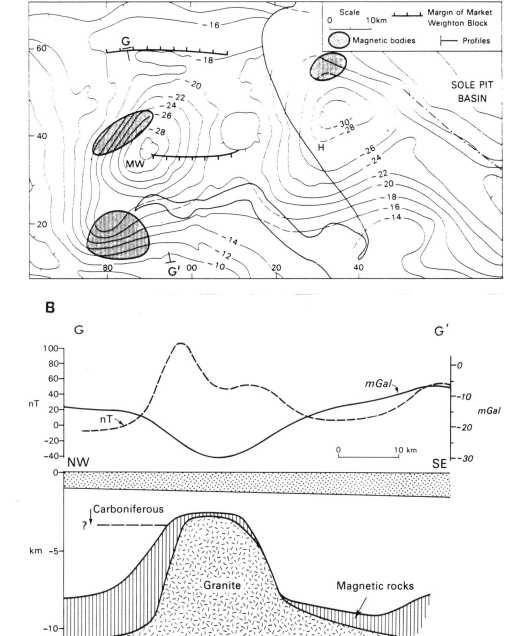

Fig. 4.18 *A*, Bouguer anomaly map of Market Weighton area with contours based on published 1:250 000 scale maps (Fig. 4.1) at 2 mGal intervals. Anomalies: MW—Market Weighton, H—Hornsea. *B*, observed Bouguer anomaly and aeromagnetic profiles (see Fig. 4.18A for location) and model for concealed granite at Market Weighton based on interpretation of geophysical data (from Rollin, 1978)

45

jority of British granites are non-magnetic, but some have flanking magnetic anomalies due to magnetic host rocks or to the formation of magnetic minerals in the metamorphic aureoles. Caledonian granites are known to occur in areas adjacent to the East Midlands, where they form the cores of basement highs beneath the Askrigg and Alston orefields (Figs. 4.2 and 4.3).

Several low-density granitic intrusions are postulated on the basis of the gravity and magnetic evidence.

Market Weighton granite

The combination of a distinct Bouguer anomaly low with flanking magnetic anomalies (Fig. 4.18) at Market Weighton can be modelled by a low-density granite that pierced and uplifted part of an elongated belt of magnetic basement rocks (Bott and others, 1978). The geophysical signature of the Market Weighton granite is reminiscent of that of the Wensleydale Granite (Wilson and Cornwell, 1982), which forms the core of the basement of the Askrigg Block. The relationship between the postulated Market Weighton granite and the subsequent history of sedimentation is complex; during Carboniferous and Jurassic to Lower Cretaceous times it appears to have been an area of relatively little subsidence, whereas in Permo-Triassic and post Lower Cretaceous times its tectonic history is similar to that of adjacent areas (Bott and others, 1978).

East of the Market Weighton granite at the northern margin of the study region another distinct Bouguer anomaly low is intersected at the coast. The centre of the low appears from marine surveys to occur about 10 km east of the coastline (Fig. 4.18), where it reaches a value of −30 mGal in an area where the pre-Mesozoic basement depths exceed −2.5 km below OD. The anomaly has a similar NE–SW elongation to the Market Weighton low and also a flanking magnetic anomaly—points taken to favour an explanation in terms of a granite rather than a sedimentary basin. An alternative interpretation in this area is the presence of thick salt deposits, but the main Permian salt diapirs are confined to the Sole Pit Basin (Day and others, 1981). As the Market Weighton granite has probably been responsible for the uplift of the basement forming the Market Weighton Block (Bott and others, 1978), it is suggested that this second postulated granite could indicate an extension of the block structure as far east as the Sole Pit Trough with the same overall E–W elongation indicated by Kent (1966).

Newark

An elongated Bouguer anomaly low swings round from an E–W trend near Rotherham to follow a northerly trend near Newark. Rollin (1978, 1982) suggested that the Newark low could be due to either a Devonian/lower Carboniferous sandstone basin or a granite. He pointed out that the presence of a granite could explain the zoned Derbyshire mineralisation (to the west), the location of small oilfields (several occur over the Bouguer anomaly low) and the thinner development of the Namurian between the Gainsborough Trough and the Edale Gulf.

The significance of the Newark anomaly is considerable in terms of understanding the deep regional structure of the East Midlands, but the ambiguity in interpreting the gravity data remains a problem. Several lines of evidence, however, seem to favour the granitic intrusion model.

(*a*) The highs on the granite model coincide with magnetic anomalies (Figs. 4.14A and 4.14B) that can be interpreted as being due to slightly magnetic igneous

rocks or to magnetite host rocks overlying the intrusion (Carboniferous igneous activity cannot be ruled out as a source of the magnetic features, however).

(*b*) The general smoothness of the contours and the positions of the maximum gradients near the centre of the anomaly are more satisfactorily explained by a granite intrusion model.

(*c*) The basin model derived with the use of a density contrast of −0.1 Mg/m³ is unrealistically deep (more than 4 km), whereas that shown in Fig. 4.14B, for a contrast of −0.35 Mg/m³, still requires an additional 2 km of sediment.

(*d*) The deep borehole at Grove (Fig. 4.14) should occur near the centre of a concealed low-density sedimentary basin. Pharaoh and others (1988a), however, noted that undated chlorite phyllites occur beneath Dinantian strata.

The evidence that favours a sedimentary basin model includes the existence, on the western flank of the Newark anomaly, of 900 m of Lower Carboniferous conglomerates and sandstones with a density of 2.58 Mg/m³ (Appendix 2) in the Eakring 146 Borehole. The similarity in the trends of the Newark–Rotherham Bouguer anomaly low and the Gainsborough Trough also tends to suggest that the geophysical feature should be related to Carboniferous structures, although it is suggested elsewhere that these may be inherited from older basement structures. In the model shown in Fig. 4.11B the main Newark anomaly is interpreted as a granite, which may have acted as a buoyant core to a concealed basement block.

A Bouguer anomaly low extends north-westwards from Newark to the Rotherham area, where another gravity minimum coincides with a pronounced magnetic anomaly. If the granite model is valid, the Rotherham anomaly must represent a cupola on an elongated and arcuate batholith. However, in this area the available deep geological evidence cannot rule out the model of a local basin of early Dinantian sediment beneath the carbonate shelf at the edge of the Gainsborough Trough. Alternative models based on the gravity and magnetic evidence are presented in Fig. 4.14, the magnetic bodies indicating the granite or ?Dinantian igneous rocks.

The Wash granites

A group of four Bouguer anomaly lows around The Wash has a more complex appearance than those described above. This may be due in part to the relatively shallow depths of the basement (above −1 km OD) in this area, but it also appears to reflect a different structural setting. The basement in The Wash area has high seismic velocities (Chroston and Sola, 1982), is often magnetic and probably comprises mainly crystalline rocks. Some of the margins of the Bouguer anomaly lows are linear and extensive faulting, in a mainly N–S direction, is indicated. The lows are also flanked frequently by magnetic highs (Fig. 4.19).

In order to determine whether the lows are due to granite or to sedimentary basins a seismic refraction profile was performed across one such feature at the eastern margin of The Wash. The relatively high velocities (5.1 to 5.8 km/s) recorded for the sub-Mesozoic basement (Fig. 4.20), together with the absence of any marked relief in the basement surface, argue against the existence of a significant thickness of low-density sediments (Chroston and others, 1987). The existence of granites in this area, suggested by these results, is consistent with the general pattern of high heat flow ascribed by Brown and others

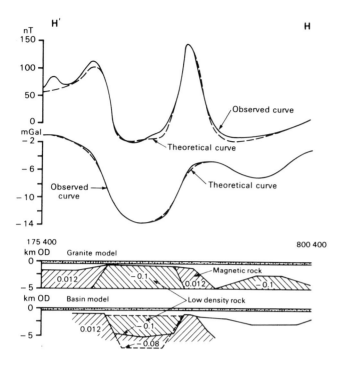

Fig. 4.19 Gravity and aeromagnetic profiles *H–H'* (Fig. 4.7) and alternative interpretations for area of relatively shallow crystalline basement

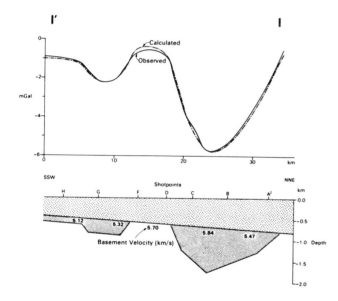

Fig. 4.20 Residual Bouguer anomaly profile and seismic refraction results for profile *I–I'* (Fig. 4.7). Observed depth and velocity of basement refractor superimposed on basin model (with density contrast of -0.18 Mg/m^3) to demonstrate that this interpretation is unlikely; a granite model is preferred as the source of the Bouguer anomaly low

(1980) to the distribution of radioelement-enriched granites.

Intrusions of intermediate composition

Intrusions of intermediate composition typically have densities that are similar to the average (about 2.73 Mg/m^3) for Lower Palaeozoic and older basement rocks. Hence they are unlikely to be identified from Bouguer anomaly data although, as they can contain

significant amounts of magnetite, aeromagnetic data are useful.

One of the most distinctive groups of aeromagnetic anomalies is that interpreted as a belt of granodioritic intrusions comparable with that at Mountsorrel in the south of the area. They do not give rise to Bouguer anomalies (Plate 7), probably because their densities are similar to those of the host rocks, and any weak feature would be difficult to distinguish in areas where variable thicknesses of Namurian sediments give rise to large anomalies. The linear nature of the magnetic anomalies suggests that their location was largely determined by a NNW/NW fracture system—which may also have determined the position of the southern margin of the Widmerpool Gulf (Fig. 4.15). The margins of the intrusions appear to slope steeply to depths in excess of 8 km. Evidence from heat flow measurements and from the geochemistry (Chapter 7) indicates that these granodiorites, although extensive, are not significant radiogenic heat sources.

In the north of the study region a deep-seated magnetic zone is interrupted by a circular negative magnetic anomaly (the Pontefract anomaly in Plate 10) the appearance of which is suggestive of an intrusion or a well-defined basin of sediments. The feature has no gravity response and its origin remains uncertain.

Basic intrusions

Rocks approaching gabbroic composition typically have sufficiently high densities (2.85–2.95 Mg/m^3) to give rise to Bouguer anomaly highs. They also frequently contain magnetite so that a combination of high magnetic and Bouguer anomalies is characteristic. No concealed basic intrusions of pre-Carboniferous age have been identified in the study area previously, but anomalies at Nocton and Spalding are thought to provide evidence of these in the basement.

Nocton

One of the clearest examples of coincident magnetic and gravity anomalies occurs in a basement high at Nocton. The combination of anomalies is sufficiently pronounced to give rise to a distinct response on images of combined data sets (Plate 10) and interpretation of profile data (Fig. 4.11) strongly suggests a narrow, steep-sided form for the intrusion.

The host rock for the proposed basic intrusion is probably the quartzites proved in boreholes in the area, including Nocton itself (Kent, 1967). These have not been included in the model but would produce, if extensive, a particularly large density contrast with basic rocks. The intrusion is elongated along a NNW-trending basement high that Kent (1966) extended to connect with the Askern–Spital high.

A series of magnetic anomalies has been tentatively interpreted as forming a lineament (Fig. 4.8) flanking this high; however, there is no convincing Bouguer anomaly evidence to support the possibility that these magnetic anomalies are due to large basic intrusions. The coincidence of the local basement high at Nocton (Fig. 4.11) with a basic intrusion is difficult to explain unless the igneous rocks formed a resistant feature at the basement surface or were intruded along a fault subsequently reactivated to form the high.

Spalding area

The Bouguer anomaly data for the area around Spalding, SW of Nocton, are characterised by high gravity values that are associated with pronounced magnetic anomalies (Plate 10). The magnetic data, in particular, indicate a complex structure consistent with the presence of a basic igneous complex, perhaps comprising several intrusions or an extensively faulted mass. The basic rocks form part of the crystalline basement around The Wash and the steepness of the Bouguer anomaly gradients in places suggests that acid and basic intrusions abut.

4.5.4 Interpretation for the Derbyshire 'Dome' Area

Unlike the block areas of the northern Pennines, the Derbyshire 'Dome' is not distinguished by any simple characteristic geophysical anomaly comparable with the gravity lows due to the concealed Weardale and Wensleydale granites. The dome, in contrast, appears to coincide with elongated anomalies that extend for considerable distances and are interpreted as lineaments (Fig. 4.8).

The images prepared by superimposing the veins and faults of the Derbyshire 'Dome' area on the Bouguer gravity and aeromagnetic data are shown in Plate 11 and Fig. 4.21 summarises the main geophysical and geological features.

Bouguer gravity data

The Bouguer anomaly data indicate that the Derbyshire 'Dome' is partly underlain by pre-Carboniferous basement rocks of higher than average density (Maroof, 1976, and Fig. 4.10), notably the southern and north-western sections of the Dinantian outcrop (Fig. 4.21). The north-eastern part of the dome, where Bouguer anomaly values are lower, includes the Eyam Borehole, where a considerable thickness (1.8 km) of Dinantian limestones was proved (Dunham, 1973). The decrease in thickness of these rocks towards the Woo Dale Borehole could explain the westward increase of Bouguer anomaly values (Fig. 4.23) if a density contrast of about 0.14 Mg/m³ with the basement rocks is assumed. In Fig. 4.24 a N–S profile, also passing close to the Eyam borehole, includes the steeper gravity gradients associated with the Matlock lineament. In order to interpret this profile it is necessary to introduce a step in the high-density basement in addition to the increasing thickness of the Dinantian, suggesting the presence of faulting in the pre-Carboniferous basement approximately along the line of the Cronkston–Bonsall fault zone (Aitkenhead and others, 1985).

Magnetics

The Derbyshire 'Dome' area contains two distinct groups of magnetic rocks, one of which occurs close to the surface, whereas the other indicates a deep-seated magnetic basement (Fig. 4.10B). The SW margin of this basement block is clearly defined by a gradual decrease of magnetic values to the SW and the form of this anomaly (the Derby lineament) indicates that the margin decreases in dip with depth (Fig. 4.22). To the NW the anomaly becomes less well defined where it coincides with the western end of the Matlock gravity lineament, and further to the west it appears to have a more E–W trend. The evidence of these trends and the depths to the source rocks suggest that the anomaly is due to Caledonian structure imposed on (?) Precambrian basement rocks.

A group of magnetic highs, interpreted elsewhere as being due to Carboniferous igneous rocks, is superimposed on the above feature. The sources of these highs occur at depths of 0.5 km, or greater, and are particularly well developed where Namurian rocks crop out, particularly near their boundary with the Dinantian (the Goyt's Moss, Eyam Moor and Calton anomalies in Fig. 4.21). The distribution of the magnetic anomalies suggests control of the igneous activity by ESE-trending fractures, including the major Grantham lineament.

General comments on the Derbyshire 'Dome'

In several parts of the Derbyshire 'Dome' the coincidence of geophysical anomalies with diverse geological features suggests, possible relationships (Fig. 4.21). The positions of the geophysical lineaments could be particularly important. Hence the Derby lineament corresponds generally with the distribution of Asbian apron reefs, suggesting that the edge of the magnetic basement might have been related to that of the Carboniferous shelf. The western part of the Grantham lineament has been defined on the basis of the linear Calton and Clay Cross anomalies (Fig. 4.21). Some evidence for its continuation to the WNW across the Dinantian outcrop is provided by the occurrence 'in line' of two volcanic vents at Monk's Dale (Stevenson and Gaunt, 1971). To the north of the Grantham lineament the Eyam Moor and Barlow Moor anomalies may represent another parallel structure, perhaps related to the westward displacement of the Dinantian outcrop and to the coincident mineral veins around Great Hucklow.

The coincidence of the Matlock lineament with the Cronkston–Bonsall fault zone is the clearest example in the area of a correlation of surface features with a deeper structure. Several alternative models are possible, but they all require a northward-dipping, higher-density basement, in addition to thickening of the Dinantian rocks (Fig. 4.24). This, together with evidence presented elsewhere (Chapter 5), suggests that the Matlock lineament formed part of a growth fault system in Dinantian times. The presence of dolomitized limestone along much of the Cronkston–Bonsall fault zone also suggests some relationship to more deep-seated structures, the dolomitisation perhaps resulted from fluids percolating upwards through these fractures rather than the downwards flow as suggested by Dunham (1952).

North of the Matlock lineament the Bouguer anomaly values decrease eastwards (Fig. 4.23), probably largely in response to the increasing thickness of Dinantian rocks. There is no evidence at the surface for any corresponding N–S-trending structural features.

There is little evidence for any direct relationship between the more deep-seated magnetic and higher-density basement rocks and the mineralisation except that the mineral veins occur generally in the area underlain by magnetic basement (Plate 11b).

Although several lineaments have been identified from the geophysical evidence and also in places from surface evidence, they do not seem to have directly affected the distribution of the mineralisation. Faults in the Cronkston–Bonsall zone, in particular, are rarely mineralised, although the margins of the dolomite in the same area were reported (Smith and others, 1967) to contain ore deposits.

The mineralisation of the Derbyshire 'Dome' is most intense near the eastern margin of the limestone outcrop; the magnetic evidence that the Carboniferous activity is also greatest to the east could be relevant as these rocks would have been associated with increased heat flow.

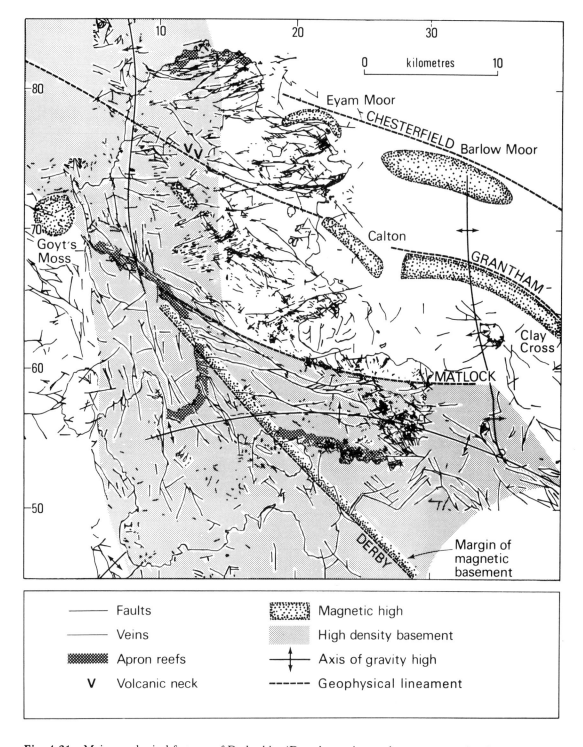

Fig. 4.21 Main geophysical features of Derbyshire 'Dome' superimposed on computer plot showing distribution of veins and faults (Map 1; cf. Plate 11)

4.6 Summary

The regional geophysical data (Bouguer gravity and aeromagnetic) for the East Midlands have been processed and interpreted to provide information on the main structural features of the area. The emphasis has been on improving the understanding of the deep geology, where the information complements and extends that obtained from seismic reflection surveys and boreholes (Chapter 5). The interpretations were carried out with conventional techniques (mainly forward modelling), but extensive use was made of an image analysing system (I²S) for data presentation and analysis. The advantages of this second approach include (1) the ability to examine large individual data sets using different methods of presentation, (2) the ability to combine data sets to examine their interrelationship, (3) the existence of computational facilities to rapidly process data and (4) improved recognition of lineaments and structural 'grain', particularly through the use of shadowgrams.

Further details on the use of the IAS are given in Appendix 1.

The presentation of regional geophysical data is a valuable contribution to the examination of combined data sets by use of the IAS, but it is important that the geological significance of the anomalies be assessed before firm conclusions are drawn concerning correlation. Within the study region, for example, Bouguer anomaly

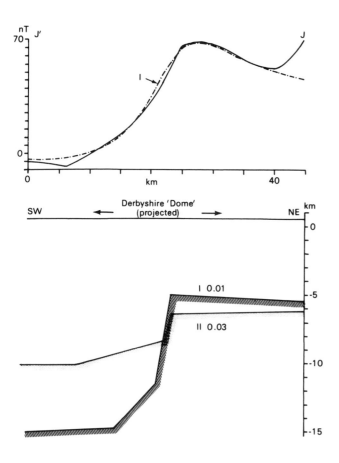

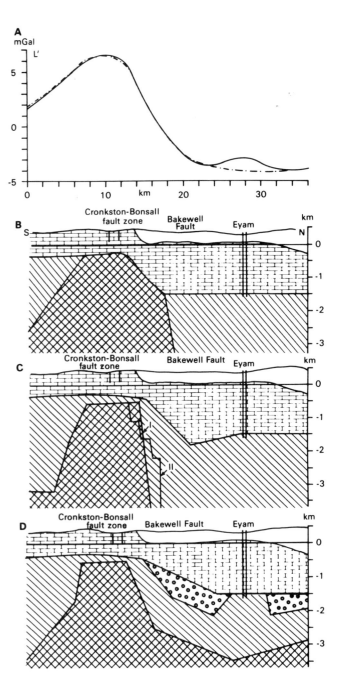

Fig. 4.22 Magnetic profile $J-J'$ (Fig. 4.7) and models (I and II) with different susceptibilities (in SI units) for margin of deep magnetic basement (part of profile $B-B'$, Fig. 4.10B). This profile does not cross the Derbyshire 'Dome' but its projected position is indicated

Fig. 4.24 A, gravity profile $L-L'$ (Fig. 4.7) and alternative models ($B-D$) for high-density basement (calculated profile shown is for model B and for model C alternatives I and II produce similar profiles)

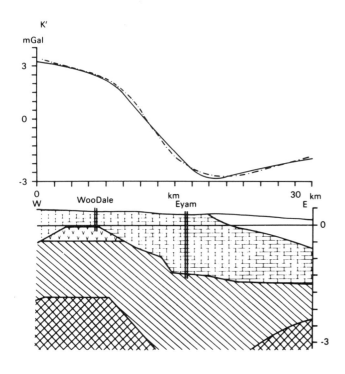

Fig. 4.23 Gravity profile $K-K'$ (Fig. 4.7) and model. V—volcanic rocks (density, 2.60 Mg/m³)

lows can indicate granite intrusions or sedimentary basins. Attempts have been made in the present study to identify the sources of major anomalies, but the ambiguity inherent in potential field interpretation means that uncertainties remain. The conclusions reached concerning the main gravity and magnetic anomalies in the East Midlands are summarised below.

The magnetic data suggest that much of the study region is underlain by magnetic basement rocks from depths of 3–4 km to more than 10 km. A particularly well-defined SW margin to this NW–SE-trending zone of (?Precambrian) magnetic rocks extends SE from the Derbyshire 'Dome'. Other magnetic anomalies exist that indicate source bodies at shallower depths (0–3 km) and three or four geological interpretations are suggested. One group of anomalies around The Wash, in the SE of the study region, is probably due to igneous or metamorphic rocks within a crystalline basement immediately beneath the Mesozoic cover rocks—this basement may be the extension of the magnetic basement that underlies much of the study region to the north. The well-defined anomaly at Nocton probably represents a basic intrusion, perhaps of Caledonian age, whereas granodiorite intrusions of the same age are thought to be responsible for a group of anomalies that, in part, flank the southern margin of the Widmerpool Gulf. Finally, it is suggested that a zone of magnetic anomalies extending to the southeast from the Derbyshire 'Dome' indicates an extensive area of Carboniferous igneous activity. Evidence from outcrops and boreholes had proved the existence of extrusive and intrusive igneous rocks ranging in age from Dinantian to late Westphalian, but the magnetic data provide evidence of their distribution and, in particular, the significance of a postulated lineament (the Grantham lineament) as an important structural feature.

Gravity stripping was carried out to remove the effect of variations in the thickness of lower-density rocks of Namurian and younger ages, but the continued existence of many anomalies on the stripped map points to sources within Dinantian or older rocks. In some cases, such as the association of gravity lows with thick Namurian basins at Gainsborough and Widmerpool, it is probable that the anomalies are due to density changes within the Carboniferous sequence. These can arise from thick sandstone conglomerate basins of early Dinantian age or thick limestone – shale sequences postulated on the basis of the tilted block model for early Carboniferous sedimentation (Chapter 5). A significant zone of gravity lows, culminating in minima at Newark and Rotherham, occupies a central position in the study region. Its origin remains uncertain, but the evidence tends to favour the existence of concealed granitic intrusions in the basement. Other gravity lows regarded as being due to low-density intrusions occur at Market Weighton and around The Wash. Additional evidence that one of the latter is due to a granite was obtained from seismic refraction surveys. In the model proposed for the structural history of Carboniferous rocks in the East Midlands, granite intrusions are considered to be relevant as being indicative of basement highs or central positions of tilted blocks.

A series of lineaments recognised from gravity and magnetic data correlates with known faults in places and provides, evidence for the existence of major structures in the basement, some of which were probably active in Carboniferous times. It is suggested that some of the ESE- or SE-trending lineaments in the Caledonian basement were reactivated in Carboniferous times.

The possible significance of some of the features recognised from the regional geophysical data is discussed in Chapter 8 in terms of their potential relevance to the formation and location of mineral deposits in the East Midlands.

5 DEEP GEOLOGY

K. Smith and N. J. P. Smith

5.1 Deep geological data

In addition to the geophysical methods outlined in the previous chapter, the deep geological interpretation of the East Midlands is based on numerous widespread deep boreholes and on a network of seismic reflection profiles.

5.1.1 SEISMIC REFLECTION

The first attempt to use seismic reflection in the East Midlands in the 1930s was abandoned when the method did not give an effective response to basement structure (Lees and Cox, 1937). As a result, early isochron maps of the area were based on the seismic refraction method, which clearly defined a deep high-velocity refractor corresponding to Dinantian limestones or pre-Carboniferous basement (Lees and Taitt, 1946; Kent, 1966; Chapter 4). Subsequent improvements in data acquisition and processing have since established seismic reflection as the premier tool of subsurface investigations in the area. Most seismic data are currently acquired on behalf of companies engaged in licensed hydrocarbon exploration.

The BGS Deep Geology Research Group occasionally acquires seismic reflection data on its own behalf. It has conducted two independent seismic reflection surveys in the study area as part of geothermal investigations on Humberside and in south Derbyshire. Alternatively, it may collaborate with an oil company and arrange to record and process deeper data from a standard seismic reflection traverse. Coherent events at lower crustal depths can be recognised on records that extend up to 15 seconds two-way travel time. Whittaker and Chadwick (1983) and Whittaker and others (1986) described several examples of data acquired by this method in the East Midlands.

Other relevant data include the Leicester University seismic reflection profile, CHARM, which extends northeastwards from Charnwood Forest into the southern margin of the study area (Maguire and others, 1983; Maguire, 1987).

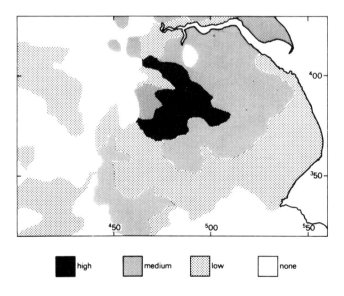

Fig. 5.1 Density of coverage by seismic reflection profiles in East Midlands

The density of seismic reflection data in the study area is shown in Fig. 5.1. Most of the lines were acquired in the surge of onshore hydrocarbon exploration activity that began in 1974 and continues to the present day. The major seismic reflectors that can be usefully interpreted in the East Midlands are listed in Table 4.2. The Mesozoic reflectors are well defined and confirm the simple structure of the Eastern England Shelf that is apparent from outcrop and borehole evidence (Chapter 2). Seismic reflection data can be used to constrain Mesozoic fault interpretations, which must otherwise be based on widely spaced exposures, drift-covered outcrop and sparse subsurface information.

On most seismic reflection profiles in the East Midlands Carboniferous rocks do not form the continuous coherent reflectors that are characteristic of the overlying Mesozoic sequence. This loss of definition is partly a function of the general increase in depth and structural complexity of the Carboniferous beneath the Eastern England Shelf and partly due to the lateral variation of Carboniferous lithologies, particularly in the Silesian. A widespread reflector within the Westphalian sequence has been correlated with the Top Hard, or Barnsley, coal seam of Westphalian B age. The correlation is broadly confirmed by numerous boreholes in the concealed East Midlands coalfield. Slight discrepancies between the reflector and the coal seam in areas of good stratigraphic control suggest that the unexpectedly strong reflection from this part of the Westphalian B sequence may be attributed to the constructive interference of several reflections from a group of closely spaced coal seams in the vicinity of the Top Hard Coal. The Westphalian B reflector provides a good link between the seismic reflection data and structural information obtained from the East Midlands coalfield.

Stratigraphically, the base of the Westphalian may be recognised by the increase in the thickness, lateral continuity and proportion of sandstones in the sequence. Mid-Silesian sandstones such as the Crawshaw Sandstone and the Rough Rock are variable, but widely developed. They give rise to an ill-defined reflector that is taken to correspond with the top of the Namurian. In many places other stratigraphic evidence must be considered before the top Namurian reflector can be interpolated through a broad zone of incoherent reflections.

The strong contrast in seismic velocity between sandstones and shales of Namurian age and Dinantian carbonates (Table 4.2) defines clearly the top of the Dinantian sequence in carbonate platform areas. In late Dinantian basins, however, the top of the Dinantian lies within a conformable sequence of Brigantian–Pendleian shales and does not correspond to a major change in acoustic impedance. In these latter areas the top of the Dinantian is interpolated by use of stratigraphic evidence.

The widespread facies change that took place in the East Midlands at the end of the Arundian stage is recognisable from seismic data. In places events are seen that separate fault-controlled early Dinantian sequences from the overlying deposits of the late Dinantian carbonate shelf.

The base of the Dinantian is not well-defined by the seismic data. In some areas a sequence of high-velocity platform carbonates rests directly on Lower Palaeozoic basement. In others early Dinantian limestones, dolomites and anhydrites overlie thick sequences of sandstones and conglomerates of late Devonian and early Carboniferous age. With the aid of the small number of boreholes that penetrate Lower Palaeozoic basement it is possible, in parts of the area, to interpolate the top of the Lower Palaeozoic. The regional variations in Dinantian thickness that arise from the interpretation of the Lower Palaeozoic surface have been incorporated in the model of Dinantian basin formation that is presented below. In the absence of borehole control, the depth conversion of seismic reflection profiles across the deep Dinantian basins remains speculative. Recent improvements in the parameters of data acquisition and processing have produced a considerable enhancement in the quality of early Carboniferous and top Lower Palaeozoic seismic reflections. A corresponding improvement in seismic interpretation would be brought about by drilling boreholes into the deeper parts of the basin to act as stratigraphic controls for the early Carboniferous sequence.

The deep seismic reflection evidence for pre-Carboniferous basement variation in the East Midlands has been described by Whittaker and Chadwick (1983) and Whittaker and others (1986). In the north stratigraphic interpretations of the short, discontinuous reflections from within the basement are not yet possible. At the south-western margin of the study area, however, a series of well-defined basement reflections probably corresponds to the relatively undeformed Lower Palaeozoic platform sediments of the Midlands Microcraton.

The potential of the seismic reflection method for the direct detection of metalliferous mineralisation has not been investigated during this study. The alternative approach, adopted here, has been to use the structural interpretations to locate suitable tectonic environments in which mineralisation may have taken place. This use of the seismic reflection method then becomes closely analogous to its application in hydrocarbon exploration, where it is used, almost exclusively, to identify potential sites of hydrocarbon accumulation, which must be subsequently tested by drilling.

The disposition of the pre-Carboniferous basement, lateral thickness changes in Carboniferous sedimentation and the pattern of Variscan faulting can all be inferred from this structural interpretation and may have acted as significant controls on carbonate-hosted mineralisation.

5.1.2 BOREHOLES

The borehole database of the East Midlands incorporates boreholes from five main sources: (1) Coal exploration, (2) Hydrocarbon exploration, (3) BGS stratigraphic boreholes, (4) Metalliferous mineral exploration and (5) Other shallow boreholes, including those drilled for water, site investigation and bulk mineral assessment.

The large data sets acquired in the course of coal and hydrocarbon exploration are most useful in regional geological analysis of the Carboniferous basins and are considered in greater detail below. The remaining categories are of lesser significance, but they provide useful information in certain areas.

Coal exploration boreholes

The exposed Yorkshire and East Midlands Coalfield is outlined in Fig. 2.5 together with that part of the concealed coalfield which has been proved solely by coal exploration. Eastwards, the extent of the coalfield has been established largely as a result of hydrocarbon exploration (Howitt and Brunstrom, 1966). Most of the Geological Survey memoirs of the area incorporate struc-

tural contour maps of the coalfield, which generally use the Top Hard coal as a structural datum (Wray and others, 1930; Edwards and others, 1940; Bromehead and others, 1933; Mitchell and others, 1947; Eden and others, 1957; Smith and others, 1973; Smith and others, 1967; Edwards, 1967; Frost and Smart, 1979).

These maps were replotted by the Deep Geology Research Group at a standard scale (1:50 000) and contour interval (25 m). Scattered borehole data from the concealed Belvoir and Selby coalfields were incorporated in the regional synthesis. Exploratory boreholes for coal are usually cored completely, and may occasionally provide substantial intersections of pre-Carboniferous basement. A few basement cores supplied by the National Coal Board (now British Coal) have been subjected to geochemical and geophysical analysis in this study (Chapters 4, 6 and 7).

Hydrocarbon exploration boreholes

The distribution of hydrocarbon boreholes in the East Midlands shows major clusters of productive wells at each of the oilfields in the area (Fig. 5.2). The older hydrocarbon boreholes were largely uncored and were logged by a combination of cuttings examination and drilling observations, such as the variation in the rate of penetration and water flow. Early geophysical logging of the boreholes was based on self-potential (SP) and resistivity logs, but the SP method was rapidly superseded as an indicator of lithological variation by the gamma (radioactivity) log. The complete suite of geophysical logs run conventionally in most modern hydrocarbon boreholes provides additional data on the dip, density, sonic velocity, temperature and porosity of the intersected formations. The main application of geophysical logging to the present study has been in regional stratigraphic correlation (Whittaker and others, 1985). Hydrocarbon boreholes are commonly drilled with a limited amount of core recovery. Cores are routinely taken at terminal depth and at the intersection of target reservoirs that, in the East Midlands, largely comprise Silesian sandstones. By themselves, sandstone cores are not stratigraphically useful in repetitive fluvio-deltaic sequences, but occasionally they incorporate parts of adjacent marine bands that can be identified palaeontologically. Marine bands can be recognised on geophysical well logs by their distinctive combination of high radioactivity and low sonic velocity. The geophysical correlation of particular marine bands between hydrocarbon boreholes, in conjunction with outcrop and cored borehole data, enables regional thickness trends to be established. Namurian and Westphalian isopachyte maps are presented in Figs. 5.8 and 5.11. Hydrocarbon boreholes are commonly terminated after proving the base of the Silesian, and rarely intersect the complete Dinantian sequence in the East Midlands. All those boreholes which penetrate the concealed Dinantian are plotted in Fig. 5.3, together with selected boreholes that have been drilled within the Dinantian outcrop. Details of many of the Dinantian intersections have been summarised by Ineson and Ford (1982).

5.2 Structural-stratigraphic model

5.2.1 INTRODUCTION

The East Midlands area lies within the Pennine Basin, a major Carboniferous basin that extends between the

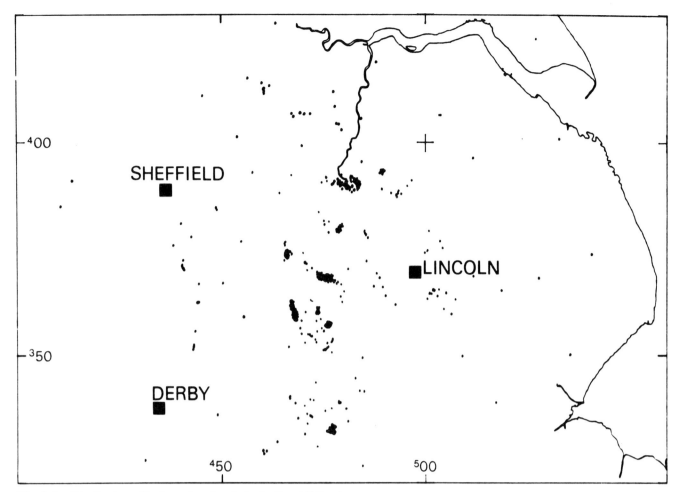

Fig. 5.2 Distribution of hydrocarbon boreholes in East Midlands

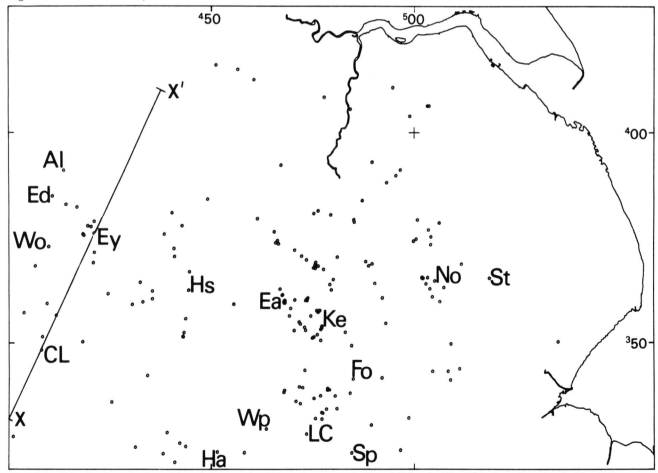

Fig. 5.3 Distribution of selected boreholes proving Dinantian strata in East Midlands (X– X' is line of section in Fig. 5.6). Boreholes: Al, Alport: CL, Caldon Low; Ea, Eakring; Ed, Edale; Ey, Eyam; Fo, Foston; Hs, Hardstoft; Ha, Hathern; Ke, Kelham; LC, Long Clawson; No, Nocton; Sp, Sproxton; St, Stixwould; Wp, Widmerpool; Wo, Woo Dale

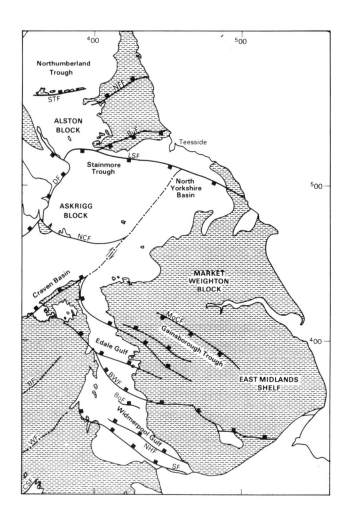

Fig. 5.4 Simplified geological map of pre-Permian surface of eastern England

Key to ornament – Westphalian, dashed; Namurian and Dinantian, blank; pre-Carboniferous, mottled. Some of the faults that affected early Carboniferous sedimentation are shown: BF, Bala; BoF, Bonsall; BuF, Butterknowle; BWF, Bakewell (Barkston in the east); CSF, Church Stretton; DF, Dent; LSF, Lunedale–Staindrop; MoCF, Morley–Campsall; NCF, North Craven; NFF, Ninety Fathom; NHF, Normanton Hills; SF, Sileby; STF, Stublick; WF, Wem. Position of concealed faults in East Midlands partly conjectural

controlled thickness and facies variation in the Carboniferous of the Pennine Basin has proved to be a useful interpretative tool (Miller and Grayson, 1982). The inlier of Dinantian carbonates, which forms the Derbyshire 'Dome', was formerly identified as a horst block, by analogy with the blocks of the northern Pennine area. After the Eyam Borehole (Dunham, 1973) proved that part of Derbyshire was underlain by a thick Dinantian sequence. Miller and Grayson (1982) attempted to explain facies variation in the area by a tilt block model (see Chapter 2). Their interpretation has been questioned by Smith and others (1985), who used geological and geophysical evidence to propose that Derbyshire was underlain by two southwesterly dipping tilted blocks. Seismic reflection profiles in the East Midlands indicate that tilted blocks or half-graben are a common feature of the area.

In this section a new structural model of the southern Pennine basin is presented that modifies and extends the interpretation of Smith and others (1985). With the aid of simplified maps and schematic sections the geological evolution of the area from the Caledonian orogeny to the present day is briefly considered. A comprehensive discussion of regional Carboniferous stratigraphy is not attempted here, although some stratigraphic details are included where they are relevant to the structural scheme. Simplified stratigraphic tables are illustrated in Chapter 2. The implications of the structural model for metallogenesis in the Pennine Basin are discussed in Chapter 8.

5.2.2 Origin of the Southern Pennine Basin

The tectonic models that have been proposed to explain the origin of Carboniferous basins in NW Europe have been critically reviewed by Haszeldine (1984). The main hypotheses all broadly support the same plate tectonic reconstruction of the Carboniferous, but differ in their assessment of the most important tectonic processes that were operating at the time. To some extent these differences have arisen because of the complexity of the European Variscides, for which a generally accepted plate tectonic scheme does not yet exist. Haszeldine (1984) has divided the various proposals into three groups, each of which has recourse to a different plate tectonic process.

(1) *Subduction models* relate basin formation on the Variscan foreland to the existence of a subduction zone in central Europe. Various mechanisms have been proposed by which tension may be created in the foreland. (*a*) Ductile creep: in this hypothesis tension is generated by slow ductile movement of the lower crust towards the subduction zone (Leeder, 1976). (*b*) Slab-pull: this force acts on a downgoing lithospheric plate. Its application in this instance depends upon an interpretation of NW Europe as forming part of a plate subducting to the south (Bott and others, 1984). (*c*) Back-arc tension: this process invokes a northward-dipping subduction zone in central France with a marginal extensional basin in SW England and an associated tensional regime extending into northern Britain (Leeder, 1982).

(2) *Oceanic spreading* This hypothesis relates widespread Carboniferous rifting in and around the British Isles to a prolonged period of tension that finally resulted in the formation of oceanic crust in the N Atlantic area during late Westphalian times (Russell and Smythe, 1978, 1983; Haszeldine, 1984).

Southern Uplands of Scotland and a conjectural Carboniferous landmass, known as St. George's Land, that crosses central England between East Anglia and North Wales. Several sub-basins, or troughs, have been identified within the northern part of this basin: these include the Northumberland Trough, the Stainmore Trough and the Craven Basin (Fig. 5.4). Troughs are areas of thick early Carboniferous basinal sedimentation. They are separated by 'blocks', such as those of Alston and Askrigg, which are contrasting areas of slow subsidence, characterised by thin Dinantian platform sediments. The original concept of blocks and troughs, as a series of horsts and graben within the Pennine Basin, has been modified in recent studies. The Stainmore Trough, for example, is now interpreted as a half-graben, the northern margin of which is defined by a major listric normal fault. The Askrigg Block, with its cover of thin Dinantian carbonates, is considered to represent the southern part of the northerly-dipping 'tilted' basement block that underlies the Stainmore Trough (Wilson and Cornwell, 1982). The idea that a series of similar tilted blocks has

(3) *Transcurrent faulting* In this model Carboniferous basins were formed as pull-apart structures in a wide zone of transcurrent movement that affected the Variscan orogen and its margins from late Devonian times onwards (Arthaud and Matte, 1977; Dewey, 1982). An assessment of the relative merits of each of these proposals is beyond the scope of this report. In the model presented here we emphasise the role of the Caledonian basement in controlling the location of the Pennine Basin, and also suggest that the orientation of the major structures within the basin is consistent with an origin in a zone of dextral shear.

Role of Caledonian basement

Leeder (1982) classified British Upper Palaeozoic basins by the varying degree to which their formation could be attributed to pre-existing Caledonian structures. He proposed that the evolution of the Carboniferous basins of central Ireland and the Northumberland Trough was strongly influenced by the position of the Iapetus Suture. Widespread Dinantian volcanism in the Scottish Borders and the extensive exhalative mineralisation of central Ireland were both related to a relict thermal regime associated with the Caledonian plate boundary.

The onset of crustal extension, south of the Iapetus Suture, divided the basement of northern England into faulted blocks, many of which contained a Caledonian granite. Parts of a major granite that forms the nucleus of the Lake District Block are presently exposed, and the Weardale and Wensleydale Granites have been proven by boreholes to underlie the Alston and Askrigg Blocks. Strong geophysical evidence (Bott and others, 1978; Chapter 4) indicates that a large granitic body also probably exists beneath the Market Weighton Block, which forms a similar area of thin Carboniferous sedimentation at the northern margin of the East Midlands shelf (Kent, 1980).

In the southern Pennine Basin known Carboniferous structures and basin isopachs commonly follow Caledonian trends. One example is provided by the inferred north-easterly trending boundaries of the Craven Basin (British Geological Survey, 1984; Gawthorpe, 1986). Elsewhere, Dinantian isopachs in the area of the Bala Fault in Wales (George, 1974) indicate that this Caledonian lineament probably acted as a Carboniferous growth fault, downthrowing to the north. In eastern England the major growth faults of the Carboniferous basin (Fig. 5.4) are broadly aligned with the main structural direction of the basement as indicated by the Aeromagnetic Map of Great Britain (Institute of Geological Sciences, 1965) (see Chapter 4).

The Mesozoic basins of southern Britain were partly controlled by growth faults, which developed by the reactivation, under tension, of thrust planes in the Variscan basement (Chadwick and others, 1983). The MOIST seismic reflection profile has shown that, in the same way, Caledonian thrusts were closely associated with the development of Mesozoic basins off the northern coast of Scotland (Brewer and Smythe, 1984). An analogous process may have influenced the formation of Carboniferous basins within the Caledonides. Deep seismic reflectors possibly related to such Caledonian basement events have been observed in northern England.

Role of strike-slip faulting

Two major groups of wrench faults were developed in the Variscan fold belt and its margins during the Carboniferous. Easterly trending and NW – SE-trending faults are typical of the fold belt itself (Badham, 1982), and to the north of the Variscan front pre-existing north-easterly trending lineaments of Caledonian age were commonly reactivated as dextral shears. In the Variscan foreland of the northern Appalachians Carboniferous strike-slip basins are associated with transcurrent faults that have dextral displacements of up to 200 km (Bradley, 1982) (Fig. 5.5b). Offsets of this scale on basement faults were not sustained along the length of the Caledonian orogen. In the British Isles, however, the effects of small-scale transcurrent movements on ENE-trending faults have been described from both central Ireland (Taylor and Andrew, 1978; Andrew and Ashton, 1985; Coller, 1984), and the Craven Basin. According to Arthurton (1983, 1984), the Ribblesdale fold belt comprises a series of *en-échelon* anticlines within the Craven Basin, which was formed by east – west-trending dextral movement along a shear zone in Dinantian–early Namurian times. The style of deformation within the Pennine Basin as a whole may be compared with that developed at the termination of a major wrench zone. In Ireland Coller (1984) has described how the Silvermines transcurrent fault dies out eastwards in a zone of WNW-trending normal faults, some of which have a dextral component. These normal faults developed in response to the extension caused by transcurrent movement along the main fault. On a larger scale it is possible that the major Dinantian growth faults of the East Midlands represent the means by which Carboniferous transcurrent movements along the main Caledonian orogen were accommodated and dispersed in the crust of eastern England. The angular relationship between the main wrenching direction and the general trend of the normal faults is consistent with that derived from the experimental evidence and observations of Wilcox and others (1973) and Sanderson and Marchini (1984). A component of dextral movement along the normal faults is to be expected where their trend broadly coincides with the direction of secondary, Riedel shears. Some of the dextral faults in the Craven Basin are of this type (Arthurton, 1984).

In plate tectonic terms, a broad east – west-trending zone of Variscan dextral strike-slip faulting marked the area where Africa, together with fragments of continental southern Europe, was translated along the Variscides during oblique continental collision (Arthaud and Matte, 1977; Badham, 1982). The transcurrent movement of Laurentia relative to the southern continents was partly accommodated in the Variscan Foreland by wrench faulting along the main Caledonian orogen. This zone of wrench faulting ended in northern and central England in a series of NW – WNW-trending normal faults. The normal faults separate a group of interconnected half-graben that make up the Pennine Basin. The bounding faults of the half-graben were possibly controlled by pre-existing basement structures, formed in an arm of the Caledonides, which extended between eastern England and northern Germany. In effect, the major components of the Variscan Foreland, which had been brought together in the Caledonian orogeny, were pulled apart during a episode of crustal extension that lasted, in the Pennine area, from late Devonian to mid-Namurian times. This episode can be seen, in part, as a simple reversal of the compression and sinistral transcurrent faulting that marked the end of the Caledonian Orogeny in late Silurian times.

In summary, the position and form of the southern Pennine Basin were strongly influenced by wrench tectonics, which controlled the gradual separation, during

a.

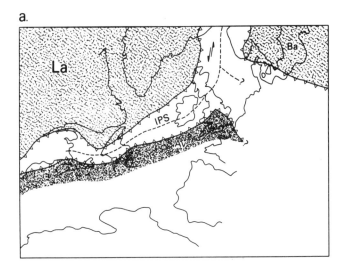

b.

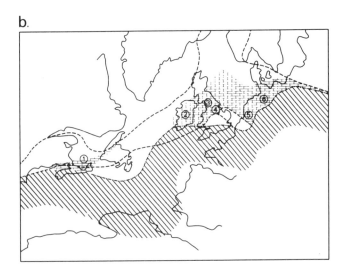

c.

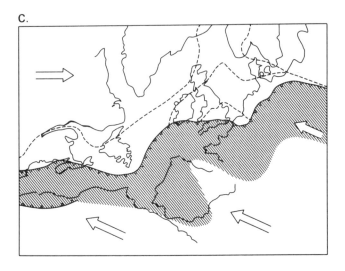

Fig. 5.5 Plate tectonic context of North Atlantic area. Modified after Haszeldine (1984)

A Late Silurian – early Devonian
Two major plates, Laurentia (La) and Baltica (Ba), are united with fragments of a third, Avalonia (Av) to form Old Red Sandstone continent. Iapetus Suture (IPS), which extends along main Caledonian orogen, forms main plate boundary. Deformation in eastern England – North Sea – North German area possibly related to late sinistral wrenching along main part of Caledonides (see text)

B Late Devonian – Tournaisian
Basin formation initiated on Variscan foreland to north of developing Variscides. Basins include (1) N. Appalachians; (2) Central Ireland; (3) Craven; (4) East Midland; (5) Dutch; (6) NW German

C Late Westphalian – early Permian
Variscides deformed during oblique continental collision. Limit of severe orogenic deformation (Variscan Front) marked by barbed line

of the paratectonic Caledonides in the north and the mildly deformed rocks of the Midland Microcraton in the south (Wills, 1973). The actual position of this boundary in the subsurface remains conjectural (Dunning, 1985). One possibility is that the northern margin of the Midland Microcraton lies immediately north of the Charnwood Forest area, and extends north-westwards towards the south-west margin of the Derbyshire 'Dome'. The Midland Microcraton, which underlies much of central England, is composed of late Precambrian basement overlain by a thick sequence of Cambrian–Tremadocian shales with unconformable Silurian platform sediments. During the Caledonian orogeny this area formed a part of Avalonia, a small plate that collided with Laurentia and Baltica. Subduction between Avalonia and Baltica generated a Caledonian fold-belt that now extends between eastern England and northern Germany, beneath the North Sea (Fig. 5.5a) (Frost and others, 1981; Cocks and Fortey, 1982). During the collision, part of the relative motion between Laurentia and Avalonia was accommodated by sinistral strike-slip faulting along the main Caledonian orogen (Soper and Hutton, 1984; Dewey and Shackleton, 1984). In late Silurian and early Devonian times the orogeny culminated with an episode of widespread granite emplacement. The Caledonides were eroded to a peneplain during the Devonian and, in places, Caledonian plutonic rocks were exposed. Thick Devonian molasse accumulated over the Midland Microcraton (Fig. 5.6a).

5.2.4 Late Devonian-Tournaisian

The legacy of the Caledonian orogeny in central and northern England was a heterogeneous crust cut by major plutonic intrusions and faults. These heterogeneities in the basement played an important role during an episode of crustal extension that began in late Devonian–Tournaisian times, when a series of tilted blocks or half-graben developed to the north of the Midland Microcraton (Fig. 5.6b). The bounding faults of the half-graben are probably related to pre-existing basement structures. In the northern Pennine area Caledonian granites formed the nuclei of tilted blocks.

In the East Midlands the basal Upper Palaeozoic sediments at the downslope end of tilted blocks commonly fail to produce coherent reflections on seismic profiles. These

crustal extension, of the Midland Microcraton from Caledonian fold belts that enclosed it to the north.

5.2.3 Late Silurian-Early Devonian

The tectonic framework of central and eastern England was established during the Caledonian orogeny. The most important structural boundary in the area at this time was the tectonic break between the moderately deformed rocks

sediments are interpreted as clastic wedges laid down on each tilt block before the first marine incursion in the area. Thick coarse-grained sandstones and conglomeratic beds have been proved at the base of the Dinantian in Caldon Low and Eakring No. 146 boreholes (Aitkenhead and Chisholm, 1982; George and others, 1976). Both these boreholes are inferred to intersect the clastic sediments deposited alongside a nearby growth fault (Fig. 5.6b) (Smith and others, 1985). In late Devonian times a marine transgression overlapped the Midland Micro-craton from the south (Butler, 1981). A subsequent trans-gression, during early Tournaisian times, extended at least as far north as Lilleshall in Shropshire (Mitchell and Reynolds, 1981). These transgressions indicate the nor-thernmost extent of the marine influence of the Variscan foredeep of southern England. Evidence for the position of the Tournaisian shoreline at this time is provided by the presence of the Hastarian anhydrites in the Hathern Borehole (Llewellyn and Stabbins, 1970). These beds were probably deposited in a sabkha environment on a coastal plain at the northern margin of the Midland Microcraton. By late Tournaisian times, as a result of ex-tension along the main Caledonian orogen, a major pull-apart basin had formed in the Craven area. Marine trans-gressions also extended into the half-graben that had developed by normal faulting in the East Midlands. The anhydrites at the base of the Eyam Borehole, which are of late Tournaisian age (Dunham, 1973; Strank, 1985), probably record the position of a contemporary shoreline on the East Midlands shelf.

5.2.5 CHADIAN – ARUNDIAN

Following the initial Tournaisian transgressions, car-bonate sedimentation was established over a wide area of the Pennine Basin. Rapid subsidence in the Craven Basin resulted in the deposition of thick shale-rich, calcareous sequences. In central and eastern England argillaceous carbonates were deposited at the downslope end of the tilted blocks. Waulsortian facies development was also widespread. Reef knolls of this age are currently exposed in the Craven Basin and in SW Derbyshire (Earp and others, 1961; Aitkenhead and Chisholm, 1982). Active growth faulting continued to control sedimentation. Con-temporaneous movement on the inferred Bakewell Fault (Smith and others, 1985) is indicated by the contrast be-tween the extremely thick early Dinantian sequence en-countered by the Eyam Borehole (Dunham, 1973) and the thin sequence proven on the upthrown side of the fault by the Woo Dale Borehole (Fig. 5.6c; Cope, 1973; Smith and others, 1985; Strank, 1985).Geophysical evidence in-dicates that on the eastern flank of the Derbyshire Dome part of the early Dinantian extension was accommodated on the Bonsall Fault (Chapter 4; Gutteridge, 1987).

Tilted blocks were buried by progressive onlap throughout this interval. The culminations of the blocks were not covered by sediment until late Arundian times.

5.2.6 HOLKERIAN – BRIGANTIAN

Some of the faults that had been active in the early Dinan-tian became inactive during the late Dinantian, whereas continued growth on the remaining faults controlled the position of late Dinantian basins, such as the Edale and Widmerpool Gulfs (Fig. 5.6d).Minor intra-Dinantian un-conformities developed on the upthrown side of the active growth faults as a result of scarp retreat. This process

occurred where sediment was removed by submarine erosion from a new block margin and was carried, by gravitational slides and turbidity currents, into the ad-jacent fault-controlled basin. Carbonate platforms, bounded by apron reefs, became established in areas of shallow basement, both at the basin margins and at the culminations of tilted blocks. These platform areas pro-graded towards the basin depocentres, across slopes com-posed of reef debris. In places, where the early Dinantian growth faults had ceased to move, the carbonate shelves extended over areas of considerably deeper basement. As a result, the pattern of thickness variation of the late Dinantian 'block' facies shows that thin sedimentation oc-curred not only at the basin margins but also over broad, elongate areas that overlie inactive early Dinantian faults. In these latter areas the carbonate platforms constitute drape structures (Smith and others, 1985) analogous to those described from the Mesozoic sequences of the North Sea (Blair, 1975).

Basic volcanism was initiated on the carbonate shelves of the East Midlands and continued intermittently up to early Westphalian times. In Derbyshire the distribution of Dinantian volcanic centres shows a broad correlation with the inferred trend of the early Dinantian Bakewell and Bonsall growth faults, which were largely inactive during the late Dinantian. The precise relationship between the volcanism and the faults and the possible significance of the lack of contemporary fault movement is not fully understood.

Gutteridge (1987) has suggested that reactivation of structures associated with the early Dinantian growth faults partly controlled the position of intra-shelf basins on the Derbyshire carbonate platform during the Brigantian. The development of these local basins restricted the sup-ply of platform-derived detritus to the contemporaneous basins that flanked the Derbyshire area and limited the progradation of the shelf.

5.2.7 NAMURIAN

Those faults which had been active during the late Dinan-tian continued to control sedimentation during the early Namurian. Southerly dipping antithetic faults, such as the Cinderhill Fault near Nottingham, may also have acted as basin-controlling faults at this time. Limestone deposition died out at the start of the Namurian as deltaic sediments entered the East Midlands, largely from the north and east. Early Namurian isopachs show a clear dif-ferentiation between basinal areas, with their thick infills largely composed of turbiditic sandstones and shales, and shelf areas characterised by thin, condensed sequences of mudstone. In Derbyshire the thick deposits of the Kinderscout and Ashover deltas (late Namurian) are separated by an area of thin early Namurian sedimen-tation that coincides with the axis of Dinantian drape about the inferred position of the Bakewell Fault (Fig. 5.9). Isopachs of the Upper Namurian stages generally in-dicate the decreasing effects of faulting on sedimentation. Deltaic lobes rapidly infilled the earlier basinal topography, and late Namurian sediments, consisting partly of extensive sheet sandstones, onlapped the basin margins. Total Namurian isopachs are illustrated in Fig. 5.8.

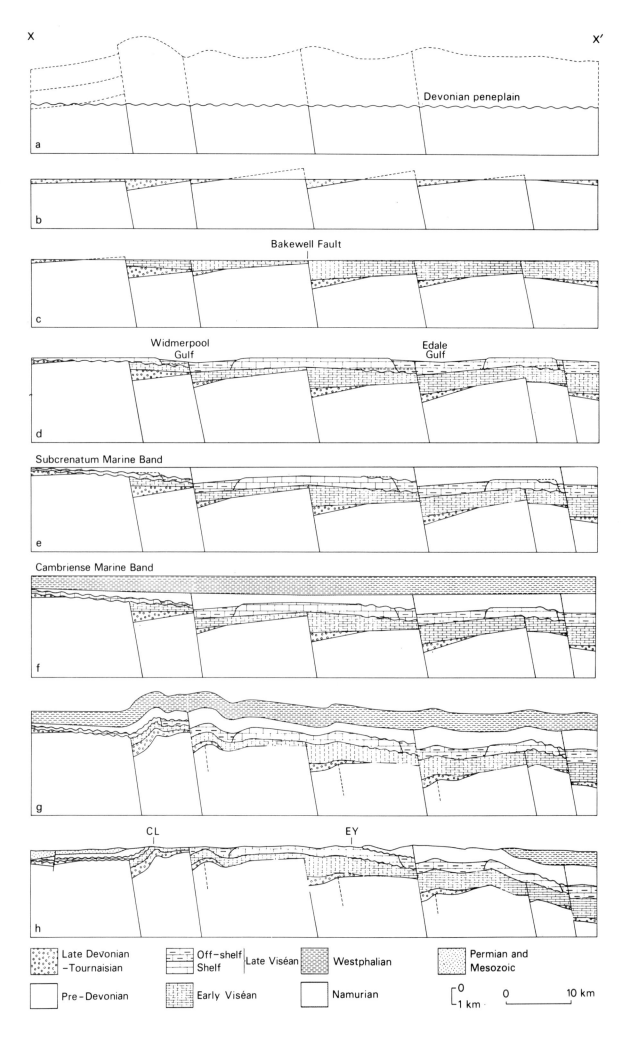

X   X′

Devonian peneplain

a

b

Bakewell Fault

c

Widmerpool
Gulf   Edale
                               Gulf

d

Subcrenatum Marine Band

e

Cambriense Marine Band

f

g

CL   EY

h

Late Devonian –Tournaisian	Off-shelf / Shelf } Late Viséan	Westphalian	Permian and Mesozoic
Pre-Devonian	Early Viséan	Namurian	0 / 1 km 0 10 km

60

Fig. 5.6 Tectonic model of southern Pennine Basin based on regional synthesis of geological and geophysical data. Conjectural tectonic evolution of SSW–NNE section (*X–X'* in Figs. 5.3 and 5.7) across East Midlands illustrates broad tectonic style of area and emphasises role of tilted blocks in controlling Carboniferous thickness and facies variation. Regional trends in Carboniferous isopachs projected onto line of section, where thicknesses shown are merely schematic

a Late Silurian-early Devonian
Deformed Caledonian basement reduced to a peneplain

b Late Devonian – Tournaisian
Tilted block structures, possibly controlled by pre-existing Caledonian fractures, initiated during crustal extension

c Chadian – Arundian
Tilted blocks continue to control thickness and facies variation

d Holkerian – Brigantian
Some early Dinantian growth faults cease to move, as crustal extension wanes, and fewer major faults control development of late Dinantian basins (or gulfs). Carbonate platforms form as drape structures across inactive faults. Intra-basin unconformities develop by scarp retreat at newly stabilised block margins

e End Namurian
Namurian deltas fill fault-controlled basins and overlap late Dinantian carbonate platforms

f Late Westphalian
Simple pattern of Westphalian thickness variation shows that major faulting was not important in controlling sedimentation

g Late Westphalian – early Permian
Inversion of the Pennine Basin and folding of East Midlands carbonate shelf during Variscan deformation. Local erosion and redeposition of late Westphalian strata and general erosion of the whole area preceded deposition of the Basal Permian Sand in Upper Permian times

h Recent
Continued uplift of Pennine axis and broad subsidence of Eastern England Shelf produced present-day outcrop pattern. Position of Caldon Low (CL) and Eyam (EY) boreholes indicated

5.2.8   WESTPHALIAN A – WESTPHALIAN C

The depositional environment that had been established in the Pennine area by the end of Namurian times continued with little change into the Westphalian. In the East Midlands more than 1000 m of shales, siltstones and sandstones with extensive coals accumulated on a widespread deltaic plain that was periodically invaded by the sea (Fig. 5.10). Although some of the controlling faults of the southern Pennine Basin continued to exhibit minor growth during the Westphalian, faulting generally was not important in controlling sedimentation. Westphalian isopachs (Calver, 1969) show that the Pennine Basin at this time had a broad saucer-like form, with a depocentre in the Manchester–Stockport area of the western Pennines (Fig. 5.11). Leeder (1982) suggested that the change from the fault-bounded 'rift'-subsidence of the early Carboniferous to the extensive 'sag'-subsidence of the late Namurian and Westphalian corresponded to the McKenzie model of basin evolution (McKenzie, 1978). In this

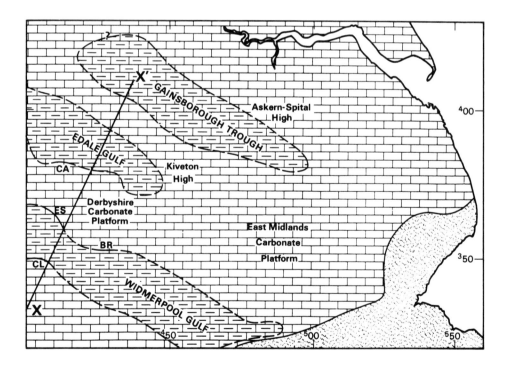

Fig. 5.7 Late Dinantian palaeogeographic map of East Midlands area showing distribution of carbonate platforms and basins. Apron reef belts in Dinantian outcrop indicated as follows: CA, Castleton; ES, Earl Sterndale; BR, Brassington; CL, Caldon Low. *X–X'* is line of section in Fig. 5.6

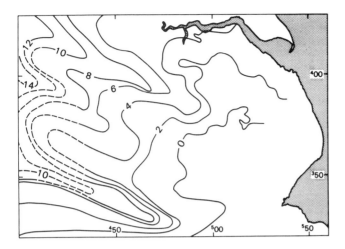

Fig. 5.8 Namurian isopach map of East Midlands (thicknesses in hundreds of metres). Dashed contours indicate estimated thicknesses in areas where Namurian incomplete or absent through erosion

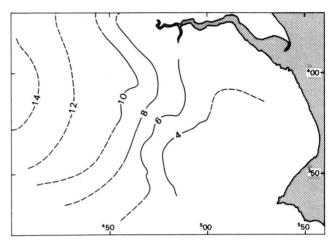

Fig. 5.11 Westphalian isopachs of East Midlands (thicknesses indicated in hundreds of metres and estimated (dashed contours) where sequence incomplete through erosion)

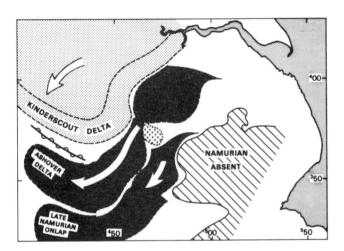

Fig. 5.9 Late Namurian deltas in East Midlands. Main areas of deposition of Kinderscout and Ashover deltas separated by the axis of drape of late Dinantian carbonate platform about inferred position of Bakewell Fault. Thin contemporaneous volcanics proved by boreholes near margin of Namurian basin

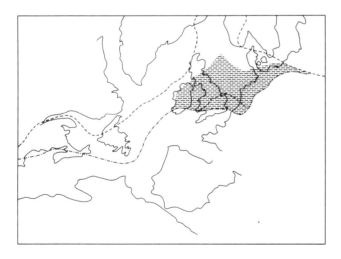

Fig. 5.10 Distribution of Westphalian paralic facies in NW Europe

model an initial period of rapid basin subsidence is related to an episode of 'stretching' or lithospheric thinning. When stretching ceases, further subsidence is caused by the cooling of the hot asthenospheric pillow that has replaced the thinned lithosphere beneath the basin. During the latter episode of 'thermal relaxation', thin sediments overlap the earlier rifted margins of the basin to form the characteristic 'steer's head' profile (Dewey, 1982). This pattern can be recognised in the East Midlands where late Namurian and Westphalian sediments overlap the southern and eastern rim of the Pennine Basin (Fig. 5.9). Bott and others (1984) have expressed doubts about the relative effectiveness of simple thermal relaxation as a subsidence mechanism in the Carboniferous basins of northern England. Subsidence in the Craven area, however, has been modelled by Dewey (1982), who derived a beta-factor of approximately 2 (equivalent to 100% stretching). A quantitative application of the McKenzie model to the whole Pennine Basin awaits geological models in which the complete Upper Palaeozoic sequence is well constrained by seismic reflection data.

5.2.9 LATE WESTPHALIAN – EARLY PERMIAN

The onset of widespread deformation in the East Midlands during the late Westphalian brought to an end the period of basin formation that had started in late Devonian times.

The gradual inversion of the Pennine Basin and the elevation of the Variscan fold belt in southern Britain generated new sources of sediment on structural highs within the former basins. In the East Midlands sediment was removed from the crests of rising anticlines and redeposited in adjacent synclinal areas (Wills, 1956; Fig. 5.6g). At the southern margin of the Pennine Basin a rapid influx of coarse-grained detritus buried the reddened sediments of Etruria Marl facies, which had been deposited on the slowly subsiding Midland Microcraton during the late Westphalian (Besly and Turner, 1983). A profound erosion of the Pennine area continued throughout Stephanian and early Permian times. At the end of the Variscan deformation the whole of the East Midlands was consolidated as part of the Eastern England shelf.

The foreland to the north of the Variscan front in Britain was deformed by a regional E–W compressive stress during the late Westphalian. The clearest illustration of the stress pattern at this time is provided by the widespread development of broadly N–S-trending compressive structures, of which the best known are those which form the western margins of the Alston and Askrigg blocks. The Dent Fault is a NNE-trending Variscan reverse fault that defines the western boundary of the Askrigg Block. It is associated, at its northern end, with a major fold, the Stainmore Monocline, which is overturned to the east (Burgess and Holliday, 1979; Dunham and Wilson, 1985). At the western margin of the Alston Block similar NNW-trending reverse faults intersect the Lower Palaeozoic rocks of the Cross Fell Inlier (Shotton, 1935; Moseley, 1972). A contemporaneous northerly trending monocline, the Burtreeford Disturbance, has also deformed Dinantian platform carbonates within the Alston Block (Dunham, 1948). Critchley (1984) has suggested that this fold may have originated by the E–W compression of a pre-existing structural weakness in the underlying Weardale Granite.

To the south of the northern Pennine blocks the Pennine Anticline is one of a series N–S-trending Variscan structures within the broad outcrop of Namurian strata in the Central Pennines. Westphalian rocks are preserved in a series of flanking synclines, such as the Goyt Trough (Stevenson and Gaunt, 1971), which are parallel with the Pennine axis.

The simple effects of E–W compression were modified in the East Midlands and in the Craven area by pre-existing basin-controlling faults. In the Craven Basin the end-Westphalian deformation reinforced the earlier episode of intra-Dinantian tectonism. Structures that had developed as a consequence of dextral shearing during basin formation were folded by renewed dextral movement during E–W compression (Arthurton, 1983, 1984). In the East Midlands, however, the different orientation of the basin-controlling, normal faults led to their re-activation as sinistral wrench faults during the later tectonism. Sinistral movement along the basement faults was partly accommodated in the cover by folding of the thick Carboniferous sequence that produced a complex series of *en-échelon* anticlines separated by broad, structurally simple synclines (Fig. 5.12). The broad pattern of NNW–NW-trending anticlines in the East Midlands is consistent with an origin in a WNW–NW-trending zone of sinistral shear. Some of the folds are directly associated with the inversion of earlier growth faults. Others may be simple anticlines or monoclines, analogous to the Burtreeford Disturbance, in which the compression of the cover sequence was accompanied by reverse faulting in the basement. In the East Midlands these anticlines often exhibit steeper western limbs associated with easterly dipping reverse faults.

The broad structural pattern in the Pennine Basin is repeated, on a smaller scale, by the orientation of veins and joints in the Pennine orefields. ENE-trending and WNW-trending veins commonly have dextral and sin-

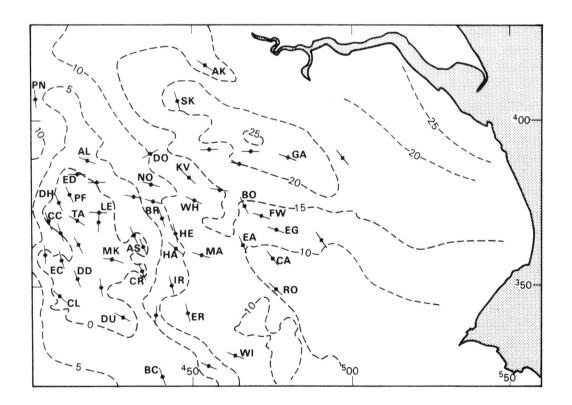

Fig. 5.12 Variscan deformation in East Midlands. Axes of major Variscan anticlines superimposed on estimated structural contours on top of Dinantian strata in hundreds of metres below sea-level. Anticlines include: AK, Askern; AL, Alport; AS, Ashover; BC, Breedon Cloud; BO, Bothamsall; BR, Brimington; CA, Caunton; CC, Countess Cliff; CL, Caldon Low; CR, Crich; DD, Dove Dale; DH, Dove Holes; DO, Don; DU, Duffield; EA, Eakring; EC, Ecton; ED, Edale; EG, Egmanton; ER, Erewash; FW, Farley's Wood; GA, Gainsborough; HA, Hardstoft; HE, Heath; IR, Ironville; KV, Kiveton; LE, Longstone Edge; MA, Mansfield; MK, Matlock; NO, Norton; PF, Peak Forest; PN, Pennine; RO, Rolleston; SK, South Kirby; TA, Taddington; WH, Whitwell; WI, Widmerpool

istral offsets, respectively (Ford, 1976; Firman, 1977; Atkinson and others, 1982; Critchley, 1984; Aitkenhead and others, 1985). These minor wrench movements developed in response to the same pervasive E–W compression that produced the main deformation to the north of the Variscan Front.

The structural development of the southern Pennine Basin may be related to events at the Variscan plate margin. Lateral movement along the Variscides during the early Carboniferous generated crustal extension in the northern plate (Badham, 1982) (Fig. 5.5b). Then, during the late Namurian and Westphalian, the foreland responded passively as thermal equilibrium was restored in the stretched lithosphere (Leeder, 1982) (Fig. 5.10). By the end of the Carboniferous, as the orogenic belt became consolidated and a series of NNW-directed nappes accumulated in southern Britain, continued lateral movement between the colliding continental fragments was expressed as a regional E–W compressive stress to the north of the Variscan Front. The change from extension to compression in the Variscan foreland, which led to the inversion of the major Dinantian basins during the late Carboniferous, was probably associated with the changing configuration of the Variscan plate margin during oblique collision (Fig. 5.5c).

5.2.10 Permian–Recent

The formation of a large Permian and Mesozoic basin in the area of the southern North Sea brought about the broad easterly tilting of the Eastern England Shelf and the further elevation of the Pennine axis. A transgression from the Zechstein Sea overlapped thin Rotliegendes sediments at the margin of the basin, and deposited the dolomites and marls that form the base of the Permian and Mesozoic sequence in eastern England. The effect of post-Triassic faulting in the area is extremely limited. Contours on the basal Permian unconformity (Fig. 2.1), which illustrate the post-Carboniferous tilting of the Eastern England Shelf, also confirm that most of the structural variation in the East Midlands area can be attributed to basin formation and deformation during the Carboniferous.

6 GEOCHEMISTRY OF SHALES

D. G. Jones and J. A. Plant

6.1 General introduction

As in the case of applied geophysical techniques (Chapter 4), geochemical exploration methods can be of considerable value in the detection of exposed or near-surface carbonate-hosted mineralisation. For example, soil sampling, was primarily responsible for the discovery of the Tynagh and Navan deposits in Ireland (Brown, 1979). In the Pennines the analysis of stream sediments and soils (e.g. Wadge and others, 1983; Bateson and Johnson, 1984) has been the main geological exploration tool, but other techniques, including the analysis of soil gas, stream water, lithogeochemistry and hydrocarbon sampling have also been applied (e.g. Ball and others, 1985; Hunting Geology and Geophysics, 1983; Ineson, 1970; Carter and Cazalet, 1984).

The effectiveness of 'direct' geochemical methods generally decreases with depth and they cannot readily be employed in the search for mineral deposits beneath a substantial thickness of cover. In these circumstances the main role of geochemistry is in constraining genetic models for the mineralisation and in the present study emphasis was therefore given to identifying potential sources of ore-forming elements and heat and to providing additional information on the sedimentary and tectonic history of the study region.

As was discussed in Chapter 3, most models for the genesis of Pennine mineralisation involve derivation of ore-forming fluids from shales* or their interaction with igneous rocks. In the case of igneous rocks, buried Caledonian granites have been assigned an important role, both tectonically and as a source of metals and heat (e.g. Dunham, 1983).

Carboniferous volcanics have also been suggested as a possible source of metals (e.g. Mostaghel and Ford, 1986).

The two most important shale formations in the study region that have been considered as sources of fluids and metals are the Lower Palaeozoic sequence, known to underlie mineralised limestone in the Eyam borehole (Dunham, 1973) and also proved in other boreholes throughout the Midlands, and the Carboniferous shales, mostly of Asbian to lower Namurian age (see e.g. Dunham, 1983; Emblin, 1978; Ineson and Ford, 1982; Mostaghel and Ford, 1986; Robinson and Ineson, 1979). Relatively few modern geological data exist for any of these rock types in the study area with the exception of the Carboniferous volcanics (MacDonald and others, 1984; Kirton, 1984) and, to a limited extent, the Namurian shales (Spears and Amin, 1981a; 1981b).

As part of this study systematic analyses (see Appendix 3) have been performed on samples of shales and igneous rocks from surface exposures and boreholes. They have been interpreted with particular reference to their potential as sources of metals, ligands, such as Cl (and F), and sources of radiogenic heat (contents of U, Th and K).

The geochemical data for igneous rocks (presented and discussed in Chapter 7) are also considered in terms of their petrogenesis in relation to the pre-Permian tectonic history of the study area, and the shales are considered (in this chapter) in relation to their depositional and diagenetic history.

*In this account the term 'shale' is used in a broad sense to cover all argillaceous rocks, including siltstones, mudstones, slates and phyllites.

6.2 Introduction to shale geochemistry

Although there is an extensive literature on shale geochemistry (see e.g. Tourtelot, 1970; Potter and others, 1980), comparatively few areas of the world have been covered systematically in terms of the range of elements analysed or of Phanerozoic stratigraphy. The most extensively studied areas are North America (e.g. Clarke, 1924; Tourtelot, 1970; Vine and Tourtelot, 1970), the USSR (e.g. Ronov and Migdisov, 1971) and Australia (Taylor and McLennan, 1985). The various estimates of 'average' shale composition (e.g. Clarke, 1924; Turekian and Wedepohl, 1961; Krauskopf, 1979; Taylor and McLennan, 1985) are thus strongly biased in favour of these regions. Modern geochemical data are generally sparse for British shales and especially for the Lower Palaeozoic interval (e.g. Leggett, 1980). Few systematic regional studies have been reported with the exception of equivalent U, and semi-quantitative XRF results for a small suite of elements, on Precambrian to Lower Palaeozoic shales (Bowie and others, 1979), and a summary account of the chemistry and mineralogy of Carboniferous mudrocks (Ramsbottom and others, 1981).

The geochemistry of shales is considered here in three main sections. First, new quantitative analyses for a large range of elements in British Lower Palaeozoic shales, from the collection of Bowie and others (1979), are presented. They provide a broad framework in which data from the scattered Lower Palaeozoic samples available in the East Midlands are considered. Lower Palaeozoic shales have been suggested to be the main source of metals in Irish-style deposits (Russell, 1978; Boyce and others, 1983) and this is supported by the Pb isotope studies of H. Mills (pers. comm.). The third major section considers the geochemistry of Viséan–Namurian shales that are coeval or contiguous with the mineralised limestones and that have been widely invoked as a source of ore fluids (e.g. Emblin, 1978; Ineson and Ford, 1982; Mostaghel and Ford, 1986; Robinson and Ineson, 1979).

Most shales comprise varying proportions of a detrital mineral fraction, a carbonate fraction and a fraction comprising organic/chemical precipitates. Sources of metals and other ligands for mineralisation are most likely to be derived from the carbonate or organic/chemical precipitate fraction. Other elements, which are concentrated in the detrital fraction, are likely to be resistant to alteration during weathering, transport and deposition, but they can be used to provide information on the provenance of the shales. The elements that are most likely to be available for leaching into hydrothermal fluids are generally those which have a high sea-water–upper crust partition coefficient (K_y^{sw}) and long calculated residence times (τ) in sea water (see e.g. Whitfield, 1979; Turner and others, 1980; Taylor and McLennan, 1985). Elements with very high K_y^{sw} and sea-water residence times (log K_y^{sw} >-3; log τ >5), including most alkaline and alkaline earth elements, B and U, are strongly partitioned into natural waters and remain in solution for long periods of time. On the other hand, elements with very low K_y^{sw} and residence times (log K_y^{sw} <-6; log τ <3), including the Ti group (Ti, Zr), the Al group, the REE (including Y) and Sc, Th, Nb and Be, are strongly excluded from natural waters and give the best information on source. Many relatively immobile elements are, however, difficult to use to deduce source. Some show marked changes in solubility as a result of changes in oxidation state (Fe, Mn) or are strongly chalcophile (Pb).

Others, such as Al, have almost constant ratios with only limited dispersion in common igneous rocks or they tend to occur in heavy mineral concentrations (e.g. Zr). The distribution of REE, Y, Th, Sc and possibly Co are probably the most useful for determining sedimentary provenance (Taylor and McLennan, 1985). The REE, which form characteristic patterns, are a special case and their content in sedimentary rocks has been used by several authors to obtain information on average upper crustal compositions (Taylor, 1964; Haskin and others, 1966; 1968; Nance and Taylor, 1976; Taylor and McLennan, 1985).

Black shales, which tend to be highly enriched in organic, carbonaceous and phosphatic material, represent a special case. Statistical analysis of chemical data from North American black shales showed that the detrital mineral fraction was characterised by Al_2O_3, TiO_2, Zr and Sc and in some cases by Be, B, Ba, Na, K, Mg and Fe (Vine and Tourtelot, 1970). The carbonate fraction commonly included Ca, Mg, Mn or Sr, whereas the organic fraction tended to be enriched in Ag, Mo, Zn, Ni, Cu, Cr, V and, less commonly, Co, Pb, Y and U. The elemental data obtained on shales are discussed in relation to groups considered to be associated with the organic fraction, the carbonate fraction and the detrital fraction, the latter providing information on provenance. In presenting the data frequent use is made of plots normalised to the average shale composition of Turekian and Wedepohl (1961). This particular composition was used because it includes data for all the elements determined in this study.

6.3 Reference collection

In order to provide a framework within which the geochemistry of Lower Palaeozoic shales from the study area could be considered, new data were obtained on a suite of Precambrian–Lower Palaeozoic shales collected as part of a BGS U exploration programme during the 1960s (Bowie and others, 1979). These data provided a regional geochemical context that was particularly important since only small numbers of samples, which are scattered both geographically and stratigraphically, were available from the East Midlands. The new data extend the range of elements determined quantitatively from that of the eleven elements, originally determined mostly by a rapid semi-quantitative XRF technique. The stratigraphic ages of the samples given in Bowie and others (1979) have also been revised, leading to several changes, particularly in NW England, the Isle of Man and Scotland. The age of some of the shale samples is still poorly known because of a lack of palaeontological con-

Table 6.1 Stratigraphy of samples in Lower Palaeozoic shale reference collection

Area	Age		No of samples
Scotland	Silurian:	Wenlock	9
		Llandovery	50
	Ordovician:	Ashgill	12
		Caradoc	27
		Llandeilo	2
		Llanvirn	6
		Arenig	14
	Cambrian:	Comley	18
NW England & Isle of Man	Silurian:	Ludlow	9
		Wenlock	10
		Llandovery	17
	Ordovician:	Ashgill	16
		Caradoc	4
		Llanvirn	5
		Arenig	80
	?Cambrian:	Ingletonian	10
N & Central Wales		Downton Group	1
	Silurian:	Ludlow	16
		Wenlock	6
		Llandovery	20
	Ordovician:	Ashgill	13
		Caradoc	10
		Llandeilo	1
		Arenig	1
		Tremadoc	1
	Cambrian:	Merioneth	3
		St Davids	18
		Comley	8
Welsh Borders & Central England		Downton Group	8
	Silurian:	Ludlow	10
		Wenlock	8
		Llandovery	6
	Ordovician:	Caradoc	27
		Llandeilo	4
		Llanvirn	9
		Arenig	4
		Tremadoc	10
	Cambrian:	Merioneth	17
		St David's	7
		Comley	6

Fig. 6.1 Location of samples analysed from Lower Palaeozoic shale reference collection

trol, so care must be taken in interpreting the data in relation to stratigraphy. For the purposes of this study only the results for samples of Cambrian–Downtonian age are considered. Descriptions of the samples are given in Table 6.1 and their locations are shown in Fig 6.1.

Before discussing the geochemical results and their interpretation the plate tectonic setting of Britain during the Lower Palaeozoic is briefly considered. A reconstruction after Phillips and others (1976), modified on the basis of subsequent work by Stillman (1981), Watson (1985) and McKerrow (1988) is shown in Fig. 6.2. Most recent models for the evolution of the British Caledonides postulate a collision orogeny related to the closing of the proto-Atlantic Ocean, Iapetus. This ocean appears to have existed throughout the Lower Palaeozoic until its closure at the end of the Silurian led to the formation of the Devonian, Old Red Sandstone continent (Phillips and others, 1976). The last events in the cycle may have involved oblique subduction and sinistral, strike-slip movements (McKerrow, 1988) that juxtaposed the American continental plate (Laurentia), of which much of Scotland formed a part, and a European continental plate (comprising Baltica and Avalonia), which included most of England and Wales (Chapter 5). The postulated end-Silurian suture follows the Solway line along the southern margin of the Southern Uplands of Scotland. To the north of this line seismic and other evidence (Bamford and others, 1977; Walton, 1983) indicates the possible presence of Lewisian-type basement, similar to the Archaean–early Proterozoic gneisses and granulites exposed in NW Scotland. Basement studies of the southern, European plate indicate that it is younger, and relatively thin, comprising calc-alkaline volcanics and sediments, ranging in age from ca 700 to 450 Ma (Hampton and

Taylor, 1983; Thorpe and others, 1984).

The reference collection includes a relatively small number of samples from the Caledonian Foreland of NW Scotland, which might be expected to reflect the geochemistry of the Laurentian continent.

Relatively large numbers of rocks were collected from the European plate, mainly from western England and Wales, together with a few samples from the English Midlands. The remainder of the collection comprises rocks from the Midland Valley and Southern Uplands of Scotland. The latter, from immediately north of the suture zone, form part of a sedimentary pile that has been interpreted as an accretionary fore-arc prism (e.g. Leggett and others, 1979) or as the infill of a back-arc basin (Stone and others, 1987). The recognition of the role of large, sinistral strike-slip faults in juxtaposing previously separated terrains makes palaeographic reconstruction of this zone particularly difficult.

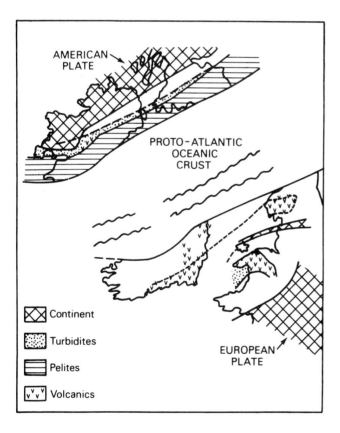

Fig. 6.2 Caledonian plate tectonic reconstruction of Britain based on Phillips and others (1976), modified after Stillman (1981), Watson (1985) and McKerrow (1988)

Table 6.2 Average abundances of elements in Lower Palaeozoic shale reference collection compared with other average shale estimates

	1	2	3	4	5
SiO_2	59.03	58.38	58.40	50.91	62.8
Al_2O_3	19.08	15.47	15.11	17.38	18.9
TiO_2	0.96	0.65	0.77	0.75	1.0
Fe_2O_3	7.52	6.75	6.64	6.62	7.2
MgO	2.61	2.45	2.49	2.32	2.2
CaO	2.00	3.12	3.09	3.50	1.3
Na_2O	1.00	1.31	1.29	1.21	1.2
K_2O	3.73	3.25	3.20	3.01	3.7
MnO	0.15		0.11	0.11	0.11
P_2O_5	0.12	0.17	0.13	0.17	0.16
As	14		13	10	
B	98		100	100	100
Ba	743		580	600	650
Be	3		3	3	
Ce	84		59	70	80
Cl	76		180	170	
Co	26		19	20	23
Cr	115		90	100	110
Cu	37		45	50	50
Li	68		66	60	75
Mo	4.4		2.6	2	1.0
Ni	57		68	80	55
Nb	18		11	15	19
Pb	25		20	20	20
Rb	134		140	140	160
S	8054		2400	2500	
Sc	14		13	15	16
Sr	126		300	400	200
Th	12		12	12	14.6
U	4.1		3.7	3.5	3.1
V	164		130	130	150
Y	34		26	35	27
Zn	89		95	90	85
Zr	184		160	180	210

1 Reference collection (this work).
2 Clarke, 1924.
3 Turekian and Wedepohl, 1961.
4 Krauskopf, 1979.
5 Taylor and McLennan, 1985.

6.3.1 COMPARISON WITH AVERAGE SHALE COMPOSITIONS

Overall element means for the reference collection are compared with average shale compositions, from a number of sources, in Table 6.2. Bowie and others (1979) indicated significant enrichment of Pb and Ba and depletion of Sr and Y compared to average shale (data of Turekian and Wedepohl, 1961, cited in Fairbridge, 1972) and the new data provide confirmation of their findings for Ba and Sr. They show only slight enhancement of Pb, however, and Y levels are within the range of average compositions. The low mean Sr figure can be attributed to low carbonate concentration, as reflected in the CaO content, although the figure is also low when compared with Taylor and McLennan's (1985) carbonate-free average. Because of the general lack of carbonate in the shales analysed major-element means are closest to those of Taylor and McLennan (1985). For the remaining trace elements the relative enrichment of Sr, and to a lesser extent of Mo, Pb and V, results from a slight sampling bias in favour of black shales, which were considered the most likely to be enriched in U. It is surprising, in the light of this bias, that the mean Cu figure is below that of average shales. Levels of Cl also appear to be significantly low.

6.3.2 TEMPORAL GEOCHEMICAL VARIATIONS

Means for each stratigraphic series are given in Table 6.3.

Variations in the major elements are largely controlled by fluctuations in quartz, carbonate and aluminosilicate contents as indicated by SiO_2, CaO and Al_2O_3 (Fig. 6.3). Silica levels peak in the Llandeilo, Caradoc and Downton. In the former two series this can be partly explained by the presence of cherty shales from southern Scotland. The Llanvirn, Caradoc and Ludlow periods are carbonate-rich, whereas aluminosilicate concentrations are greatest in the Upper Cambrian (Merioneth Series) and the Arenig. TiO_2, Fe_2O_3, MgO, Na_2O and K_2O contents mirror changes in Al_2O_3 levels to varying degrees. The TiO_2 maximum in the Llanvirn, associated with enhanced MgO, Cr and Ni (Fig 6.3), indicates a significant basic/ultrabasic igneous component. This component is present only in sediments from the Barr Group of SW Scotland, and is almost certainly derived from the underlying Ballantrae Igneous Complex, which has been interpreted as an ophiolite (e.g. Church and Gayer, 1973; Bluck, 1978). High TiO_2 in the Downton, linked to elevated levels of SiO_2, P_2O_5, Ce, Nb, Th, Y and Zr (Fig. 6.3), clearly reflects relatively arenaceous rocks, enriched in heavy minerals and phosphatic vertebrate remains, represented, for example, by samples from the Ludlow Bone Bed. The Tremadoc Series is also phosphatic, but lacks the suite of elements (Ce, Nb, Th, Y, Zr) associated with heavy minerals (Fig. 6.3). Manganese shows a marked enrichment in the Middle Cambrian (St David's Series), which reflects the inclusion of samples from in and around a Mn 'ore bed' from the Harlech Dome.

Table 6.3 Reference collection: element means for major stratigraphic groups

	L Cambrian	M Cambrian	U Cambrian	Tremadoc	Arenig	Llanvirn	Llandeilo	Caradoc	Ashgill	Llandovery	Wenlock	Ludlow	Downton
SiO_2	59.03	58.41	58.90	48.60	58.40	52.28	64.78	63.83	58.29	58.74	58.05	58.98	62.73
Al_2O_3	18.78	19.99	20.57	18.87	20.94	18.91	17.05	17.54	19.49	19.60	18.09	15.93	17.64
TiO_2	0.85	0.96	1.00	0.92	1.03	1.15	0.72	1.01	0.96	0.97	0.85	0.83	1.07
Fe_2O_3	6.03	8.81	6.96	7.37	9.34	7.30	5.63	6.58	7.68	7.87	6.81	5.79	6.92
MgO	2.59	2.00	2.24	1.79	1.80	3.65	1.82	2.28	2.53	2.95	4.26	3.65	2.97
CaO	1.96	0.57	0.29	8.80	0.58	5.53	2.12	1.81	2.72	0.90	3.67	5.00	0.96
Na_2O	0.78	0.96	0.76	0.67	1.07	0.81	0.82	0.96	1.08	1.09	1.02	1.29	0.98
K_2O	6.03	3.92	3.93	2.97	3.57	2.97	3.55	3.07	3.43	3.78	3.73	3.48	3.21
MnO	0.08	0.67	0.16	0.23	0.17	0.09	0.06	0.06	0.16	0.11	0.08	0.11	0.16
P_2O_5	0.13	0.11	0.07	0.30	0.18	0.10	0.11	0.13	0.09	0.05	0.07	0.16	0.47
As	6.1	18.6	24.7	4.7	22.9	4.2	1.9	8.2	11.3	23.7	7.7	3.3	5.6
B	125.5	83.3	87.2	68.4	127.1	85.4	87.1	84.5	108.5	97.3	93.3	67.7	76.8
Ba	671	1010	1865	1057	707	578	873	763	750	639	523	526	647
Be	2.5	3.3	3.3	2.6	3.2	2.8	3.1	2.5	2.6	2.5	2.5	2.4	3.6
Ce	73.8	95.5	94.3	75.8	88.0	83.4	86.9	74.8	81.1	83.5	81.4	88.7	106.8
Cl	51.0	45.4	113.6	108.9	77.2	72.0	93.0	164.2	25.8	50.7	96.0	39.4	36.7
Co	15.4	26.6	44.0	27.5	27.7	25.8	16.1	21.6	24.9	35.2	20.8	15.9	17.7
Cr	77.0	90.1	108.7	93.5	116.3	171.6	71.1	115.9	122.9	127.5	117.9	113.5	117.1
Cu	27.6	37.1	65.9	26.8	34.9	32.9	20.9	29.5	35.4	51.4	31.4	27.0	21.9
Li	35.3	43.6	61.3	96.5	95.0	93.7	50.0	73.9	80.3	62.2	55.2	40.9	47.0
Mo	3.2	5.0	23.3	9.7	2.8	1.3	3.1	6.9	1.6	3.8	2.2	2.7	0.7
Ni	31.3	41.3	91.6	44.8	48.7	103.0	33.4	47.0	62.2	70.5	68.2	59.1	55.1
Nb	16.1	18.1	19.5	16.9	19.2	19.6	19.0	16.8	18.4	16.9	18.2	18.0	20.7
Pb	15.0	23.6	46.1	21.7	27.3	19.8	13.7	26.8	24.0	31.2	18.8	18.2	16.4
Rb	140.9	136.0	141.7	111.2	148.6	121.5	137.0	121.2	134.1	135.7	131.6	116.3	115.1
S	891	4553	14374	1442	1927	2546	868	5380	5994	12712	3430	778	345
Sc	12.7	16.7	15.7	16.9	12.9	16.6	12.3	15.1	14.3	14.2	13.3	12.6	13.1
Sr	114.8	85.2	74.0	147.5	110.1	120.1	124.7	114.4	271.5	97.3	125.8	179.3	90.6
Th	10.6	11.0	13.6	11.0	11.9	11.8	13.4	9.8	10.8	11.9	13.5	11.9	15.8
V	127.1	174.8	454.9	228.9	144.4	121.3	76.4	182.4	153.7	185.5	125.3	92.4	84.7
Y	29.5	35.2	36.5	31.5	34.3	32.5	38.7	36.5	31.4	30.0	31.6	39.2	53.8
Zn	67.5	81.6	92.4	91.1	108.4	100.2	72.3	79.7	96.2	86.7	85.1	81.0	85.7
Zr	207.4	185.2	139.6	116.7	164.6	178.0	182.1	171.9	165.6	176.9	200.0	262.3	404.4
U	2.9	3.4	21.4	2.2	3.6	2.7	2.2	4.2	2.9	3.7	3.1	3.2	3.7

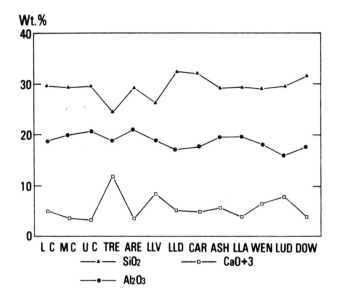

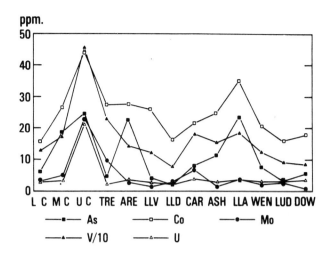

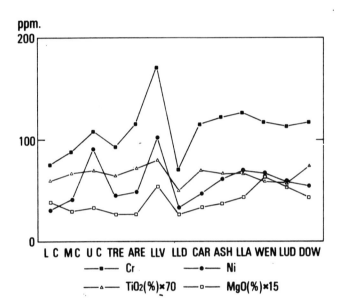

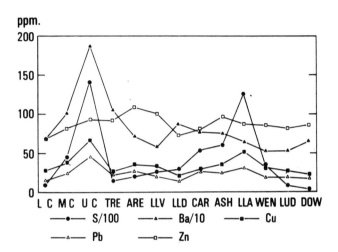

Fig. 6.3 Variations in mean concentrations of elements through Lower Palaeozoic for samples from reference collection

LC, L. Cambrian (Comley Series); MC, M. Cambrian (St David's Series), UC, U. Cambrian (Merioneth Series); TRE, Tremadoc; LLV, Llanvirn, LLD, Llandeilo, CAR, Caradoc, ASH, Ashgill; LLA, Llandovery; WEN, Wenlock, LUD, Ludlow, DOW, Downton

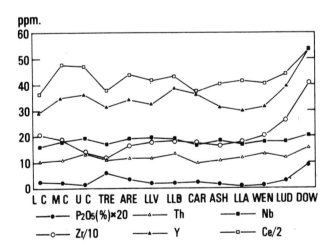

Potash is most abundant in the Lower Cambrian (Comley Series) because of exceptionally high levels (8–12%) in the Fucoid Beds of NW Scotland, first described by Bowie and others (1966). The K_2O occurs in authigenic feldspars, thought to have formed in an evaporite sequence (Allison and Russell, 1985).

Several trace elements, including As, Co, Cr, Cu, Mo, Ni, Pb, S, U and V, show maxima coincident with the occurrence of black shale units through the Lower Palaeozoic sequence (Leggett, 1980) with peaks in the Upper Cambrian, Caradoc, Llandovery and to a lesser extent in the Arenig–Llandeilo (Fig. 6.3). Black shales of Caradoc and Lower Llandovery age are particularly widespread in Britain and can be related to eustatic rises in sea-level (Leggett, 1980). The greatest enrichment of many of these elements (As, Co, Cu, Mo, Pb, S, V and U) occurs, however, in the Upper Cambrian. This series includes the Monk Park Shales of the Nuneaton inlier and the White Leaved Oak and Shoot Rough Road Shales of the Welsh Borders, which are of the same age as the trace metal-enriched Alum Shales of Scandinavia (see e.g. Armands, 1972; Andersson and others, 1982; Bjørlykke, 1974).

6.3.3 REGIONAL GEOCHEMICAL VARIATIONS

The mean concentration of elements in stratigraphic series may conceal considerable variation, both locally and regionally. This can be illustrated by examining data for the Caradoc, for example—a period for which a good sample suite is available with reasonably good

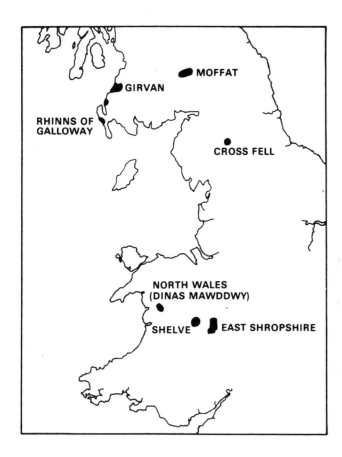

Fig. 6.4 Locations of analysed samples of Caradoc age from Lower Palaeozoic shale reference collection

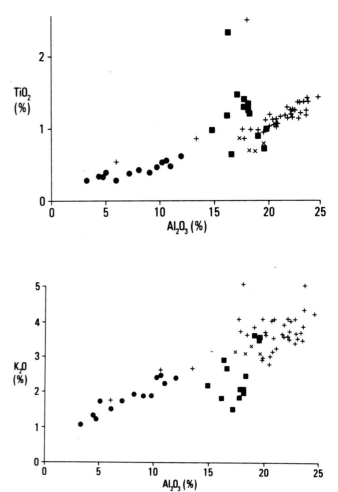

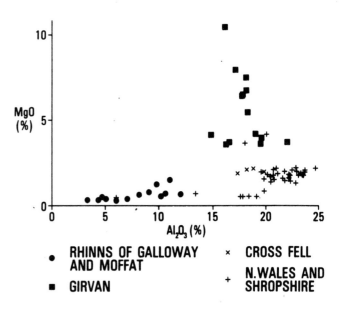

Fig. 6.5 Plots of $TiO_2 - Al_2O_3$, $K_2O - Al_2O_3$ and $MgO - Al_2O_3$ for groups of Caradoc samples from reference collection. For locations see Fig. 6.4

Table 6.4 Stratigraphic groupings of Caradoc shales (numbers of samples in parentheses)

Shelve inlier		
	Whittery Group	(4)
	Hagley Group	(1)
	Aldress Shales	(3)
	Rorrington Beds	(3)
East Shropshire		
	Onny Shales	(3)
	Acton Scott Beds	(3)
	Cheney Flags	(2)
	Chatwall Flags	(1)
	Chatwall Sandstone	(1)
	Harnage Shales	(3)
	Hoar Edge Grits	(3)
North Wales (Dinas Mawddwy)		
	Ceiswyn Beds	(5)
	Nod Glas	(5)
Cross Fell		
	Dufton Shales	(4)
Rhinns of Galloway and Moffat		
	Lower Hartfell Shales	(9)
	Glenkiln Shales	(2)
Girvan area		
	Lower Whitehouse Group	(4)
	Ardwell Group	(7)
	Balclatchie Group	(1)
	'Glenkiln Shales'	(2)

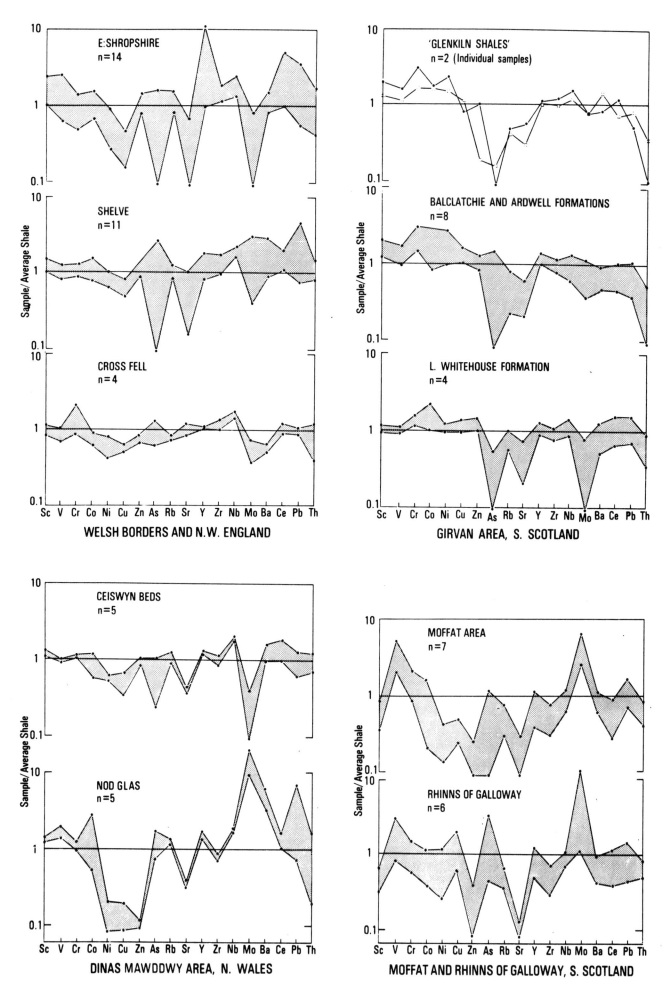

Fig. 6.6 Average shale normalised trace-element diagrams for groups of Caradoc shales from reference collection. For locations see Fig. 6.4

71

biostratigraphical control and with a fairly even geographic spread to the north and south of the postulated Iapetus suture (see Fig. 6.4 and Table 6.4).

Plots of the major elements TiO_2, K_2O and MgO against Al_2O_3 (Fig. 6.5) are related to the aluminosilicate mineral content. Samples from the Rhinns of Galloway and the Moffat area of south Scotland form a distinct group, characterised by very low Al_2O_3 content and high SiO_2 concentrations, reflecting the association of the shales with cherts of possible biogenic origin. The $K_2O-Al_2O_3$ plot is principally indicative of the illite + K-feldspar content. In the case of shales from the Moffat area the good correlation of K_2O and TiO_2 with Al_2O_3 reflects the predominance of illite in the clay mineral assemblage (Stephens and others, 1975). The limited scatter about these trends suggests that a similar, relatively uniform composition occurs generally in Caradocian shales, particularly in the Anglo-Welsh plate south of the Iapetus suture.

Samples from North Wales and the Welsh Borders form a fairly coherent group with much greater levels of TiO_2, K_2O and Al_2O_3, whereas the Cross Fell samples are transitional in composition between this group and rocks from the Rhinns of Galloway and Moffat.

With a few exceptions, the samples that deviate most from the illite trend are those from the Girvan area. The most pronounced difference between these rocks and other Caradocian shales, however, is their marked enrichment of MgO relative to Al_2O_3 (Fig. 6.5). The differences are most simply explained by the presence of a significant component of basic igneous detritus in the Girvan shales, which form an onlapping, unconformable cover to the Ballantrae Igneous Complex. Shales from the higher parts of the Girvan sequence have major-element characteristics more closely comparable with those of other Caradoc shales, probably reflecting the gradual burial of the complex during a north-westerly directed marine transgression (Ingham, 1978). Plots of trace elements, normalised to the average shale values of Turekian and Wedepohl (1961), indicate enhanced levels of Cr and Ni, relative to both average shale and to other Caradoc samples, in the lowest members of the sequence (Fig. 6.6), which are also consistent with a large component of basic detritus.

Samples from the Rhinns of Galloway and Moffat have generally similar trace-element characteristics, the normalised plots showing 'spikes' of V, As, Mo and Pb. Vanadium and Mo are particularly enriched in the Moffat rocks, and relatively high levels of Ni and Cu occur in the Rhinns of Galloway. The patterns are typical of black shales, in which they are considered to be associated with the high content of organic matter and processes such as sulphate reduction and pyrite formation. Enrichment of Pb, Zn, Cu and Fe in Ordovician–Silurian shales from the Moffat area has also been attributed to similar factors by Stephens and others (1975). The normalised trace-element pattern for the Nod Glas black shales from North Wales is comparable (Fig. 6.6), but, in addition to the V, As, Mo and Pb 'spikes', there is slight Co enrichment, a

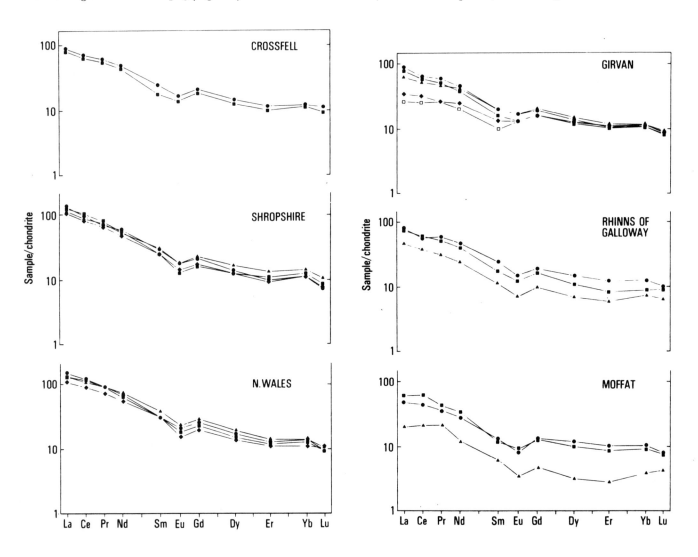

Fig. 6.7 Chondrite-normalised REE plots for groups of Caradoc shales from the reference collection. For locations see Fig. 6.4

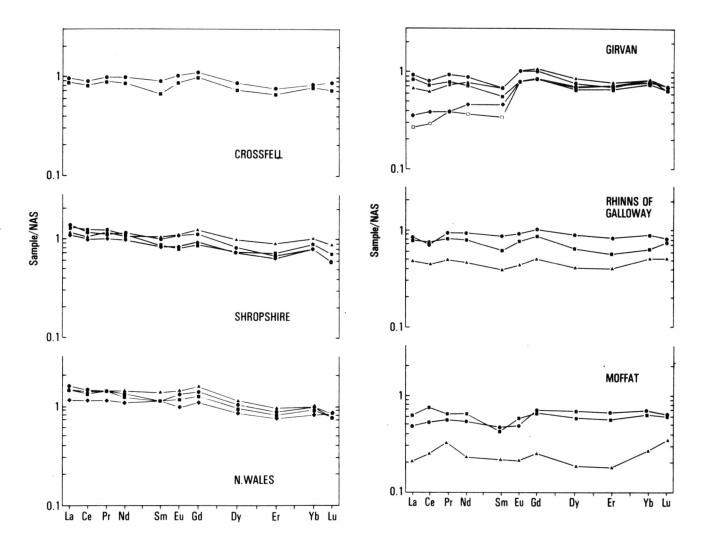

Fig. 6.8 Normalised REE plots (normalised against North American Shale composite, NAS) for groups of Caradoc shales from reference collection

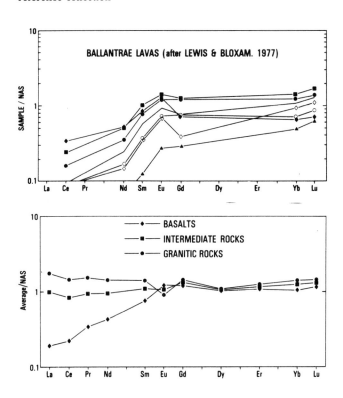

Fig. 6.9 Normalised REE plots (normalised against North American Shale composite, NAS) for Ballantrae lavas and for means of major igneous rock groups

high Ba content and markedly low levels of Ni, Cu and Zn. High levels of V, Co, Mo and Ba do not occur in the easterly outcrops of the Nod Glas, near Welshpool, where Ni contents are much closer to those of average shales (Cave, 1965). Such lateral variations are consistent with the contrast between the rich, benthic fauna near Welshpool (Cave, 1965) and the lack of benthos further to west (Leggett, 1980), where bottom conditions appear to have been anoxic.

Although the contents of organically bound metals in the Nod Glas shales are comparable with those of similar age from the Southern Uplands, the significant differences in other elements, such as Sc, Co, Rb, Ba and Ce, suggest different provenances. Patterns of the least mobile trace elements (e.g. Sc, Cr, Y, Zr, Nb, Ce and Th) in the Nod Glas are typical of the Anglo-Welsh samples generally, whereas the Southern Uplands rocks are broadly similar to those from Girvan, if allowance is made for the basic igneous component present there. The existence of these two groups of shales with contrasting detrital components is consistent with their palaeographic setting on opposite sides of the Iapetus Ocean. Samples of Dufton Shale from Cross Fell appear to have an Anglo-Welsh signature, except for the relatively high Sr levels, which reflect a 15–20% carbonate content.

Results of REE analyses are summarised in Figs. 6.7 and 6.8. The chondrite-normalised diagrams are typical of shales in general, with enrichment of LREE relative to

HREE and a small negative Eu anomaly. Sedimentary rocks are, however, generally normalised against the composite of 40 North American shales (NAS) after Haskin and others (1968). Samples from Girvan, especially from the lower part of the sequence, are clearly shown to be low in LREE compared to NAS (Fig. 6.8), providing further evidence of the presence of basic igneous detritus, such patterns being typical of the underlying Ballantrae lavas (Fig. 6.9) and of basaltic rocks in general (Fig. 6.9). Shales from the Southern Uplands generally have REE patterns similar to NAS, but with lower total REE levels, especially in samples from the Moffat area. This appears to reflect, at least partly, SiO_2 dilution of the fine, clay fraction of the sediments, which generally contains the REE (Fleet, 1984). Anglo-Welsh Caradocian shales have REE patterns more closely comparable with NAS, although samples from North Wales and Shropshire have a slight enrichment of LREE and depletion of HREE relative to NAS. The general increase in total REE southwards, from the Southern Uplands to North Wales, broadly follows an increase in Al_2O_3 (e.g. Fig. 6.5).

6.3.4 STRATIGRAPHIC VARIATION OF THE ORE–FORMING ELEMENTS (BA, CU, PB, ZN AND S)

The stratigraphic variation of the elements most directly relevant to Mississippi Valley type mineralisation (Ba, Cu, Pb, Zn and S) is shown in Fig. 6.3. Copper, Pb and S enrichments are generally associated with black shales. Lead and Cu levels are highest in the Upper Cambrian and Llandovery with slight relative enrichment of Pb in the Arenig and Caradoc. Sulphur contents also peak in the Upper Cambrian and Llandovery, at levels greater than 1%, and mean values exceed 0.5% in the Caradoc and Ashgill. Barium levels are greatest in the Middle Cambrian to Tremadoc shales with maximum values in the Upper Cambrian. High levels also occur in the Llandeilo. Very high Ba concentrations (up to 5–7%) have been reported from the Cambrian–Lower Ordovician of Norway by Bjørlykke (1974). The Ba is associated with the Ba feldspar, hyalophone, which is believed to be a diagenetic replacement of baryte (Bjørlykke and Griffen, 1973). Barium is also enriched in the Precambrian of Scotland, both in the Lewisian and its arkosic cover formations, particularly the Moine, where it occurs in detrital K-feldspars (Plant and others, 1984).

The distribution of Zn differs from that of the other ore-forming elements, with maximum mean values in the Arenig–Llanvirn and Ashgill, suggesting a different control on its distribution. The mean Zn/Pb ratio for the reference collection is 5.0, values greater than 10.0 occurring in about 10% of the samples.

From the stratigraphic means of the elements normalised against the average shale values of Turekian and Wedepohl (1961) it is clear that the Upper Cambrian and Llandovery shales are relatively enriched in all of the ore-forming elements with the exception of Zn. Sulphur, Pb and Ba levels are also significantly greater than those of average shales in the Middle Cambrian, Arenig, Caradoc and Ashgill (Fig. 6.10). Vine and Tourtelot (1970) defined 'minimum enrichment values' (ppm) of metals in North American black shales as Ba, 1000; Cu, 200; Pb, 100; Zn, 1500 (no data for S).

In the Cambrian–Downtonian shales studied mean values exceed these minimum enrichment figures only in the case of Ba, in samples of the Middle Cambrian to Tremadoc age (Fig. 6.3). Only two samples have Cu

values greater than the minimum enrichment level, 11 samples, of Middle–Upper Cambrian, Arenig, Caradoc and Llandovery age, have Pb contents greater than 100 ppm, and no individual sample has a Zn content as high as 1500 ppm.

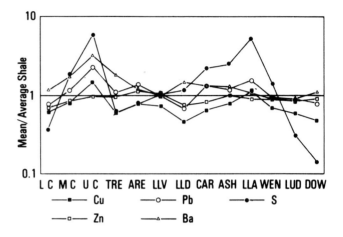

Fig. 6.10 Temporal variations in mean Cu, Pb, S, Zn and Ba normalised to average shale (see Fig. 6.3 for full key)

In summary, it is clear that certain periods of widespread black shale deposition are associated with trace-element enrichment. At the stratigraphic intervals studied the Upper Cambrian and Llandovery have a much greater potential as metal source rocks than strata of other periods.

6.3.5 HEAT PRODUCTION

The mean heat production value from the reference collection of $2.2\,\mu W\,m^{-3}$ is closely comparable with the figure of $2.0\,\mu W\,m^{-3}$ computed from the K, U and Th concentrations of average shale. These values are 50% higher than average crustal heat production and therefore thick shale-filled basins can represent zones of enhanced crustal heat production increasing heat flow. Mudrocks also have lower thermal conductivities than most other common lithologies (see e.g. Wheildon and Rollin, 1986) such that thick flat-lying sequences might also be expected to act as thermal blankets maintaining high geothermal gradients.

Heat production is largely a function of the U content of rocks and thus the uraniferous Upper Cambrian (Merioneth Series) has by far the greatest value (Table 6.5). The remaining periods of widespread black shale deposition (Arenig, Caradoc and Llandovery) have slightly higher heat production figures than those of most other series (Table 6.5). The value for the Comley Series is enhanced by the very high K_2O contents in the Fucoid Beds (section 6.3.2) and the Downtonian figure reflects U associated with phosphates, and Th-rich heavy minerals. None of the formations with relatively high heat production is sufficiently thick, however, to significantly affect the prevailing thermal regime.

6.4 Lower Palaeozoic shales in basement of East Midlands

A total of 218 samples of Lower Palaeozoic shale from the basement of the East Midlands, and adjoining areas to the south and east, was studied using core from BGS stratigraphic boreholes and a small number of oil exploration wells. Some samples were taken from outside the study area because of a general lack of available material and in order to sample older Lower Palaeozoic basement that is exposed near Nuneaton. The position of the boreholes is shown in relation to pre-Devonian subcrop in Fig. 6.11 and the stratigraphy and lithology of the samples is given in Table 6.6. Information on the age of the samples collected is highly variable. A fairly detailed biostratigraphy is available for the Cambrian–Tremadoc Stockingford Shales, drilled in the Merevale boreholes on the Nuneaton inlier (Taylor and Rushton, 1971). On the other hand, a whole-rock Rb–Sr date of 470 Ma provides the only indication of the age of an unfossiliferous sequence of shales at the base of the Thorpe-by-Water borehole (A. H. Bath, cited in Richardson and Oxburgh, 1978).

The samples can be subdivided into two suites on the basis of their deformation histories. The first includes those of the Merevale, Twycross and Home Farm boreholes, from the south-west corner of the study region, which are uncleaved and of low density (ca 2500–2600 kg m⁻³). The second group, which covers the remainder of the area, is cleaved and more highly indurated, with saturated densities around 2700 kg m⁻³, approximating to grain densities. The two groups fall on either side of a NW–SE line that extends between the Rotherwood and

Table 6.5 Mean heat production figures for Lower Palaeozoic shale reference collection (using a density of 2700 kg m⁻³)

Stratigraphic unit	Heat production (μW m⁻³)	No. U analyses
Downton Group	2.4	3
Ludlow	2.0	16
Wenlock	2.1	13
Llandovery	2.1	37
Ashgill	1.8	17
Caradoc	2.0	29
Llandeilo	1.8	3
Llanvirn	1.8	9
Arenig	2.1	31
Tremadoc	1.6	3
Merioneth	6.9	7
St David's	2.0	12
Comley	2.0	16

Twycross boreholes, through Huntingdon, to the coast south of Ipswich. This reflects the boundary between the Midlands Microcraton to the SW and a zone of Caledonian deformation to the NE. The size of the microcraton is much reduced compared to that shown by Wills (1978). The boundary here is probably much closer to that of Turner (1949) than the south-western edge of the fault-bounded trench or aulacogen that was proposed by Evans (1979). If the aulacogen model is accepted, its north-eastern edge must be moved into the North Sea as strongly cleaved and indurated Lower Palaeozoic

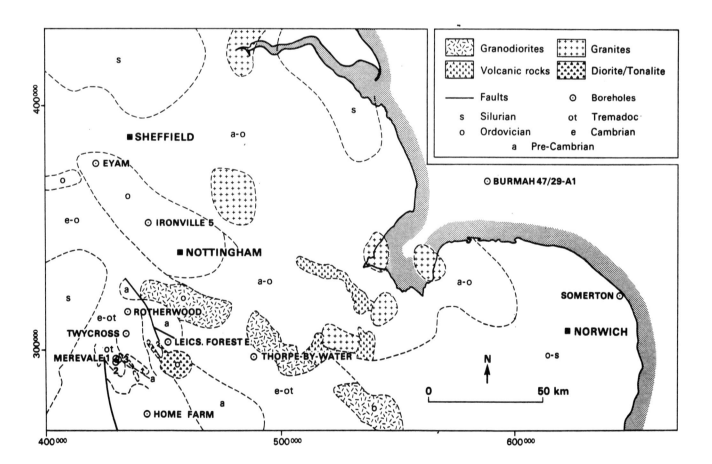

Fig. 6.11 Locations of boreholes in study area sampled for Lower Palaeozoic shales in relation to pre-Devonian subcrop (from an interpretation by N. J. P. Smith)

75

mudrocks are present in the Burmah 47/29-A1 and Somerton boreholes (see Fig 6.11). Recent detailed studies of the basement of eastern England (Pharaoh and others, 1988a) favour Turner's (1949) model with the strike of the Anglo-Welsh Caledonides swinging around beneath the Pennines, from a NE–SW orientation in the west to a NW–SE alignment in the east and joining the continental Caledonides in the Ardennes.

The Merevale boreholes provide the most detailed stratigraphical section through Lower Palaeozoic shales in the study region and their geochemistry is described first. These data are then compared with the shale sequences of the Home Farm and Twycross boreholes, at the edge of the postulated Midland Microcraton, and with the group of more indurated shales, of mainly Ordovician–Silurian age, from boreholes to the north and east.

One hundred samples from nine boreholes were analysed and compared with the Merevale boreholes. More than half of the samples came from the relatively long basement cores of Eyam and Thorpe-by-Water. Elsewhere only short basement intersections were made or, in the case of oil exploration wells, only a short terminal core was taken. Few samples were available from these bore-holes, but they nevertheless provide a general indication of the nature and chemical composition of argillaceous Lower Palaeozoic basement in the East Midlands. The boreholes at Eyam and Ironville provide information on basement close to the South Pennine Orefield.

6.4.1 SAMPLES FROM THE MIDLANDS MICROCRATON

The Merevale Boreholes

The stratigraphy of the Nuneaton inlier is presented in Fig. 6.12. The Merevale boreholes sample a section from the Middle Cambrian Abbey Shales to the basal Tremadocian Merevale Shales. The latter are overlain unconformably by Old Red Sandstone and the top 1.2 m are reddened. The sequence mainly comprises light grey-black mudstones with some purple (Purley Shales) and greyish green varieties (Merevale Shales) together with minor sandstone, limestone and dolomitic layers. Arenaceous units at the top and base of the Outwoods Shales sequence are termed the Moorwood Flags and Shales and the Mancetter Grits and Shales, respectively (Taylor and Rushton, 1971).

Table 6.6 Details of Lower Palaeozoic shale samples from East Midlands

Borehole/sample type	No.	Stratigraphy	Age
SOMERTON Bluish grey to very dark grey cleaved mudstone/siltstone	6	?L Silurian	
BURMAH 47/29-A1 Dark grey well-cleaved mudstones	6	Llanvirn/Llandeilo (Wills, 1978)	
EYAM Red-brown, grey-green and dark grey siltstone and mudstone, cleaved	27	Arenig or Llanvirn (Dunham, 1973)	
IRONVILLE NO 5 Dark and light grey cleaved mudstone	5	Early Ordovician	?Arenig or Tremadoc
THORPE-BY-WATER Purple-brown and grey mudstone and siltstone	35	?Cambro-Ordovician (Bath in Richardson and Oxburgh, 1978)	
HOME FARM Purple-grey, red-brown and dark grey mudstone	4	Merevale Shales (IGS, 1979)	Tremadoc
LEICESTER FOREST EAST Dark grey and purple cleaved mudstone	8	Probable Tremadoc (IGS, 1979)	
ROTHERWOOD Dark grey well-cleaved	5	Upper Cambrian (pre-Tremadoc) (IGS, 1979)	
TWYCROSS Dark purple silty mudstone and pale grey-green mudstone	4	Presumed Cambrian (IGS, 1979)	
MEREVALE NOS 1–3 Mainly dark grey mudstones (overall range pale grey-black)	118	Stockingford Shales (Taylor and Rushton, 1971)	M Cambrian to Tremadoc

The overall geochemistry of the Stockingford Shales and of individual units in the sequence is summarised in Table 6.7 and presented in Figs. 6.13, 6.14 and 6.15. As a whole the concentrations of major elements and elements normally associated with the detrital fraction of shales approximate those of average shales. SiO_2, CaO and Na_2O are slightly below and Al_2O_3 and Fe_2O_3 slightly above the values of average shales. The sequence is strongly enriched, however, in the trace elements As, Ba, Cu, Mo, Pb, S and V (Fig. 6.15) with levels comparable with those of black shales elsewhere in Britain, particularly those of Cambrian age (section 6.3.2). At Nuneaton the enrichment in these trace elements, found in the Upper Cambrian of Britain generally, also extends down into the Middle Cambrian. Levels of Fe_2O_3 are also high and together with elevated S concentrations reflect the abundance of pyrite.

Within the Stockingford Shale sequence there is considerable variation in the concentration of As, Cu, Mo, Ni, Pb, S, U and Ba. Particularly high levels of these elements and of Fe_2O_3 and S occur in the Monks Park Shale, which is the equivalent of the uraniferous and metal-enriched Upper Cambrian Alum Shales of Sweden. Levels of Cr, Co, Y, Zr, Nb and Th, which are normally associated with the detrital fraction of shales, are, however, at or below average levels, although Ce values are up to twice those of average shales, which suggests a degree of LREE enrichment. The Monks Park Shales are one of the most pyritic units of the Merevale sequence with a total S content up to 10.8% (Taylor and Rushton, 1971), equivalent to 20.2% pyrite. The trace-element and pyrite levels are thought to reflect an initially high content of organic matter, although the C content of only 2.42% (Taylor and Rushton, 1971) is not particularly high. It is likely that much of the organic material was oxidised during diagenesis, with the associated reduction of SO_4^{2-} producing pyrite. According to the SO_4^{2-} reduction equations and the organic composition (the Redfield composition) given in Froehlich and others (1979), the average 13% pyrite in the Monks Park Shales would have required the destruction of 14% organic matter, equivalent to an initial C content of 5.1%. The division of the Monks Park Shales into a lower series of grey micaceous shaly mudstones with lower trace-element contents and an upper series of more metal-enriched black mudstones (Fig. 6.14) is also consistent with an organic association for As, Cu, Mo, Ni, Pb, S and V. Ba and Zn levels show no significant differences between the two series. Baryte has been reported in dolomitic nodules in the Monks Park Shales (Taylor and Rushton, 1971) comparable with that found in the Cambro-Ordovician of Norway (Bjørlykke, 1974; Bjørlykke and Griffin, 1973).

The Monks Park Shales pass upward by sedimentary transition into the Merevale Shales, the basal part of which has levels of the above trace elements approaching those of the Monks Park Shales. Above this horizon these trace-element levels generally decline. The top of the sequence, which is reddened below the ORS unconformity, has particularly low levels of Mo and Cu, values for other trace elements, except Ba, approaching those of average shales. The reddening is also associated with increased Fe_2O_3, but S and MnO contents are lower, probably reflecting destruction of pyrite or siderite during Old Red Sandstone erosion and surface weathering. These phases are concentrated in the basal Merevale Shales, particularly in burrow infills. The dolomitic bands noted by Taylor and Rushton (1971) in the Merevale Shales are not represented in the samples collected.

The Monks Park Shales grade downwards into the Moorwood Flags and Shales, the Outwoods Shales and the Mancetter Grits and Shales, all of which have a similar geochemistry characterised by markedly lower contents of certain trace elements than those of the Monks Park Shales; Ni, Pb, V and Zn concentrations are close to those of average shale, but As, Ba, Cu, Mo, S and U levels remain appreciably higher (Figs. 6.14 and 6.15). Geochemical variations through this part of the sequence generally reflect gradational changes in sedimentation. The base of the Mancetter unit is erosive (Taylor and Rushton, 1971) and increased levels of several trace elements (As, Cu, Mo, V, Zn) above the erosion surface may reflect the incorporation of detritus from the underlying Abbey Shales.

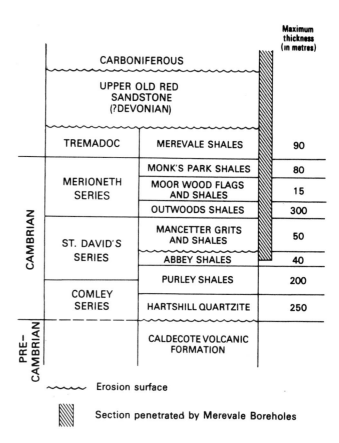

Fig. 6.12 Stratigraphic summary of Stockingford Shales of Nuneaton inlier

The Abbey Shales are distinctive bluish grey mudstones with thin limestones and glauconitic sandstones. They have higher As, Cu, Mo, Pb, V and Zn levels than most of the Stockingford sequence; their As and Cu levels match those of the Monks Park Shales, Pb and Zn contents are even higher and Mo and V levels are intermediate between those of the Monks Park and other shales. This unit is also distinguished in having lower Ba concentrations than the remainder of the Stockingford sequence, although they are, nevertheless, approximately three times those of average shales, and only average S levels (1.4%). The relatively high mean CaO content of 2.47% is biased by a value of 16.2% in one of the samples analysed. The mean for the remaining samples drops to 0.94% when this sample is excluded. The decrease in Ba content from the Upper Cambrian to the Middle Cambrian through the Mancetter Grits and Shales precisely follows that identified in rocks from the reference collec-

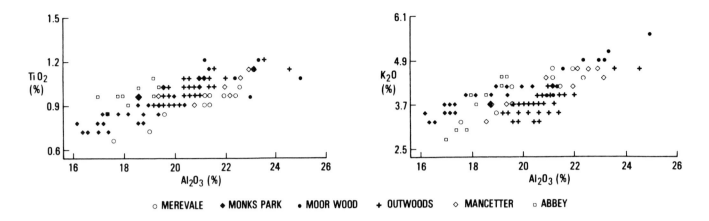

Fig. 6.13 $TiO_2-Al_2O_3$ and $K_2O-Al_2O_3$ plots for Stockingford Shales of Nuneaton inlier. For stratigraphy see Fig. 6.12

tion covering the same time period elsewhere in Britain (see section 6.4).

Twycross

The Twycross samples are approximately the same age as those of Merevale, whereas the rocks from Home Farm are slightly younger. Both boreholes are situated near to the edge of the Midlands Microcraton to the SW of the zone affected by late Caledonian deformation.

Samples from the Twycross borehole, which are

presumed to be of Cambrian age (Institute of Geological Sciences, 1979), comprise dark purple silty mudstone and pale greenish grey siltstones, reddened beneath the Triassic unconformity. Geochemically, the samples can be divided into two groups, one comprising samples that are fissile and contain higher Fe_2O_3, MgO, Na_2O, Co, Ni, Cu and Zn, and a second group of banded shales with higher SiO_2, Al_2O_3, TiO_2, CaO, K_2O, As, Ce, Pb and Y contents (Fig. 6.17). The high Ba content of all of the samples is consistent with a Cambrian age, but the

Table 6.7 Means for individual units of Stockingford Shales, Merevale boreholes (M. Cambrian–Tremadoc)

	Merevale Shales	Monks Park Shales	Moorwood Flags and Shales	Outwoods Shales	Mancetter Grits and Shales	Abbey Shales
SiO	59.95	55.98	56.34	56.54	56.56	58.88
Al_2O_3	20.71	18.34	21.09	20.48	21.05	17.32
TiO_2	0.88	0.84	1.03	0.99	1.01	0.92
Fe_2O_3	10.61	10.62	8.16	9.62	9.94	9.31
MgO	2.30	2.05	2.37	2.44	2.34	2.79
CaO	0.43	0.79	0.92	1.21	0.61	2.47
Na_2O	0.42	0.65	0.86	0.95	0.89	1.11
K_2O	4.18	3.83	4.27	3.72	4.10	3.51
MnO	0.24	0.06	0.26	0.57	0.37	0.59
P_2O_5	0.06	0.09	0.12	0.13	0.13	0.14
As	51	92	28	33	56	90
Ba	3147	2564	2634	2186	1557	1668
Ce	97	91	103	101	104	102
Cl	16	43	104	57	28	49
Co	24	19	17	19	18	29
Cr	94	102	105	102	106	99
Cu	60	106	49	63	58	112
Mo	8.8	45.6	8.0	6.5	8.6	12.6
Ni	88	144	60	61	72	69
Nb	15	15	17	17	17	17
Pb	50	48	18	22	29	64
Rb	146	129	146	137	145	116
S	28051	67685	22236	25308	24845	14066
Sc	12	10	12	12	12	13
Sr	149	109	104	105	103	124
Th	12	13	12	13	13	13
U	5.9	23.8	9.8	5.3	3.4	5.4
V	145	431	170	156	148	218
Y	26	32	33	33	32	40
Zn	76	141	79	108	105	178
Zr	95	92	142	128	129	136

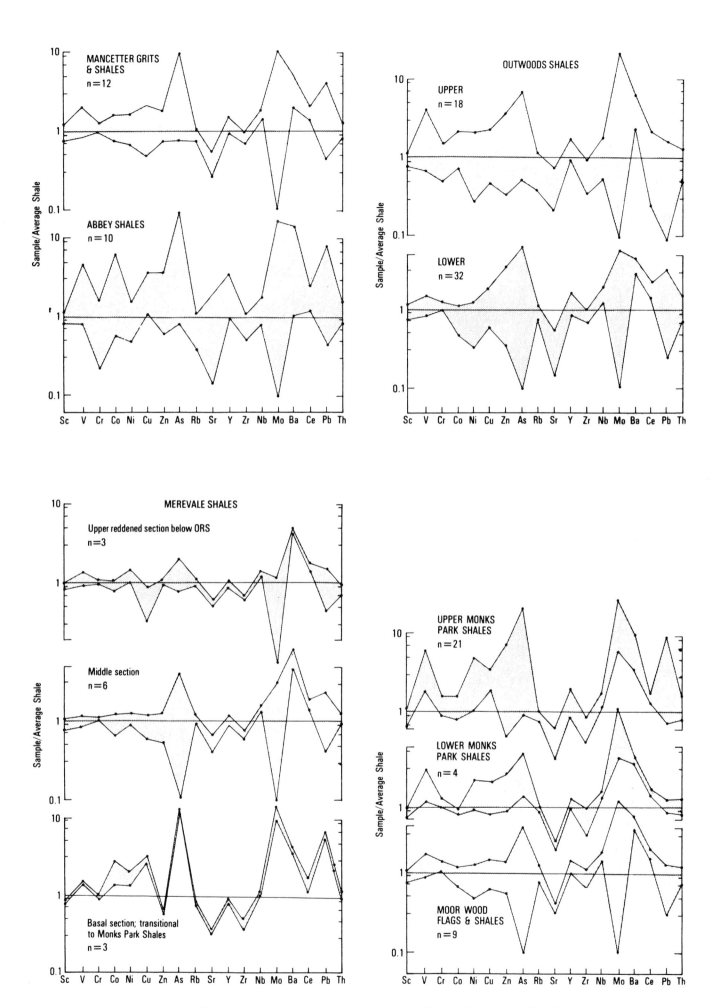

Fig. 6.14 Average shale normalised trace-element diagram for Stockingford Shales of Nuneaton inlier. For stratigraphy see Fig. 6.12

lithology and trace-element geochemistry of the fissile samples is generally more like that of the Tremadocian rocks of the Merevale boreholes.

The trace-element contents of the banded rocks differ markedly from those of Merevale, possibly reflecting leaching during Triassic times, which removed mobile elements and increased the proportion of relatively immobile elements such as Al_2O_3, TiO_2, Ce and Y.

Home Farm

The Home Farm samples are from higher in the Tremadocian sequence than those of Merevale and they differ geochemically, particularly in their lower contents of Ba (Table 6.8; Figs. 6.16, 6.17, 6.18). The decrease in Ba and other metals of organic association (V, Cu, etc.) in the higher parts of the Tremadocian sequence of Home Farm relative to the Cambrian–Tremadoc sequence of Merevale follows a similar temporal trend that can be identified throughout England and Wales in the reference collection (see section 6.3.2). Levels of Sc, Co and other elements normally held in detrital minerals are higher in the Home Farm borehole than those at Merevale, but their association with increased Al_2O_3 suggests that this reflects, at least in part, an increased proportion of clay minerals such as illite. The Home Farm samples have been oxidised to a purple-brown colour immediately beneath the Triassic unconformity, but in comparison with reddened samples from beneath the ORS unconformity at Merevale they have lower contents of V, Cu, As, Sr, Mo and Pb and higher levels of Sc, Co, Y, Zr, Nb and Th. As outlined above, many of these differences are probably of primary origin, but an overprint of more intense weathering cannot be discounted.

6.4.2 Shales from Tract of Caledonian Deformation

These rocks are generally distinguished from the sequences to the south by their low trace-element contents, including metalliferous and heat-producing elements, and by their higher TiO_2/Al_2O_3 ratios.

Leicester Forest East

This borehole intersected rocks of probable Tremadocian age (Institute of Geological Sciences, 1979) that are purplish brown immediately beneath the Triassic unconformity and grey siltstones with oxidised cleavage surfaces lower in the sequence. The geochemistry of these rocks is generally comparable with that of the younger Tremadoc rocks of the Home Farm borehole (although Zn levels are lower and As levels higher) rather than with the basal Tremadoc sequence represented at Merevale (Figs. 6.16, 6.17, 6.18).

Rotherwood

The lithology and fauna of the Rotherwood samples indicate an Upper Cambrian age, although geochemically they most closely resemble the Middle Cambrian Abbey shales of the Merevale boreholes with enrichment in Co and As. Their low Mo contents differ markedly from those of the Monks Park Shale, but their general enrichment in Ba follows that of the Upper Cambrian shales elsewhere in Britain (section 6.3.4) and the Middle–Upper Cambrian shales of Nuneaton.

Thorpe by Water

A Rb–Sr date of 470 Ma (Llanvirn, Harland and others, 1982; or Arenig–Llanvirn, Odin and others, 1982) has been reported for shales from this borehole. The core shows evidence of considerable oxidation beneath the Triassic unconformity, however, so this age can be regarded only as a minimum: an older age would be more consistent with the geochemical evidence. The major- and trace-element geochemistry of the samples is most closely comparable with that of Tremadocian rocks of Leicester Forest East or Home Farm. The TiO_2/Al_2O_3 ratio, for example, averages 0.045 (Fig. 6.16), which is within the range of the Merevale–Abbey Shales succession (0.042–0.053) and considerably lower than that of known Ordovician shales—for example, those from Eyam, which average 0.062. The radiometric date could reasonably be interpreted as recording a diagenetic or metamorphic event, such as cleavage formation.

Burmah 47/29-A1

Of the remaining boreholes, the samples from the Burmah well are the most distinctive, with high levels of Cr, Ni, MgO and TiO_2 (Figs. 6.16, 6.17 and Table 6.8), consistent with an important component of basic igneous rock detritus (cf. samples from Girvan in the reference collection (section 6.3.3). By use of TiO_2–Al_2O_3 and K_2O–Al_2O_3 plots, however, samples from this borehole clearly fall into the Ordovician–Silurian field with higher ratios than Cambrian–Tremadoc samples (Fig. 6.16). The slightly elevated Ba levels of 860 ppm are comparable with those found in Tremadoc–Arenig or Caradoc age rocks in the reference collection. The high mean S content, approaching 4000 ppm, reflects the presence of fine sulphides in the samples.

Ironville 5 and Eyam

The age, location and major- and trace-element content (Table 6.8; Figs. 6.16–6.18) of the Ironville 5 and Eyam samples are closely comparable with similar trends and ratios of TiO_2/Al_2O_3 and K_2O/Al_2O_3. Ba, Rb and K are higher in the Ironville 5 borehole than that of Eyam, however, with Ba levels almost twice those of average shales—a feature more typical of Tremadoc than Arenig shales in the reference collection (section 6.3.2). The lithology differs markedly from Tremadoc shales to the south, however, dark grey argillites being the typical rock type. The Eyam borehole is affected by reddish staining cross-cutting cleavage and bedding, but this shows no obvious relationship to the overlying red bed sequences.

Somerton

The Somerton borehole has a higher carbonate content than any of the other boreholes sampled; this is reflected by elevated CaO and MgO and lower Al_2O_3 levels. It also has the highest K_2O/Al_2O_3 ratios. These distinctive major-element characteristics suggest a different provenance for these shales, and perhaps those of Burmah 47/29-A1. The high S value reflects the presence of scattered, fine-grained pyrite in the samples.

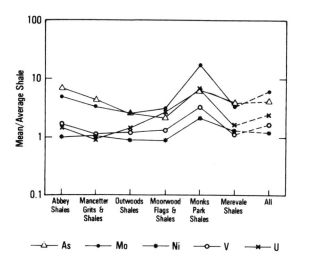

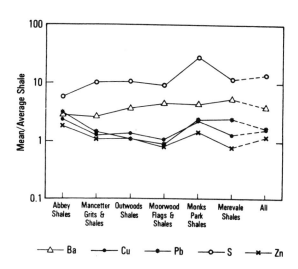

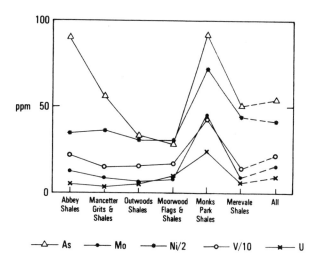

Fig. 6.15   Temporal variations in selected trace elements for Stockingford Shales of Nuneaton inlier. For stratigraphy see Fig. 6.12

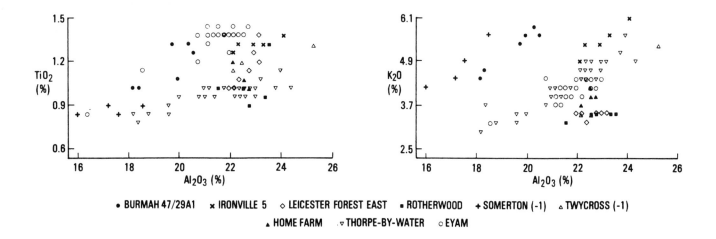

Fig. 6.16   TiO$_2$–Al$_2$O$_3$ and K$_2$O–Al$_2$O$_3$ plots for Lower Palaeozoic shales from east central England. For borehole locations see Fig. 6.11

Table 6.8 Means from borehole samples of Lower Palaeozoic shales

	Merevale	Home Farm	Leicester Forest East	Rother-wood	Twycross	Thorpe-by-Water	Burmah 47/29 A1	Ironville No. 5	Somerton	Eyam
SiO_2	56.65	56.02	56.34	55.57	59.65	58.33	53.92	55.67	57.04	56.68
Al_2O_3	19.88	22.54	22.67	22.82	24.17	21.83	19.57	23.10	16.26	21.59
TiO_2	0.94	1.10	1.17	1.03	1.25	0.98	1.18	1.31	0.86	1.34
Fe_2O_3	9.82	8.67	9.05	9.24	5.70	7.58	8.72	7.05	6.55	7.92
MgO	2.36	2.42	2.16	1.96	1.77	2.14	5.30	2.46	4.30	3.36
CaO	1.06	0.26	0.37	0.52	0.31	0.36	0.59	0.58	4.24	0.35
Na_2O	0.83	0.99	1.23	1.39	0.95	1.08	1.02	0.43	0.87	0.97
K_2O	3.86	4.00	3.41	3.43	4.12	4.35	5.30	5.47	4.40	4.09
MnO	0.39	0.04	0.11	0.92	0.02	0.06	0.09	0.05	0.09	0.06
P_2O_5	0.11	0.10	0.09	0.11	0.11	0.09	0.16	0.10	0.14	0.11
As	54	6	11	49	19	3	12	6	9	10
Ba	2290	593	725	1650	2291	949	860	1089	563	645
Ce	99	99	99	97	64	84	94	95	81	74
Cl	50	148	17	26	74	48	277	44	381	384
Co	20	21	23	37	11	14	26	20	17	21
Cr	101	99	110	103	87	89	172	113	124	118
Cu	74	11	19	34	12	20	34	10	29	22
Mo	15.9	0.8	0.3	0.0	1.3	0.2	0.1	0.8	1.5	0.5
Ni	83	61	67	76	57	44	131	64	85	62
Nb	16	20	19	17	20	19	20	21	18	19
Pb	34	12	7	17	43	7	19	8	22	11
Rb	136	155	146	140	130	162	199	170	165	128
S	33331	150	1281	1484	495	330	3887	317	9469	2941
Sc	12	18	20	19	18	16	22	20	15	22
Sr	111	115	131	144	277	144	51	53	115	111
Th	13	13	12	13	11	12	13	10	12	11
U	9.2	2.6	2.4	3.0	4.2	3.7	3.4	2.6	4.3	3.0
V	219	112	119	114	123	102	141	125	114	135
Y	33	31	29	29	42	32	35	34	33	30
Zn	115	110	65	90	44	80	96	55	94	78
Zr	119	142	130	112	179	181	196	200	204	217

6.4.3 ORE-FORMING ELEMENTS

The levels of the elements Pb, Ba, Zn, S and Cu in the Stockingford Shale sequence, and particularly in the Monks Park Shale, are all higher than those of average shales, enrichment relative to the average decreasing in the order S, Ba, Pb, Cu, Zn (Fig. 6.15). However, the only average concentration that exceeds the minimum enrichment value with which Vine and Tourtelot (1970) defined a 'metal-rich' black shale is that of Ba. The mean Ba contents from individual units of the Stockingford sequence also exceed this figure. Individual samples of basal Merevale Shales and Upper Monks Park and basal Abbey Shales exceed the minimum enrichment value of 100 ppm for Pb, but levels of Zn, S and, particularly, Cu are considerably lower. The Stockingford Shales represent an excellent potential source of Ba, and of Pb, Zn, S and Cu for the formation of mineral deposits, but they are poorly indurated and have probably not lost all the water bound in clays. Moreover, it is difficult to envisage that they provided the source of fluids for the Pennine mineralisation since they are separated from the orefields by an extensive tract of shales, which were indurated and cleaved during late Caledonian tectonism.

Elsewhere in the study region Ba concentrations are above average in all the borehole samples studied, except Somerton, with the highest values (approximately twice the average) in Middle–Upper Cambrian and Tremadoc samples from the SW of the region, including samples from Rotherwood and Twycross (Fig. 6.18). Similar values are found in rocks of this age from elsewhere in England and Wales (section 6.3.2). The Cu content is, in contrast, generally low throughout the basement shales of the East Midlands with values consistent with those found in shales of similar age. The overall average value of 37 ppm for Cu in the reference collection is appreciably higher than in any of the borehole samples except those from Rotherwood, Burmah and Somerton. Pb values are high to the SW of the main study region, the above-average values of the Merevale borehole extending NE as far as Twycross. Values are close to the 20 ppm average for Pb at Rotherwood, Somerton and Burmah, but elsewhere they fall well below this value. Zn is highest (just above the 95 ppm average) at Rotherwood, Home Farm, Burmah and Somerton, but values elsewhere are below average. S is enriched in the Eyam, Burmah and Somerton boreholes where sulphides are visible to the naked eye, but it falls to exceptionally low levels in the Home Farm, Thorpe-by-Water and Ironville 5 boreholes.

Most of the metals that make up the ore mineral assemblage in the Pennine orefields reach their highest concentrations in the Middle–Upper Cambrian and lowest Tremadocian rocks. Generally lower levels occur in the younger (Ordovician–Silurian) sequences that form the basement in the vicinity of the South Pennines. The extent to which this is related to the depositional environment or to the post-depositional history of the shales is not clear. Most of the samples from boreholes near to the South Pennine Orefield are cleared and indurated and may have lost quantities of mobile elements as a result. The Caledonian cleavage development predates the mineralisation in the South Pennines, however, by at

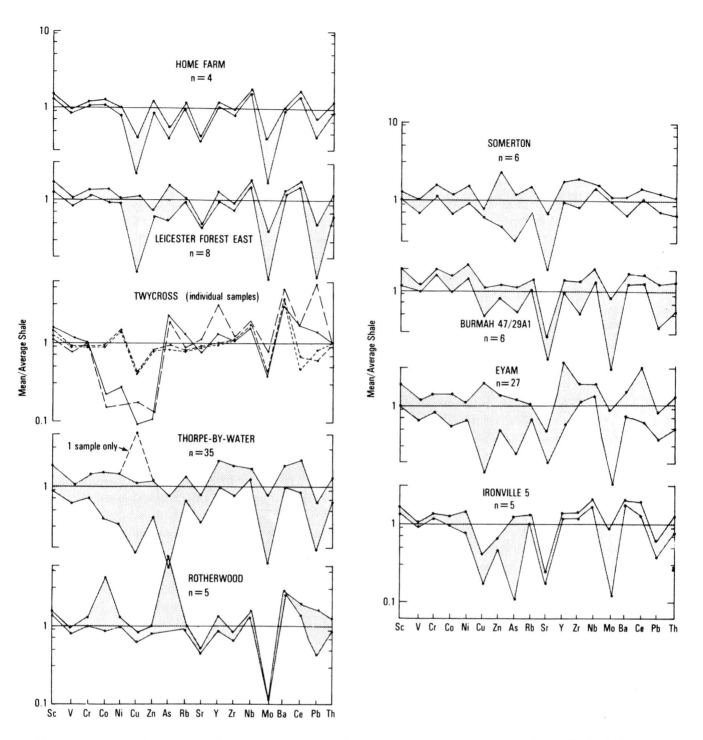

Fig. 6.17 Average shale normalised trace element diagrams for Lower Palaeozoic shales from east Central England. For borehole locations see Fig. 6.11

least 100 Ma (Chapter 8) and it would thus appear highly unlikely that these rocks could have been a source of mineralising fluids. The distribution of metals in the East Midlands, however, fits reasonably well with the age-related distribution shown by the reference collection, which mainly comprises cleaved mudrocks, with enrichment of metals in, for example, the Upper Cambrian.

6.4.4 Heat Production

The high heat production observed in the Upper Cambrian shales from the reference collection (section 6.3.5) is also apparent in the Stockingford Shales from Nuneaton (Table 6.9), particularly in the Monks Park Shales, where the mean uranium content is 23.8 ppm (Table 6.7). The Monks Park Shales, however, form less than 10% of the total Cambrian–Tremadoc sequence; the thickest unit sampled, the Outwoods Shales, has only a slightly greater heat production ($2.5\mu W\ m^{-3}$) than average shale. The overall mean for the Stockingford Shales, weighted for the thickness of each unit, is $3.2\ \mu W\ m^{-3}$, 50% greater than average shale.

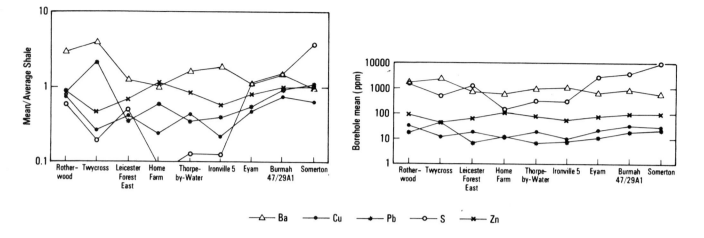

Fig. 6.18 Temporal variations in Ba, Cu, Pb, S and Zn for Lower Palaeozoic shales from east Central England (B normalised to average shale). For borehole locations see Fig. 6.11

Table 6.9 Heat production figures for Lower Palaeozoic shales in and around study area (densities based on Appendix 2)

Stratigraphic unit or borehole name	Heat production ($\mu W\ m^{-3}$)	No. of U analyses	Density used (kg m^{-3})
Somerton	2.5	2	2750
Burmah 47/29 A1	2.3	2	2750
Eyam	1.9	5	2730
Ironville 5	1.9	2	2750
Thorpe-by-Water	2.2	5	2740
Home Farm	1.8	2	2540
Leicester Forest East	1.7	2	2690
Rotherwood	2.0	2	2760
Twycross	2.1	2	2510
Mean, excluding Stockingford Shales	2.0	24	
Merevale Shales	2.6	3	2600
Monks Park Shales	7.2	5	2600
Moorwood Flags and Shales	3.6	2	2600
Outwoods Shales	2.5	10	2600
Mancetter Grits and Shales	2.1	3	2600
Abbey Shales	2.5	2	2600
Stockingford Shales: mean	3.5	25	2600
thickness weighted mean	3.2	25	2600

The average heat production for the remaining samples from the East Midlands (excluding Merevale boreholes) is the same as that of average shale (2.0 $\mu W\ m^{-3}$) with a fairly restricted range of values from 1.7 to 2.5 $\mu W\ m^{-3}$). With the exception of the Cambrian at Nuneaton, heat production figures are therefore close to the average shale for the Lower Palaeozoic argillites from the East Midlands, but generally 50% greater than average crust.

6.5 Carboniferous (Viséan–Namurian) shales

The Viséan–Namurian shales, unlike those of the Lower Palaeozoic, which were deposited in the oceans and seas of the developing Caledonian orogenic belt, were laid down in continental basins and shelf seas formed as the crust was extended and thinned during the Lower Carboniferous (Chapter 5). There are thus marked local facies changes with Viséan shelf limestones and later thin, condensed sequences of Namurian shale developed on basement highs, and thick basinal sequences in the deeper parts of half-graben. The stratigraphy of the Peak District of Derbyshire and Staffordshire is summarised in Fig. 6.19. The off-shelf sequence consists primarily of turbiditic facies, intercalated with pelagic mudstones; carbonate turbidites predominate in the Viséan and terrigenous clastic turbidites (protoquartzites) in the Namurian.

Approximately 400 Viséan–Namurian shale samples were obtained from core from thirty boreholes drilled in the region (Fig. 6.20) for a variety of purposes (Table 6.10). In many cases the core had been retained for palaeontological studies and only small samples (of the order of 100 g) were available for analysis.

Larger samples were prepared from the Duffield, Birchover, Ipstones Edge, Lees Farm and New Mixon Hay 2 boreholes, however. The samples include shales from the Asbian (B_2) and Brigantian (P_1 and P_2) zones of the Viséan through the Pendleian (E_1), Arnsbergian (E_2), Chokierian (H_1), Alportian (H_2), Kinderscoutian (R_1) and Marsdenian (R_2) zones of the Namurian.

The basement block, which underlay the Viséan carbonate platform, remained as a stable positive feature during the Lower Namurian and thus the E_1–H_2 portion of the section is greatly condensed. In the North Staffordshire Basin the early turbidites were probably derived from the west with later members from the south. The Kinderscout turbidites, in the Central Pennine Basin to the north, were derived from the north and the Ashover Grits represent a facies change from turbidite to delta-top deposits. The highest portion of the sequence sampled was from below the Ashover Grit horizon, so the samples studied mainly represent the turbidite/ hemipelagic sequences, or their equivalents overlying the blocks.

In the Namurian (and possibly to a lesser extent in the Viséan) cycles of faunal change (Ramsbottom and others, 1962), which probably reflect variations in salinity

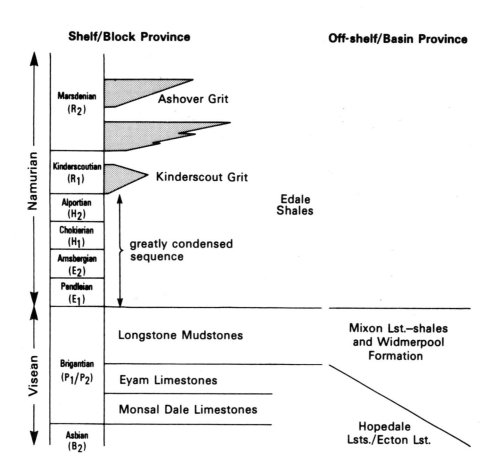

Fig. 6.19 General Viséan–Namurian stratigraphy for Peak District of Derbyshire and Staffordshire (after Aitkenhead and Chisholm, 1982)

Table 6.10 Summary of Viséan–Namurian shales analysed

Borehole name	No. of samples	Stratigraphic interval sampled
Alport	33	Brigantian (P_1)–Kinderscoutian (R_1)
Ballhaye	5	Arnsbergian (E_2)
Birchover	10	Kinderscoutian (R_1)–Marsdenian (R_2)
Bretton Clough	21	Arnsbergian (E_{2a})–Kinderscoutian (R_{1c})
Buskey Cottage	5	Brigantian (P_2)
Castleton 1	6	Brigantian (P_2)–Pendleian (E_{1c})
Duffield	107	Asbian (B_2)–Chokierian (H_{1a})
Field House	1	Brigantian (P_2)
Glutton Bridge	11	Pendleian (E_1)–Arnsbergian (E_2)
Gun Hill	4	Brigantian (P)
Highoredish	19	Brigantian (P)–Kinderscoutian (R_1)
Hope Cement Works	8	Brigantian (P)–Arnsbergian (E_{2b})
Hucklow Edge 1	14	Pendleian (E_{1a})–Kinderscoutian (R_{1a})
Ipstones Edge	28	Arnsbergian (E_2)–Kinderscoutian (R_1)
Ironville 3	2	Brigantian (P) and Kinderscoutian (R_1)
Lees Farm	21	Brigantian (P_{1b})–Pendleian (E_{1a})
Mam Tor 2	4	Chokierian (H_{1a})
Mam Tor 4	3	?Arnsbergian (E_{2a})
Mam Tor 6	3	Chokierian (H_{1a})
Mam Tor 8	3	Arnsbergian (E_{2b})
New Mixon Hay 2	12	?Asbian (B_2)
Stanton T1	2	Chokierian (H_{1a})
Stanton T5	15	Brigantian (P)–Kinderscoutian (R_{1b})
Stoop Farm	10	Arnsbergian (E_2)–Kinderscoutian (R_1)
Uppertown	31	Brigantian (P)–Marsdenian (R_2)
Wardlow Mires 1	13	Brigantian (P_2)–Arnsbergian (E_{2a})
Warslow Hall	4	Late Asbian and Brigantian
Waterfall	4	Late Asbian and Brigantian
Wetton	5	Late Asbian and Brigantian
Widmerpool 1	4	Brigantian (P_2)–?Alportian (H_2)

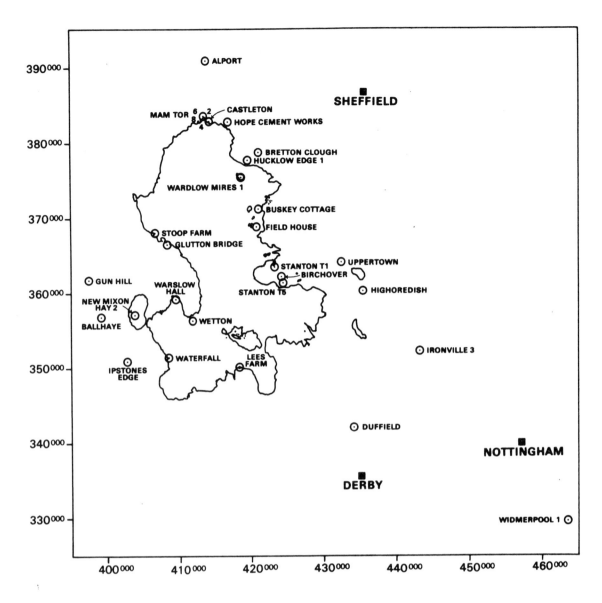

Fig. 6.20 Locations of boreholes sampled for Viséan–Namurian shales in relation to outcrop of Viséan–Namurian boundary (Viséan inside the boundary)

(Aitkenhead, 1977) can be identified. Only the fish faunal phases (indicative of low salinity) are represented in Namurian turbidites from Duffield, whereas goniatite phases (indicative of higher salinity) are represented by pelagic interbeds. The sequence probably reflects the formation of deltas, which built out to the edge of basins and gave rise to turbiditic sediments that were followed by marine transgressions marked by marine bands.

The stratigraphy of the samples collected is normally very well constrained, unlike the Lower Palaeozoic shales described in previous sections, and it is possible in the case of almost all of the samples to assign them to a particular stage and in many cases to a single zone or subzone by use of faunal indices—in particular, goniatites. Hence geographical and temporal changes can be identified readily and sediments of shelf and off-shelf environments (Viséan) or stable block and basin facies (Namurian) compared.

Lithologically, the samples collected are mudstones, silty mudstones and siltstones, which are commonly micaceous and which range from medium grey to black in colour, although most of the samples are dark grey (GSA rock colour chart). They range from mudstones and siltstones, which produce no acid reaction, to argillaceous

limestones, which produce a strong reaction with dilute acids. The samples are frequently fossiliferous, containing, for example, plant, goniatite, bivalve and fish remains and some have nodules of siderite or phosphate. Pyrite occurs fairly commonly and many samples have calcite veinlets. The samples range from fissile and laminated to massive.

The location of samples, in relation to the Dinantian outcrop of the South Pennines, is shown in Fig. 6.20. Coverage is good, except in the NW between Stoop Farm and Mam Tor, although many samples are close to the Dinantian outcrop since the limestone was often the target of drilling operations.

6.5.1 GEOCHEMISTRY OF VISÉAN SHALES

Of the major elements, CaO contents are the highest, compared to average shale, and are equivalent to 11.05–19.73 % $CaCO_3$, with an average of at least 15 %, after allowing for CaO in aluminosilicate. P_2O_5 is also high, reflecting the presence of phosphatic nodules in the rocks. Other major elements, especially SiO_2, TiO_2, Fe_2O_3, Na_2O, K_2O and MnO, are all low compared to

Table 6.11 Summary of the geochemistry of Viséan shales

	All Viséan (n = 137)		Asbian (B₂) (n = 23)		Brigantian (P₁) (n = 50)		P₂ (n = 64)	
	Mean	sd	Mean	sd	Mean	sd	Mean	sd
SiO_2	54.44	13.50	57.53	9.53	52.78	12.30	54.84	15.40
Al_2O_3	14.19	5.59	16.66	5.62	15.25	5.76	12.47	4.95
TiO_2	0.55	0.29	0.69	0.33	0.56	0.28	0.48	0.27
Fe_2O_3	4.67	3.32	4.36	1.93	4.63	3.53	4.81	3.58
MgO	2.33	2.67	2.76	3.62	1.66	0.99	2.69	3.07
CaO	11.05	10.00	7.91	9.29	12.29	10.60	11.21	9.77
Na_2O	0.65	0.71	0.96	1.56	0.71	0.35	0.49	0.29
K_2O	2.25	1.00	2.81	1.39	2.30	0.91	2.00	0.80
MnO	0.07	0.13	0.04	0.01	0.05	0.09	0.09	0.16
P_2O_5	0.45	1.24	0.13	0.13	0.27	0.57	0.71	1.72
As	32	32	46	41	28	34	30	26
Ba	369	1128	321	169	263	166	470	1642
Ce	67	28	80	25	72	31	60	24
Cl	646	650	421	589	1004	642	434	552
Co	18	12	22	11	17	11	19	13
Cr	119	62	139	46	120	56	111	71
Cu	55	41	88	81	44	22	52	22
Mo	23	25	16	24	17	25	30	24
Ni	97	53	100	46	86	47	105	60
Nb	11	5	14	5	12	5	10	4
Pb	31	29	48	38	22	13	32	32
Rb	91	45	118	60	97	44	77	35
S	33242	28627	20123	9525	32180	32219	38787	28899
Sc	15	3	17	4	15	3	14	3
Sr	439	368	340	379	530	394	404	331
Th	7	4	10	4	8	4	7	3
U	9.4	(n = 16)	10.7	(n = 1)	7.9	(n = 5)	10.0	(n = 10)
V	174	113	188	87	154	96	185	131
Y	27	13	24	10	27	13	28	15
Zn	113	198	143	419	97	73	115	136
Zr	112	64	128	55	115	57	103	71

average shale, reflecting the effect of carbonate dilution (Table 6.11; Fig. 6.21). The trace elements As, Cl, Mo, Ni, S, V, U and Pb (all of which are typically associated with the organic fraction of shales) and, to a lesser extent, Cr, Cu, Zn and Sr, are higher than in average shales (Fig. 6.21). The content of As is more than twice the average, Mo almost ten times and U approximately three times the content of average shale. A few trace elements, including Ba, Rb, (cf. K_2O), Th and Zr (Fig. 6.21), are low relative to average shales. Although the Sr content is approximately 50% higher than in average shales, it is low when the level of CaO (three times that of the average shale of Turekian and Wedepohl, 1961) is considered. The Al_2O_3 content is close to the average, whereas TiO_2, Fe_2O_3, Na_2O and K_2O contents are low, which suggests that kaolinite, which is the only aluminosilicate that does not contain appreciable quantities of other elements, is the most important clay mineral. Mineralogical data, mostly for Namurian shales, confirm the presence of kaolinite (Perrin, 1971).

Comparison of shelf and off-shelf facies

Shales (the Longstone Mudstones) developed on the shelf only in P_2 times, so samples from the $B_2 - P_1$ zones are all from off-shelf environments. The shelf samples generally have slightly higher SiO_2 levels, but with a large standard deviation, whereas off-shelf samples have higher Al_2O_3 contents and higher contents of TiO_2, Fe_2O_3, Na_2O and K_2O associated with aluminosilicates (Table 6.13). The variation between boreholes is almost as great, however, as that between shelf and off-shelf facies. In terms of trace elements the shelf facies generally has slightly higher Ba, Cu, Zn, Ni and Sr levels with higher As, Cl, Nb, Pb, S, Th, Zr, Pb, (Cr and Nb) levels in samples of off-shelf facies.

6.5.2 Geochemistry of Namurian Shales

Comparison with average shales

The contents of major elements in the Namurian shales are close to those of average shales (Table 6.12; Figs. 6.21), although CaO (with a mean of 4.94%) and P_2O_5 contents are slightly higher and the Na_2O and K_2O contents are lower. The values are, nevertheless, much closer to those of average shales than in the case of the Viséan samples and other major elements, including SiO_2, Al_2O_3, TiO_2, Fe_2O_3 and MnO, which are low in the Viséan, are at average levels in the Namurian. The trace elements As, Co, Cu, Mo, Ni, S, V, Zn and U and, to a lesser extent, Cr, Pb and Y are higher than average shales (Figs. 6.21) and show a generally similar pattern to that of Viséan shales, although Co and Sr are higher in Viséan samples and the content of Ce, Zn and Y is higher in the Namurian samples; Mo, S, V and U contents are also

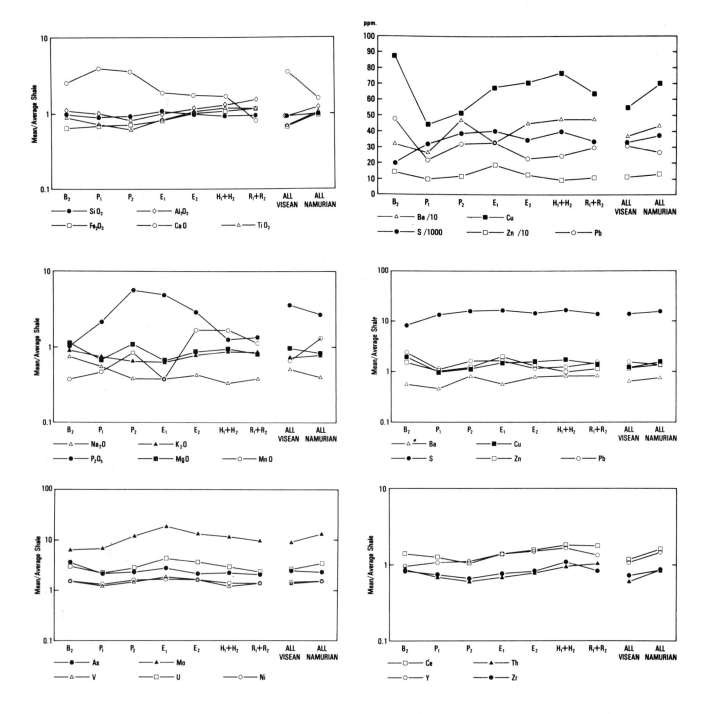

Fig. 6.21 Variations in mean concentrations of elements from Asbian (B₂) to Marsdenian (R₂) (data normalised to average shale). For stratigraphy see Fig. 6.19

higher in Namurian samples. Sr and Ba levels are low relative to average shales in the Namurian, but to a lesser extent than in the Viséan.

Temporal trends

The means of the zonal groups (E₁, E₂, H₁ + H₂, R₁ + R₂) represented by the Namurian samples indicate some general time trends. The clearest geochemical evidence of change is from E₁ through E₂ to H₁ + H₂; the composition of R₁ + R₂ shales is similar to those of the H₁ and H₂ zones. The trend mainly reflects the reduced importance of carbonate and the increasing influx of terriginous clastic material in the Namurian with the deposition of siliciclastic turbidites and deltaic deposits.

SiO₂ levels decline with time, the highest values occurring in the E₁ zone and the lowest values in the H₁ and H₂

zones, whereas the Al₂O₃ content increases with time together with the levels of Fe₂O₃, TiO₂, MgO (although this is low in the R₁ and R₂ zones) and K₂O (Table 6.12; Fig. 6.21). These trends suggest that the proportion of clay minerals (e.g. kaolinite and illite) increased relative to silica (quartz) during much of the Namurian. The CaO and Sr levels decrease with time with a marked drop from the H to the R zones and P₂O₅ decreases from the E₁ to R zones, reflecting the decrease in both carbonate and organic detritus. MnO is highest in the E₂ and H zones with lower contents in the R and E₁ zones.

K₂O–Al₂O₃ and TiO₂–Al₂O₃ plots (Figs. 6.22, 6.23) generally show a good correlation and trend, which probably reflect the composition of illite. Certain samples, particularly those of R₂ and R₂ age, plot away from this trend, however, and have higher Al₂O₃/K₂O ratios, which probably reflect an increased proportion of clay

Table 6.12 Summary of geochemistry of Namurian shales

	All Namurian (n = 271)		Pendleian (E₁) (n = 66)		Arnsbergian (E₂) (n = 89)		Chokierian & Alportian (H₁ & H₂) (n = 58)		Kinderscoutian & Marsdenian (R₁ & R₂) (n = 55)	
	Mean	sd	Mean	sd	Mean	sd	Mean	sd	Mean	sd
SiO_2	58.25	8.52	63.15	7.62	58.10	9.06	54.95	9.35	55.82	4.25
Al_2O_3	18.61	5.97	15.12	5.78	17.35	6.12	19.84	4.19	23.36	3.93
TiO_2	0.82	0.38	0.65	0.41	0.80	0.41	0.91	0.35	0.90	0.22
Fe_2O_3	6.80	3.36	5.52	2.99	6.78	4.53	7.25	1.83	7.92	2.23
MgO	2.01	0.95	1.61	0.99	2.15	0.96	2.30	1.14	2.02	0.36
CaO	4.94	5.77	5.83	5.39	5.78	6.04	5.20	6.80	2.50	3.75
Na_2O	0.51	0.20	0.49	0.18	0.55	0.22	0.44	0.18	0.52	0.20
K_2O	2.40	0.69	1.95	0.66	2.40	0.77	2.71	0.58	2.61	0.36
MnO	0.14	0.28	0.04	0.03	0.21	0.43	0.18	0.22	0.11	0.08
P_2O_5	0.34	0.73	0.62	1.19	0.37	0.67	0.16	0.09	0.16	0.16
As	30	26	36	33	27	25	30	22	29	20
Ba	484	205	328	155	465	288	477	127	470	95
Ce	92	32	79	43	89	30	104	24	101	17
Cl	122	186	173	205	179	222	40	62	30	31
Co	25	13	22	14	23	13	30	14	25	8
Cr	119	62	119	40	133	65	121	33	144	95
Cu	70	33	68	35	71	23	77	39	64	37
Mo	33	35	47	52	34	30	29	24	23	23
Ni	100	52	110	58	105	66	91	35	90	29
Nb	15	5	13	5	15	6	17	5	15	3
Pb	27	24	33	41	22	11	25	18	28	14
Rb	100	33	81	30	98	38	112	29	112	18
S	37294	25321	40313	29323	34977	25590	40501	24434	35257	20177
Sc	18	4	14	4	18	4	19	3	20	2
Sr	206	140	274	91	214	127	187	112	134	56
Th	10	4	8	3	9	4	11	3	13	3
U	12.4	(n = 45)	15.5	(n = 13)	13.1	(n = 14)	10.4	(n = 12)	8.3	(n = 6)
V	194	125	231	149	209	133	150	53	173	120
Y	37	18	35	26	37	19	41	9	34	7
Zn	129	216	187	422	123	81	94	35	107	36
Zr	134	66	120	60	125	59	168	90	124	36

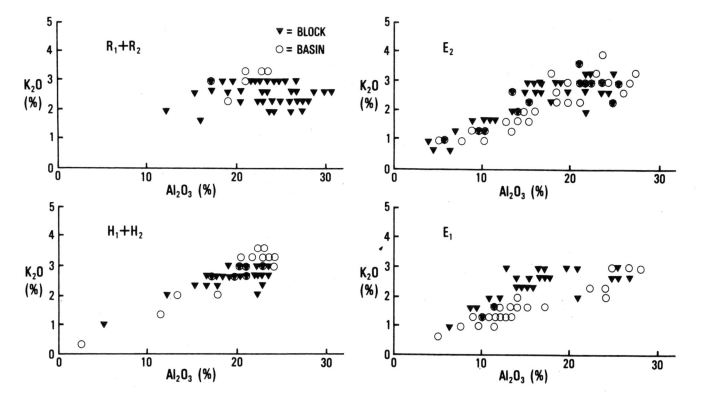

Fig. 6.22 K_2O–Al_2O_3 plot for 'Block' and 'Basin' facies of E_1 to R_1 and R_2 zones of Namurian

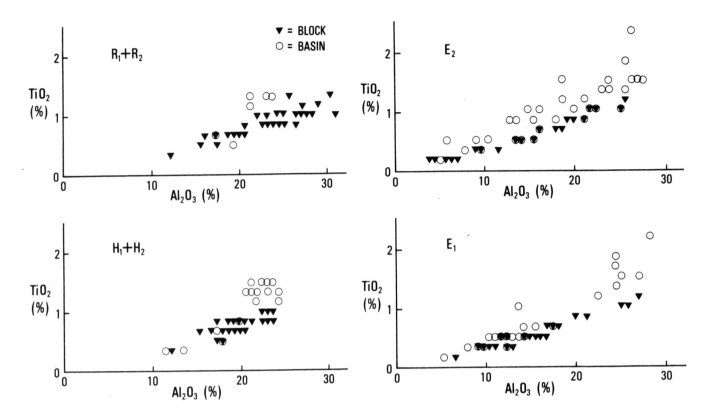

Fig. 6.23 TiO₂–Al₂O₃ plot for 'Block' and 'Basin' facies of E_1 to R_1 and R_2 zones of Namurian

minerals, such as kaolinite, which contains Al_2O_3 but only insignificant amounts of K_2O. Evidence of an increased kaolinite content in R_1 and R_2 zone samples is consistent with the change from turbidite deposition to delta-top sandstones and probably reflects differential sedimentation of relatively large kaolinite grains or flocculation of kaolinite close to the shoreline, and transport of illite and other fine-grained/less readily flocculated clays away from the shore. A similar pattern is seen off the mouth of the Amazon at the present day (Gibbs, 1977).

Ba levels (Fig. 6.21) approximate to those of average shales, possibly reflecting a feldspar component, although shales from the E_1 zone have values below the average. Ce, Rb, Sc, Th, Zr and Y contents, except in the R zones (Table 6.12; Fig. 6.21), increase from E_1 to R and H zones, reflecting the increasing proportion of aluminosilicates. As, Cl, Mo, Ni, V, Zn and U decrease (Fig. 6.21) from the E_1 to R and H zones, probably reflecting the declining amounts of organic material in the shales.

Comparison of Block and Basin facies shales

Samples from 'Block' and 'Basin' areas generally show the same general geochemical trends with time as the average trends described above, particularly for major elements (see Table 6.13). Co and Cr levels increase with time, although Ni levels show considerable fluctuation in basin facies shales. P_2O_5, As, Ba, Ce, Co, Cu, Mo, Ni, S, V, Y and Zn levels are consistently higher on the block in the E to H zones, reflecting the highly condensed nature of the sequence, with little dilution of reactive organic matter by a detrital component. The increase in thickness of the E_1–H_1 zones, from the sequence overlying the Derbyshire block to the Widmerpool Gulf, is about a factor of five (Frost and Smart, 1979). Hence phosphates and metals associated with the organic fraction, together with Ce and Y (which may also be concentrated in

phosphates) and Ba are higher in shales on the block, whereas SiO_2, TiO_2, Nb and Zr are consistently higher in basin facies shales, reflecting their content of coarse terrigenous turbidites.

Previous studies of geochemistry of Namurian in study area

The chemistry of the sandstone and shale of the Mam Tor Beds (R_{1c}) and marine and non-marine black shales of 'Block' facies from the Tansley borehole has been described by Spears and Amin (1981a, b).

Spears and Amin (1981a) found the best discriminator of marine and non-marine shales from the Tansley borehole to be a plot of Pb–Cu; all of the non-marine samples plotted relatively near to the origin and there was no overlap with marine samples, which had consistently higher contents of Pb and Cu. All of the non-marine (fish phase) samples studied by Spears and Amin (1981a) were R_2 in age, however, and all of marine samples were from the E_1 to R_1 zones, except for five samples from the R_2 zone (overall only two E_1 samples, no E_2, two H_1 and four R_1, whereas the remainder (36) are of R_2 age).

The greater proportion of samples for the present study were from the lower parts of the sequence; samples from the Uppertown and Highoredish boreholes, very near to Tansley, failed to show the clear relationship described by Spears and Amin (for detailed borehole descriptions see Ramsbottom and others, 1962). In the Uppertown borehole Pb levels are much lower than those reported by Spears and Amin. Although discrimination of marine and non-marine shales can be made by use of a Pb–Cu plot for E, H and R_2 zones, it does not apply to the R_1 zone, which has low levels, particularly of Cu, in marine samples (Fig. 6.24). In the Highoredish borehole nearly all of the samples collected are marine, but the non-marine samples contained higher Pb and Cu than all but three of these samples. Most samples are from the E–H

Table 6.13 Summary of geochemistry of shelf/block versus off-shelf/basin environments for shales of Brigantian (P_2) to Marsdenian (R_2) age. (n.b. mean for Ba in Brigantian shelf samples excludes 1 sample with 1.3% Ba)

| | Brigantian (P_2) | | Pendleian (E_1) | | Arnsbergian (E_2) | | Chokierian & Alportian (H_1 & H_2) | | Kinderscoutian & Marsdenian (R_1 & R_2) | |
	Shelf ($n = 20$)	Off-shelf ($n = 44$)	Block ($n = 30$)	Basin ($n = 36$)	Block ($n = 48$)	Basin ($n = 41$)	Block ($n = 34$)	Basin ($n = 24$)	Block ($n = 49$)	Basin ($n = 6$)
SiO_2	58.54	53.16	60.61	65.25	57.73	58.54	52.53	58.39	55.57	57.85
Al_2O_3	10.27	13.48	16.14	14.27	16.84	17.96	19.77	19.94	23.67	20.84
TiO_2	0.32	0.55	0.59	0.69	0.66	0.96	0.76	1.14	0.88	1.10
Fe_2O_3	3.09	5.59	5.75	5.32	6.62	6.98	8.19	5.92	8.03	6.97
MgO	1.92	3.04	1.89	1.38	2.10	2.21	2.31	2.28	2.01	2.13
CaO	12.87	10.46	5.68	5.94	6.24	5.26	5.86	4.27	2.35	3.70
Na_2O	0.37	0.55	0.47	0.51	0.49	0.61	0.50	0.35	0.56	0.25
K_2O	1.80	2.10	2.32	1.63	2.45	2.34	2.60	2.86	2.55	3.12
MnO	0.16	0.06	0.05	0.04	0.23	0.19	0.18	0.17	0.11	0.09
P_2O_5	1.25	0.47	0.92	0.38	0.45	0.27	0.16	0.15	0.16	0.12
As	16	36	43	30	33	21	42	13	29	26
Ba	303	249	410	260	505	418	485	465	469	475
Ce	54	62	88	72	91	87	107	99	101	98
Cl	67	602	45	280	53	326	33	52	31	21
Co	19	19	28	17	25	20	36	21	24	30
Cr	92	119	129	111	132	135	112	132	140	178
Cu	66	45	94	46	79	61	100	46	64	65
Mo	28	30	67	30	41	25	42	11	24	16
Ni	135	91	145	81	117	91	103	75	89	102
Nb	8	11	13	13	14	17	15	19	15	18
Pb	22	36	38	29	22	22	22	29	26	46
Rb	62	84	91	73	97	100	105	121	110	132
S	25516	44819	44419	36892	41507	27331	56586	17714	35740	31316
Sc	15	14	16	13	18	18	20	17	20	19
Sr	474	373	248	295	236	187	211	150	134	132
Th	6	7	8	7	9	9	11	11	13	11
U	8.9	10.8	29.6	9.2	17.6	9.7	11.4	9.1	9.8	5.2
V	203	176	309	166	238	176	162	134	165	232
Y	34	26	41	30	39	35	43	40	33	35
Zn	167	91	282	107	132	112	103	80	100	164
Zr	63	122	106	131	107	146	123	231	119	165
No. of U analyses	4	6	4	9	6	8	7	5	4	2

Table 6.14 Heat production of Viséan–Namurian shales (using density of 2510 kg m^{-3})

Stratigraphic unit	Heat production (μW m^{-3})	No. of U analyses
Viséan–Namurian	3.6	61
Namurian	3.9	45
Visean	2.9	16
Namurian: Block facies	4.8	21
Basin facies	2.9	24
Namurian		
R_1 and R_2 zones	3.1	6
H_1 and H_2 zones	3.5	12
E_2 zone	4.0	14
E_1 zone	4.5	13
Viséan		
P_2 zone	3.1	10
P_1 zone	2.6	5
B_2 zone	3.5	1

zones and fall in a similar area of the plot to those from Uppertown (Fig. 6.25).

6.5.3 Ore-forming Elements (Ba, Cu, Pb, S and Zn)

Ba levels in the Viséan and Namurian shales studied are generally below those of average shales (Figs. 6.21). The values increase up the stratigraphic column, reaching a maximum in samples of block facies from the E_2 and H_2–H_2 zones, and in block and basin samples from the R_1 and R_2 zones, probably reflecting an increased content of feldspar. Overall, Ba levels are less than 20% of those in the Cambrian shales from Nuneaton (see section 6.4.1).

Average Pb values in the Viséan are generally about 50% higher than in average shales (Fig. 6.21) and the zonal averages decrease in the order B_2, P_2, P_1; in the Namurian, the averages decrease in the order E_1, R_{1+2}, H_{1+2}, E_2. The highest concentrations of Pb are in off-shelf samples of the P_2 zone. In the Namurian, differences in the Pb content between block and basin facies are not clearcut, although there are generally higher contents in block facies samples for the E_1 and E_2 zones with higher contents in the basin facies samples of the H and R zones.

Sulphur contents are generally many times those of average shales, partly reflecting the high content of pyrite

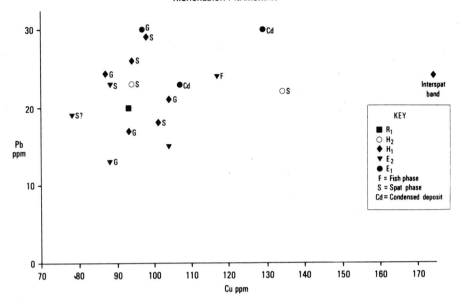

Fig. 6.24 Pb–Cu plot for samples from Highoredish borehole (faunal phases after Ramsbottom and others, 1962)

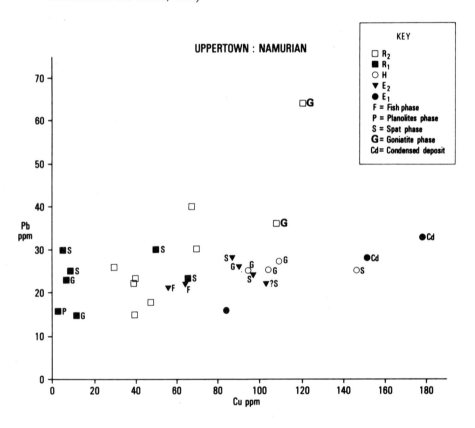

Fig. 6.25 Pb–Cu plot for samples from Uppertown borehole (faunal phases after Ramsbottom and others, 1962)

in the samples. In the Viséan the zonal means increase from the B_2 to the P_2 zones. There is a greater degree of fluctuation throughout the Namurian but, overall, S levels are higher than, in the Viséan.

Zinc levels are slightly above those of average shale in both the Viséan and Namurian samples. In the Viséan zones Zn decreases in the order B_2, P_2, P_1 and in the Namurian in the order E_1, E_2, R_{1+2}, H_{1+2}.

Copper levels are generally somewhat higher than the average for shales, values in the Namurian being generally higher than those in the Viséan. In the Viséan Cu

levels are highest in the B_2 zone; they are slightly below average in the P_1 zone and above average in the P_2 zone. In the Namurian the highest average values are in the E_2 and in the H_{1+2} zones. In the lower Namurian (E–H zones) block facies shales the content of Cu is twice that of shales of basin facies.

With the exception of Ba, concentrations of the key ore-forming elements are therefore above average in these shales and, on this ground alone, they have reasonable potential as a source. More significantly, they occur in precisely the right location, immediately adjacent to the

92

mineralised limestones, and the tectonic model presented in Chapter 5 suggests that their structural position would have facilitated movement of fluids up-dip into the carbonates. Thermal modelling of a Viséan–Namurian basin (the Edale Gulf) shows that fluids of the correct temperature, as deduced from fluid inclusions studies, could have been generated during the Upper Westphalian to Lower Permian period (Chapter 8). If the volume of argillaceous sediment in the Edale Gulf is considered (assuming a conservative thickness of 2 km), it is necessary to remove only very small percentages of the Pb, Zn and Ba currently present in the rocks (aproximately 2%, 0.01% and 0.03%, respectively) to provide the quantities of these elements extracted from the South Pennine Orefield. In the case of F, if the average Carboniferous mudrock figure of 480 ppm of Ramsbottom and others (1981) is used, about 0.1% of this amount would provide the recovered output. Clearly, the production figures from the orefield are an underestimate of the total tonnages of these elements present, but the calculation ignores any contribution from the larger volume of mudrocks present in the Widmerpool Gulf, which abuts the southern end of the orefield. It would therefore seem entirely feasible to provide all of the ore-forming elements in the South Pennines by dewatering the adjacent basins without recourse to more distant sources, such as the North Sea Basin (as proposed by e.g. Ford, 1976; Ineson and Ford, 1982; Mostaghel and Ford, 1986).

Zn/Pb ratios

Low Zn/Pb ratios are a feature of the Pennine orefields (see Chapters 3 and 8) and values in the shales have therefore been calculated. The overall mean for the Carboniferous shales is close to that of average shale and average continental crust, although some zones contain lower values (see Fig. 6.26). For example, the mean of the B_2 zone is only 1.88 (with a standard deviation of 2.84) when one outlying sample with a ratio of 73 is excluded. This is still very different from the figure 0.1 or less found in the ores. The other average values for the different zones are summarised in Fig. 6.26 for rocks of different facies.

It is apparent that the highest Zn/Pb ratios are in the highly condensed sequences E_1 to E_2 of block facies; although there is a slight possibility that these could have been enriched by mineralising fluids, it is more likely that the high ratio reflects conditions of sedimentation. Spears and Amin (1981b) correlated Zn levels with illite content, whereas Pb appeared to be associated with organic matter and pyrite, and it therefore seems likely that the proportion of detrital clays relative to organic matter most readily explains the varying ratios in the shales.

6.5.4 Heat Production

Heat production figures for Viséan–Namurian shales (Table 6.14) are generally significantly higher than those of Lower Palaeozoic shales (Tables 6.5, 6.9) with a mean of 3.6 μW m^{-3} as compared to 2.0–2.1 μW m^{-3}. This may reflect a loss of U as a result of Caledonian low-grade metamorphism of the Lower Palaeozoic samples. The Namurian generally has higher heat production than the Viséan. Within the Namurian there is a marked variation between block and basin facies, with much higher levels in the former (4.8 μW m^{-3}) than in the latter (2.9 μW m^{-3}). The highest values, stratigraphically, occur in the E_1 and E_2 zones. Cosgrove (Appendix B in Ramsbottom and others, 1962) reported relatively high levels of radioactivity from beds of E_1 age, which he attributed to uranium in phosphatic material. He also observed that marine shales generally had higher radioactivity than non-marine shales. No simple relationship between P_2O_5 and U contents has been noted in the present study.

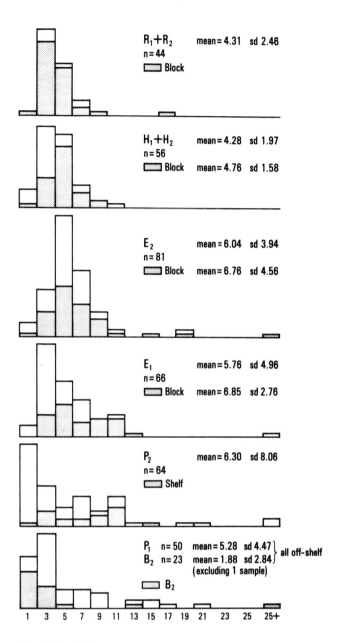

Fig. 6.26 Histograms of Zn/Pb from Asbian (B_2) to the Marsdenian (R_2)

Higher levels of a number of trace elements (including Pb, Cu, V, Ni and Zn) in Namurian marine shales from the Tansley borehole, compared to non-marine shales, have been attributed to reactions that involve organic matter and oxyhydroxides controlled by salinity and slow sedimentation rates (Spears and Amin, 1981a). A similar association might be invoked for U, the higher sedimentation rates in basinal environments leading to reduced U concentrations.

The heat production for basinal Namurian facies is, however, twice that of average crust. The thickest Viséan–Namurian shale sequences are probably of the order of 2.5–3.0 km (K. Smith, pers. comm.) and the largest East Midlands sub-basin (the Widmerpool Gulf) has an area greater than 2500 km^2. Such sequences could therefore have provided a significant contribution of heat and, owing to their low thermal conductivity, have helped to some extent in maintaining elevated temperatures near their base; this may have been particularly important in the generation of hot mineralising fluids under the higher geothermal gradients (50°C + /km) postulated for the late Carboniferous to early Permian period (Chapter 8).

6.6 Summary

Geochemical data for the Lower Palaeozoic shales from most parts of Britain (the reference collection) indicate that periods of widespread black shale deposition, at discrete stratigraphic intervals, have greater potential as sources of the main ore-forming elements for MVT deposits than mudrocks of other ages. The most notable of these intervals is the Upper Cambrian (Merioneth Series), which includes strata equivalent to the Alum Shales of Scandinavia—a unit well known for its exceptionally high levels of many trace metals.

Just south of the study area, at Nuneaton, the Stockingford shales, of Middle–Upper Cambrian age, also have higher than average contents of trace elements, including Ba, Cu, Pb, S and Zn. These esentially undeformed and poorly indurated rocks on the Midland Microcraton are, however, separated from the South Pennine Orefield by 30–40 km of cleaved and indurated Lower Palaeozoic shales, deformed by late Caledonian events. Levels of the key ore-forming elements in these argillites are lower than, or similar to, values for average shale and they have clearly lost most of their contained fluids a considerable time (> 100 Ma) prior to the onset of mineralisation.

Thick basinal sequences of Viséan to Namurian shales have the greatest potential as a source of fluids and metals for mineralisation. They generally have higher than average contents of the key elements, they are geographically and stratigraphically in the right position and it can be demonstrated that the appropriate fluid temperatures could have been attained during basin evolution. It is necessary to extract only a very small amount of the elements from the shales from local basins, such as the Edale Gulf or Widmerpool Gulf, to provide the total quantities present in the South Pennine Orefield. The shales have heat production values twice those of average crust and low thermal conductivities and could, therefore, have helped to maintain an already elevated geothermal regime.

7 GEOCHEMISTRY OF PRE-MESOZOIC IGNEOUS ROCKS

P. C. Webb and G. C. Brown

This section describes the geochemistry of igneous rocks from the Caledonian/Precambrian basement and the Carboniferous cover sequence of the East Midlands and Pennine regions. Information that the igneous rocks provide on the tectonic evolution of the region is considered and particular reference is made to their role as sources of radiogenic heat and ore-forming metals. In the study region, basement rocks are concealed and few boreholes intersect basement rocks of igneous origin. Consequently, studies have been extended to central and northern England, where igneous basement with similar geophysical signatures (Chapter 4) to that found in the study area may be sampled from boreholes and at outcrop. Although Carboniferous igneous rocks are exposed in the west of the study region, they too are concealed in the east and are sampled from boreholes. These rocks may be considered within the following groups.

(1) The Charnwood late Precambrian inlier: Charnian dacitic and andesitic lavas (porphyroid) and the Charnwood Northern and Southern diorite plutons are exposed in Leicestershire
(2) Concealed basement of uncertain age: borehole samples, largely lacking age controls, include acid and basic volcanics and intrusive rocks (the boreholes include Upwood, Warboys, Glinton, Great Osgrove Wood, Woo Dale and Beckermonds Scar)
(3) Caledonian plutons: the South Leicestershire diorites (Croft and Enderby) and the Mountsorrel Granodiorite are exposed, whereas the Wensleydale and Weardale granites are accessible only through boreholes
(4) Carboniferous basalts: basic lavas and sills of Dinantian age are exposed in Derbyshire and representatives of Westphalian age are found in boreholes in the Vale of Belvoir.

The nature of the igneous units sampled and their ages (where known) are listed in approximate order of increasing age in Table 7.1. The petrology and geochemistry of each group are described in sections 7.1–7.4. Tectonic environments are investigated with the use of geochemical discrimination diagrams. However, results from individual discrimination diagrams should be treated with caution, as they are usually defined only for restricted ranges of rock composition (e.g. basalts, granites) and are used here to consider compositional ranges of basalts to andesites and diorites to granites, respectively. They are also known occasionally to give ambiguous results—for example, when mixed magma sources are involved.

Heat production characteristics, of significance when considering potential heat sources for driving hydrothermal convection systems, are covered in section 7.5. Element mobilities, although described along with the geochemistry of each rock unit in sections 7.1 to 7.4, are summarised and tabulated in section 7.6. Finally, a synthesis of tectonic and genetic interpretations of the igneous units is presented in section 7.7.

Although there is extensive literature on the pre-Mesozoic igneous rocks of the East Midlands and northern England, existing geochemical studies involve mainly individual suites. One of the aims of this study is to prepare a systematic data set for igneous rocks of the region to identify important temporal and spatial changes in chemistry. Details of analytical techniques and methods of sample preparation are the same as for shales and are given in Appendix 3.

7.1 Charnwood late Precambrian inlier

The location of the late Precambrian inlier, which projects from beneath Permo-Trias cover in the Charnwood Forest area, is illustrated in Fig. 7.1. In structure, it is a south-easterly plunging antiform, comprising a sequence, more than 2.5 km thick, of bedded volcaniclastics, greywackes and slates. Minor dacitic and andesitic volcanics and high-level intrusions (the porphyroid sequence) occur in the lower part of the Charnian succession, which is intruded by diorites, local in extent in the north (Charnwood Northern Diorite), but more extensive in the south (Charnwood Southern Diorite). A similar sequence occurs at Nuneaton in Warwickshire where the Caldecote volcanics are cut by dioritic intrusions (Evans and others, 1968; Thorpe, 1982). The Charnian volcanics have been dated by K–Ar (Meneisy and Miller, 1963) at 694 ± 29 Ma, which may be regarded as a minimum age. The Charnwood Southern Diorites have a Rb–Sr whole rock age of 540 ± 57 Ma (Cribb, 1975), which is believed to be a magmatic age (but the error is large). The corresponding date for the Charnwood Northern Diorite of 304 ± 90 Ma (Cribb, 1975) is of doubtful significance; Thorpe and others (1984) noted that some data points fall on the isochron for the Southern Diorite and suggested that a similar magmatic age might be appropriate. The other samples may have been affected by hydrothermal activity (Le Bas, 1981). The Charnian volcanic and sedimentary succession was deposited, and probably tectonised, prior to emplacement of the diorites. Similar diorites at Nuneaton are unconformably overlain by Cambrian quartzites, indicating a late Precambrian–early Cambrian age for these intrusions. Although samples from Nuneaton have been obtained for geochemical analysis, the data are not yet available for inclusion in this report. The following account therefore concentrates on the Charnwood rocks, previously described by Thorpe (1972) and discussed in a regional framework by Thorpe (1974), Le Bas (1981, 1982) and Thorpe and others (1984).

7.1.1 PETROLOGY AND MINERALOGY

Charnwood Porphyroid

A variety of andesitic to rhyodacitic lavas and pyroclastics was sampled from the 'porphyroid' unit, largely from Bardon Hill and Whitwick quarries (Fig. 7.1). Although some intrusive relationships have been recognised previously (Thorpe, 1974), the fine-grained groundmass of the porphyritic rock, which is predominant at Whitwick, and the irregular bounding surfaces of texturally distinct, small-scale units suggest an extrusive origin. Lavas also occur at Bardon Hill, but pyroclastics, including coarse agglomerate and tuff, are dominant and possibly form part of a vent complex. Some lavas are more uniformly fine-grained and lack the quartz and feldspar phenocrysts of the typical porphyroid more common at Whitwick. These phenocrysts are often fragmented and angular, but also show the effects of corrosion. The groundmass is usually dark with a dusting of oxides. Some pseudomorphs of ferromagnesian minerals are preserved by outlines of opaque oxide. Products of

Table 7.1 Pre-Mesozoic igneous units sampled in central, eastern and northern England

Unit	Age (Ma)	Characteristics	Occurrence
CARBONIFEROUS VOLCANICS			
Westphalian	315–310	Alkaline and tholeiitic within-plate basalt	Concealed sills and lava flows in Nottinghamshire
Viséan–Namurian	335–315	Tholeiitic (locally alkaline) within-plate basalts	Exposed sills, lava flows, vents, tuffs and ashes in Derbyshire
CALEDONIAN PLUTONICS			
Weardale Granite (Rookhope Borehole)	394 ± 34 (Rb–Sr)	Foliated calc-alkaline granite with subduction and arc-collision related signatures	Concealed batholith beneath northern Pennines (Alston Block)
Wensleydale Granite (Raydale Borehole)	400 ± 10 (Rb–Sr)	Calc-alkaline granite with subduction and within-plate related signatures	Concealed pluton beneath central Pennines (Askrigg Block)
Mountsorrel Granodiorite (+ Kirby Lane Borehole)	430 ± 7 (Rb–Sr) 423 ± 17 (Rb–Sr)	Calc-alkaline subduction-related granodiorite	Igneous complex exposed N. of Leicester
South Leicestershire Diorites	452 ± 8 (U–Pb, zircon)	Calc-alkaline subduction-related diorite	Igneous complex exposed SW of Leicester
UNCERTAIN AGE, VARIABLE CHARACTER BY BOREHOLE			
Beckermonds Scar	(Post-Ordovician?)	Calc-alkaline microdiorite	Concealed sill/dyke in central Pennines
Woo Dale	(Ordovician?)	Andesite	Concealed lavas and tuffs near Buxton
Warboys	(Post-Ordovician?)	Altered diorite	Concealed pluton N. of Huntingdon
Upwood		?Basaltic andesite	Concealed lava and agglomerate; N. of Huntingdon
Glinton, Great Osgrove Wood, Orton, Oxenden Hall, North Creake	?	Dacite and rhyolite	Concealed flow-banded lavas and tuffs in eastern England
LATE PRECAMBRIAN PLUTONICS/VOLCANICS			
Charnwood S. Diorite	540 ± 57 (Rb–Sr)	Calc-alkaline subduction-related diorites	Plutons exposed in Charnwood Forest N. of Leicester
Charnwood N. Diorite	?304 ± 90 (Rb–Sr)		
Charnian 'Porphyroid'	684 ± 29 (K–Ar)	Calc-alkaline subduction-related andesite to rhyodacite	Lavas and high-level intrusives exposed in Charnwood Forest

chloritisation, sericitisation, epidotisation and carbonatisation are common. Some of the fine-grained rocks are cleaved and fibrous alteration products often occupy strain shadows. Whether the cleavage is of late Precambrian or of Caledonian age is uncertain (Evans, 1979).

Charnwood Northern Diorite

The most accessible exposures of the Northern Diorite are in the Longcliffe and Newhurst quarries at Shepshed (Fig. 7.1). The elongate mass of diorite has chilled margins and is in contact (sometimes faulted) with bedded and massive volcaniclastics in both quarries. Typically, the rock is dark grey-green, reflecting the fine-grained strongly chloritised and epidotised matrix (sometimes also sericitised and carbonated), and has creamy pink patches of partially sericitised and carbonated plagioclase. Primary ferromagnesian minerals have not been found. Some outlines and cleavage traces, possibly of amphibole, are pseudomorphously outlined by opaque oxide and, rarely, chloritised pseudomorphs of amphibole are visible in hand specimen. Originally, the rock was a basic diorite, but alteration or, possibly, low-grade metamorphism has

been extensive. Carbonate and epidote veining is common and orange-coloured haematised zones are often found adjacent to the latter. Deformation of the veins clearly demonstrates the highly tectonised nature of the intrusion. This makes it difficult, especially at the southern margin, to distinguish in hand specimen between the diorite and the coarser and more massive units of tuffaceous material.

Charnwood Southern Diorite

The outcrop of distinctively mottled green and orange-pink granophyric diorite, known as Markfieldite, is more extensive than the Northern Diorite and is quarried at numerous sites along a NW-trending linear tract between Leicester and Coalville. Sampling was mainly from quarries near Groby Pool and at Cliffe Hill (Fig. 7.1). The original mineralogy is better preserved than in the Northern Diorite, although alteration of feldspars to sericite and of ferromagnesian minerals to chlorite is common. Well-formed plagioclase and amphibole crystals (sometimes with cores of clinopyroxene) are set in a granophyric intergrowth of quartz and alkali feldspar.

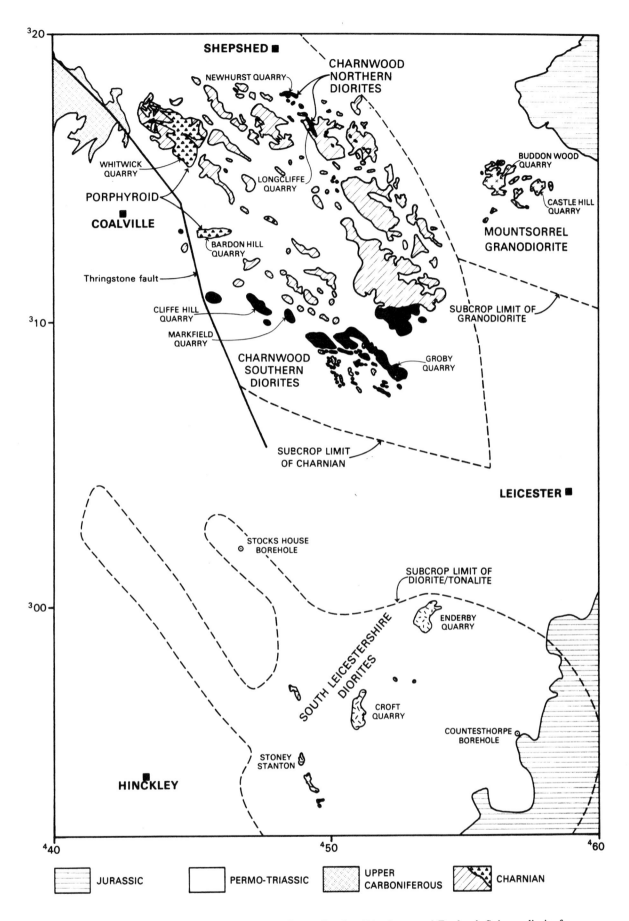

Fig. 7.1 Distribution of basement outcrops and sampling localities in central England. Subcrop limits from an interpretation by N. J. P. Smith

Ferromagnesian constituents are less abundant than in the Northern Diorite. Epidote and calcite veining are common and are associated with different types of wall-rock alteration. Orange, haematised zones are found next to epidote veins, and homogeneous dark green zones are associated with calcite veins. Sulphide and baryte mineralisation is more evident than in the Northern Diorite.

7.1.2 GEOCHEMISTRY

In order to provide a geochemical framework for examination of the Precambrian igneous basement, the mean compositions of representative suites of samples are listed in Table 7.2. Rocks from dykes, veins and alteration zones have been excluded.

Large standard deviations may result either from regular primary variations of elements that tend to be highly fractionated or to scatter produced by mobilisation during hydrothermal alteration/mineralisation processes. A wide range of variation diagrams has been studied to distinguish these types of variation, identify the primary geochemical characteristics of each unit and examine possible interrelationships between units. Selected diagrams for major and trace elements are shown in Fig. 7.2, and for base metals are shown separately in Fig. 7.3. The three Charnwood igneous suites exhibit clear magmatic fractionation trends for most major and trace elements. Some elements, however, have been remobilised and alteration characteristics are superimposed on magmatic trends. In particular, K_2O, Rb, Ba and Zn are

enriched and Fe_2O_3, MgO, Na_2O and CaO are depleted in alteration zones associated with calcite veins in the Southern Diorite (e.g. Fig. 7.2 and Fig. 7.3). Although SiO_2 contents are probably also affected, this is minor and does not prevent the use of SiO_2 variation diagrams for examination of the effects of fractionation. Cu, Pb, Na_2O and Sr variation diagrams for the diorites show large degrees of scatter (e.g. Fig. 7.3, b and d), probably reflecting both enrichment and depletion as a result of alteration. It is not possible to tell whether there was a net gain or loss of these mobile base metals, but some elements (e.g. Ba) are enriched in mineralised samples.

Many variation diagrams (see Figs. 7.2 and 7.3) show almost continuous primary variation through the less evolved Northern and the more evolved Southern Diorite suites with increasing contents of incompatible elements K_2O, Rb, Ba, Th, Zr and Y and decreasing contents of MgO, Fe_2O_3, CaO, Co, Zn and V. These patterns are consistent with fractionation of ferromagnesian minerals and plagioclase feldspar. Although some elements, e.g. V, Co, Fe_2O_3 (see Fig. 7.4) show a small compositional gap, there are enough similarities to suggest that the Northern and Southern Diorites are related and possibly were derived from the same parental magma.

The Charnwood Porphyroid also forms a geochemically coherent suite, but it is of andesitic to rhyodacitic composition, and generally more evolved in terms of its SiO_2 content than the Charnwood Diorites (Table 7.2). They are chemically distinct in many respects: lower in K_2O, Rb, Ba, Th, Na_2O, Zr and Sr than diorites with equivalent SiO_2, but higher in MgO, CaO, Fe_2O_3, Co and V (see Figs. 7.2 and 7.3). Few elements in the por-

Table 7.2 Geochemistry of igneous rocks from Charnwood inlier

	Porphyroid Mean(19)	$2 \times sd$	Northern Diorites Mean(14)	$2 \times sd$	Southern Diorites Mean(16)	$2 \times sd$
SiO_2	63.53	12.10	50.47	2.70	57.41	4.70
TiO_2	0.44	0.20	1.02	0.30	0.60	0.20
Al_2O_3	14.03	1.70	14.75	2.40	16.14	1.60
Fe_2O_3	8.03	5.60	14.55	10.00	9.43	3.00
MgO	4.92	4.60	6.08	1.50	3.70	2.15
CaO	4.25	4.30	8.49	3.10	5.20	3.00
Na_2O	3.44	3.50	2.16	1.00	3.10	0.92
K_2O	0.79	1.40	1.19	0.80	2.58	1.19
MnO	0.13	0.10	0.21	0.00	0.19	0.08
P_2O_5	0.07	0.10	0.20	0.10	0.35	0.10
Total	99.63		99.12		98.70	
As	9	10	16	17	7	7
Ba	234	292	253	424	1014	435
Ce	19	19	31	16	31	18
Cl	55	111	127	71	176	159
Co	17	16	34	6	18	8
Cr	36	95	42	22	66	173
Cu	34	107	100	88	78	67
Mo	0	1	0	1	1	1
Ni	8	10	15	7	5	4
Nb	2	3	2	2	3	3
Pb	9	6	15	13	14	9
Rb	16	28	29	17	85	53
S	141	142	178	257	191	140
Sc	25	14	40	9	24	11
Sr	91	102	277	114	268	130
Th	3	5	4	5	7	4
V	144	129	490	122	141	93
Y	28	10	23	8	25	5
Zn	61	38	97	29	92	74
Zr	64	24	53	17	75	21

phyroid show the degree of scatter that is seen in the diorites, post-magmatic mobilisation being generally less intense. The effect on Na₂O and Sr contents is variable, whereas MgO and CaO contents are usually depleted. Primary abundances of most base metals (except Zn) were lower than for the diorites (Fig. 7.3), and subsequent mobilisation effects minor, with relatively little scatter of data away from primary trends. Some samples, however, show depletion of Pb and possibly Ba.

Magmatic enrichment of incompatible elements with respect to SiO₂ variation in the porphyroid is not as significant as for the diorites (Fig. 7.2 a, c), which suggests that there were different controls of fractionation. Additionally, the most basic of porphyroid compositions are depleted in incompatible elements compared to the diorites, but they are enriched in CaO and MgO. These characteristics suggest a more primitive source composition for the porphyroid suite or separation from an amphibole-rich residue, in contrast to a pyroxene-rich residue in the case of the diorites. Alternatively, there could have been a greater degree of crustal contamination of the diorite parental magma. Like the diorites, the porphyroid exhibits a calc-alkaline trend on an AFM diagram (Fig. 7.4), but at a higher MgO/FeO ratio. Their chondrite-normalised REE plots are flatter than those of the diorites (Fig. 7.5), which are relatively LREE enriched.

Normalised multielement plots (Fig. 7.6) also show much greater large ion lithophile (LIL) element enrichment of the diorites than the porphyroid, but all of the

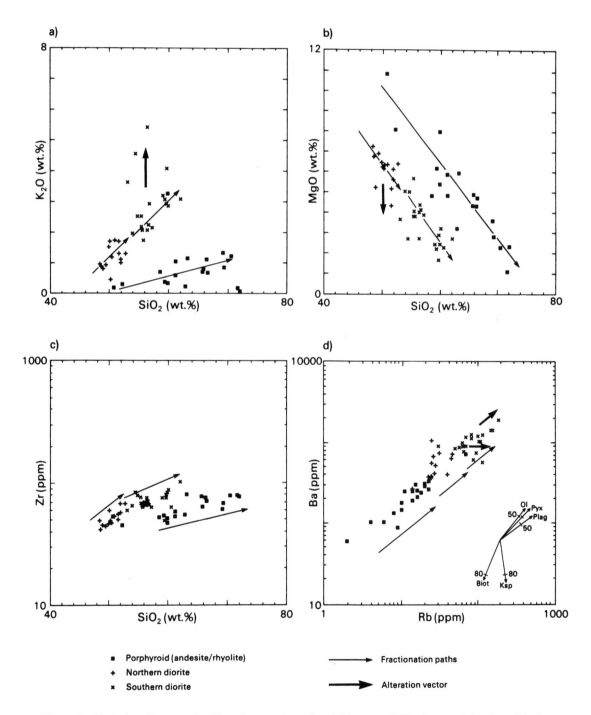

■ Porphyroid (andesite/rhyolite)
+ Northern diorite
× Southern diorite

———▶ Fractionation paths
━━━▶ Alteration vector

Fig. 7.2 Variation diagrams for Charnian porphyroid and Charnwood Northern and Southern Diorites. Mineral vectors in Rb–Ba diagram serve to show possible minerals controlling fractionation. Vectors show theoretical effects on melt composition of crystallising single mineral phases (assuming Raleigh fractionation) and are annotated according to proportion (%) of melt remaining. After Tindle and Pearce (1981)

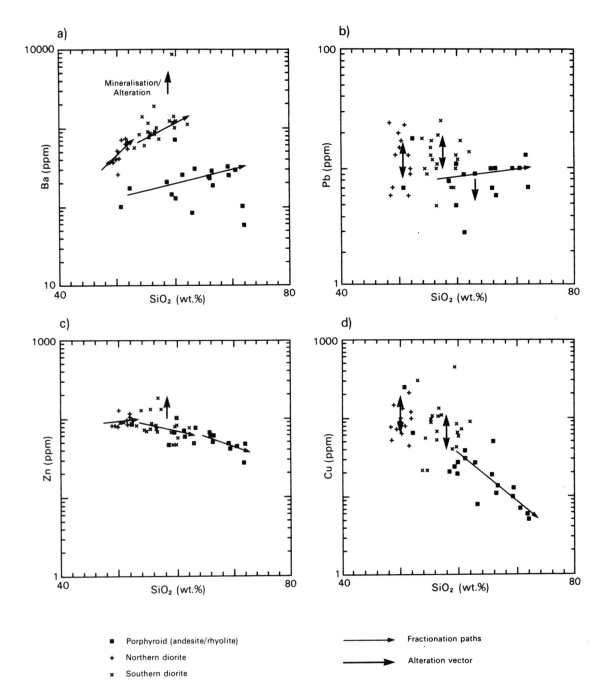

Fig. 7.3 Base-metal variation diagrams for Charnian porphyroids and Charnwood Northern and Southern Diorites

samples, especially the diorites, exhibit low Ta–Nb abundances—a feature of subduction-related igneous rocks generally (Pearce, 1982). Considering the tectonic setting discrimination diagrams for granites of Pearce and others (1984), the more acid members of the Charnian igneous suites all plot in the 'volcanic arc' fields on both Y–Nb and Y + Nb–Rb diagrams (Fig. 7.7 a, b). The more basic members also plot in the 'volcanic arc' field of the Ti–Zr–Y discriminant diagram developed for basalts (Pearce and Cann, 1973).

Although the porphyroid is generally more acid in composition than the diorites, it appears to reflect a less mature arc environment. Given that the diorite magma formed after a time gap (poorly defined) of at least 30 Ma, and perhaps as much as 100 Ma, the source may have become enriched in LIL elements or contamination may have been more extensive. These are both features of a more mature arc environment.

7.2 Concealed basement of uncertain age

Material has been examined from boreholes that have penetrated basement at Glinton, Great Osgrove Wood, Cox's Walk, North Creake, Orton, Oxendon Hall, Upwood, Warboys, Sproxton, Wittering, Woo Dale (see Fig. 7.8) and Beckermonds Scar (Fig. 3.1). Material recovered includes cleaved sediments, volcaniclastics, volcanics and intrusive rocks that have been variously assigned to the Precambrian and Caledonian. No reliable radiometric dates are available at present and unconformably overlying sediments are usually Carboniferous or Permo-Trias. Cox's Walk, Sproxton and Wittering boreholes appear to contain only sedimentary rocks and have not been considered further, whereas North Creake, Orton and Oxendon Hall boreholes contain igneous rocks and were available for observation but not analysis. The remainder contain igneous rocks that were examined

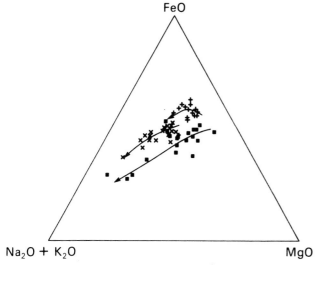

FeO

Na₂O + K₂O   MgO

- ■ Porphyroid (andesite/dacite)
- + Northern Diorite
- ✕ Southern Diorite
- ⟵ Fractionation trends

Fig. 7.4 AFM diagram for Charnian porphyroids and Charnwood diorites showing their typically calc-alkaline fractionation trends

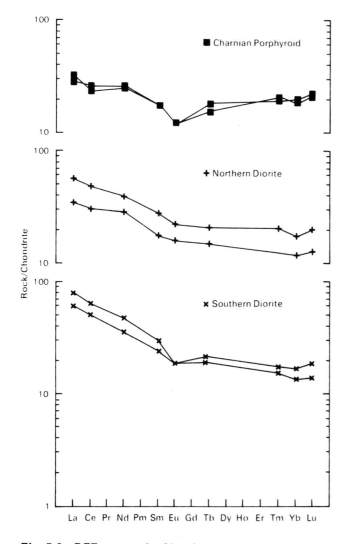

Fig. 7.5 REE patterns for Charnian porphyroids and Charnwood diorites (normalisation factors of Evensen and others, 1978)

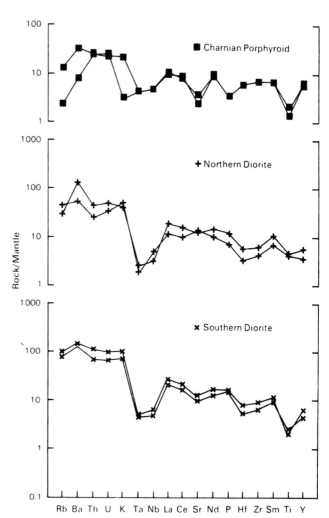

Fig. 7.6 Mantle normalised multielement diagrams for Charnian porphyroids and Charnwood Diorites (normalisation factors of Wood, 1979)

petrographically and analysed geochemically.

7.2.1 PETROLOGY AND MINERALOGY

Glinton

Approximately 30 m of acid volcanics, overlain by Triassic sediments, and forming a 'seismic basement', was described by Kent (1962) as devitrified tuffs. The presence of fine compositional banding suggests, however, that they may be flow-banded rhyolites, with corroded crystals of quartz and plagioclase set in a fine groundmass. Whichever origin is preferred, the Glinton volcanics represent extrusive acid magmatism. Sericitisation has been the most significant effect of alteration.

North Creake

Approximately 60 m of agglomerate and tuff, which occurs at the base of the N. Creake boring and is overlain by Triassic sediments, was considered by Kent (1947) to be Precambrian in age. Petrographically and mineralogically, the material is similar to samples from Glinton. However, quartz is more angular, which suggests the possibility of explosive volcanism. The rock is highly sericitised, and, unlike the samples from Glinton, epidote and chlorite are also present as alteration products.

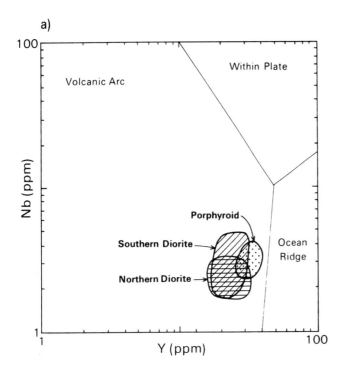

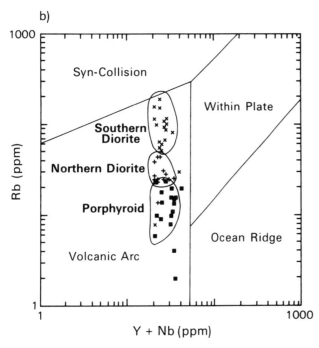

Fig. 7.7 Tectonic setting discrimination diagrams for granitic compositions (fields from Pearce and others, 1984) showing affinities of Charnwood inlier

Orton

The presence of cleaved ignimbrite beneath Triassic sediments was reported by Poole and others (1968) and compared to Charnian welded tuffs. Petrographic examination suggests, however, greater similarity with the Glinton volcanics.

Oxendon Hall

A welded tuff, also beneath Triassic sediments, was reported by Poole and others (1968) as being similar to that at Orton. Microscopic examination again indicates similarities to the Glinton volcanics. Corroded quartz crystals are scattered in a finely banded partially sericitised groundmass.

Upwood

Approximately 25 m of coarse volcanic agglomerate was penetrated in the Upwood boring (Institute of Geological Sciences, 1966). The large clasts are of similar material to the matrix, and comprise variably altered plagioclase–phyric amygdaloidal basic lava. The groundmass is chloritised and strongly carbonated. Some haematisation is present particularly towards the upper surface, which is probably weathered. Amygdales are filled with quartz and zeolites. No epidotisation is apparent.

Warboys

Nearly 50 m of coarse-grained altered basic diorite was drilled in the Warboys Borehole (Institute of Geological Sciences, 1966). Plagioclase phenocrysts are zoned and least altered. Pyroxenes and amphiboles are mainly chloritised, and in some cases serpentinised. Sericite and epidote are also common alteration products. One sample contains interstitial biotite, zoned clinopyroxene and evidence of both carbonate alteration and late silicification. Le Bas (1972) has suggested a Caledonian age in view of presumed intrusion into Ordovician sediments and a minimum age of $311/305 \pm 10$ Ma by K–Ar (Fitch and others, 1970)

Beckermonds Scar

The quartz-microdiorite from the Beckermonds Scar Borehole was described in detail by Berridge (1982) and believed to post-date the Wensleydale Granite. The unit is about 15 m thick, intruded into Ordovician low-grade metasediments. Sericitic and carbonate alteration of plagioclase and chloritic alteration of amphibole is extensive, only quartz, opaque oxide and accessory apatite representing the primary mineralogy.

Woo Dale

The presence of haematized lavas, volcanic breccias and pyroclastics at the base of the Woo Dale Borehole was reported by Cope (1973). These rocks vary from fine-grained lavas to fragmental tuffs, and are strongly haematised. Cope (1979) gave a minimum age of 383 ± 6 Ma from a K–Ar determination, but microfossils have given an Ordovician age (Evans, 1979).

Great Osgrove Wood

Here, fine-grained and finely banded rocks, described by Burgess (1982) as acid ash-flow tuffs, are partly sericitised, and bands/lenses of quartz mosaic suggest silicification. They are similar to the Glinton volcanics, although generally finer-grained and more strongly banded.

7.2.2 Geochemistry

Mean compositions of these basement units are presented in Table 7.3. The range of compositions (reflected by standard deviations) for each set of borehole samples is often large where post-magmatic mobilisation has occurred, but primary variations are generally small. Zeolitisation, chloritisation, sericitisation, carbonatisation and silicification of lavas and agglomerate at Upwood have affected most major elements, whereas the chloritisation, sericitisation and carbonatisation at Warboys have affected Ca but few other elements. Although the acid volcanics have also been affected by sericitisation, carbonatisation and silicification, only at Great Osgrove Wood has the bulk geochemistry been strongly affected, with enrichment of K_2O at the expense of Na_2O, and higher K/Rb compared to other volcanics. These samples also contain high and variable levels of Sr, despite extremely low CaO contents. In view of these observations, emphasis is placed on the use of immobile elements, such as Zr and Y, in making comparisons and characterising these borehole samples.

Variation diagrams (e.g. Figs. 7.9 and 7.10) are used to identify coherent suites of samples and possible relationships between them. The geochemistry of the basic rocks from Upwood and Warboys suggests that they are related in spite of their different appearance and environments of crystallisation. Together they appear to form coherent fractionation sequences for many elements, including K_2O, Rb, Ba, Na_2O, Sr, Zr, Y and Co, although some

primary patterns, especially Fe_2O_3 and CaO at Upwood, are affected by alteration. This close relationship is not surprising in view of their geographical proximity (see Fig. 7.8). Base-metal contents are similar, except for much higher Cu contents in Warboys, matched only by two Upwood samples (Fig. 7.10), suggesting, perhaps, Cu depletion at Upwood or enrichment at Warboys. Scatter in the Pb data (Fig. 7.10b) may reflect analytical difficulties at low concentrations rather than the effects of mobilisation.

On the tectonic setting discrimination diagrams of Pearce and Cann (1973) for basalts, Upwood and Warboys samples plot in the 'volcanic arc' field of the Ti–Zr–Y diagram (Fig. 7.11a). Although it may not be strictly valid to use these diagrams for rocks that do not represent liquids, and this could be the case for Warboys, the agreement with Upwood data is remarkable. Warboys samples, however, plot on a convincing calc-alkaline trend in the AFM diagram (Fig. 7.11b). Hence it seems likely that these basic to intermediate rocks were formed in an arc environment. In terms of many incompatible elements, K_2O, Rb, Ba, Na_2O and Y, they show remarkable similarity to the Charnwood Northern Diorite (cf. Tables 7.2 and 7.3), but they are enriched relatively in MgO and Zr and depleted in Fe_2O_3.

The acid volcanics from the Glinton and Great Osgrove Wood boreholes also have some features in common—in particular, the depletion of K_2O, Rb, Ba and Zn with increasing SiO_2, an effect that, arguably, could have resulted from K-feldspar plus biotite fractionation (Fig.

Table 7.3 Geochemistry of concealed basement of uncertain age

	Upwood Mean(8)	2 × sd	Warboys Mean(10)	2 × sd	Beckermonds Scar Mean(5)	2 × sd	Glinton Mean(11)	2 × sd	Gt. Osgrove Wood Mean(8)	2 × sd
SiO_2	48.80	2.20	52.15	1.70	54.45	3.90	70.18	9.20	74.43	3.30
TiO_2	0.94	0.68	1.00	0.40	1.48	0.50	0.44	0.30	0.18	0.10
Al_2O_3	18.91	4.90	13.92	1.50	16.87	2.10	16.12	6.00	15.94	2.20
Fe_2O_3	9.64	5.60	10.60	1.00	8.55	3.00	2.53	1.90	0.58	0.50
MgO	7.52	3.00	9.11	2.20	4.83	1.70	0.64	0.54	0.31	0.10
CaO	8.62	4.10	8.05	4.10	5.30	3.50	1.82	1.30	0.07	0.10
Na_2O	1.78	1.30	2.61	1.30	3.48	2.70	2.96	1.60	0.49	0.40
K_2O	1.27	0.90	1.71	1.60	2.49	1.80	4.74	1.60	7.63	1.30
MnO	0.19	0.10	0.16	0.03	0.16	0.08	0.08	0.10	0.00	0.00
P_2O_5	0.14	0.20	0.26	0.30	0.17	0.10	0.08	0.00	0.13	0.10
Total	97.81		99.57		97.78		99.59		99.76	
As	9	9	5	6	15	18	10	7	10	8
Ba	183	172	322	249	529	542	989	350	710	145
Ce	31	34	36	23	52	27	89	29	138	107
Cl	300	545	238	126	43	21	38	53	162	98
Co	31	19	49	21	12	4	25	21	16	20
Cr	143	183	251	196	56	120	9	16	12	19
Cu	40	72	178	153	6	3	7	3	6	2
Mo	0	1	1	2	2	4	0	1	0	1
Ni	47	66	90	59	2	3	2	4	9	5
Nb	7	6	9	4	14	3	18	6	12	3
Pb	6	5	8	4	10	6	23	6	9	77
Rb	25	16	39	38	99	67	190	73	190	35
S	930	2946	483	543	445	989	125	184	209	167
Sc	28	8	26	8	27	6	10	5	5	2
Sr	170	137	229	163	227	123	148	43	434	430
Th	3	3	5	8	6	4	14	6	14	4
V	234	54	284	78	108	75	18	15	13	6
Y	20	7	25	11	33	19	51	17	40	33
Zn	69	60	81	13	74	39	69	45	10	3
Zr	93	41	116	32	156	29	380	175	182	45

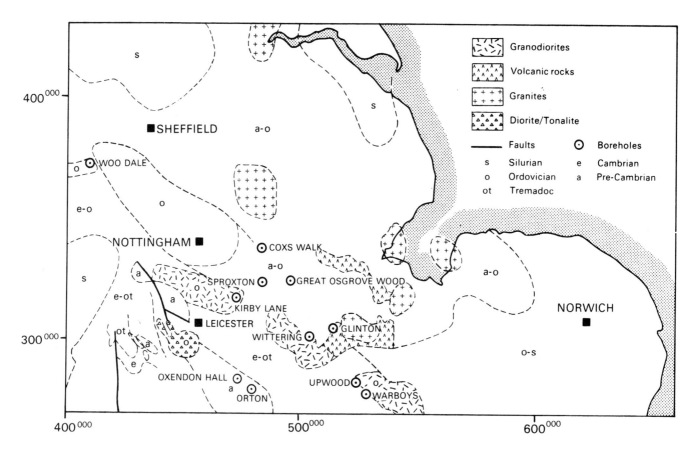

Fig. 7.8 Distribution of sampled boreholes penetrating Pre-Carboniferous basement in East Midlands (sub-Carboniferous surface from an interpretation by N. J. P. Smith)

7.9). The relatively high levels of the immobile incompatible elements, Zr, Y and Nb at Glinton suggest differences in their source compositions. Indeed, these elements are depleted with fractionation at Glinton, but are either unchanged or enriched, in the case of Y, at Great Osgrove Wood, suggesting different accessory mineral controls. Base-metal contents vary little in both suites and appear not to have been mobilised (Fig. 7.10). Cu and Ba are at similar concentrations in both, whereas Zn and Pb are lower at Great Osgrove Wood.

Examining these elements in relation to tectonic setting diagrams for granites, to which they approximate in composition, the Great Osgrove Wood data plot largely in the 'volcanic arc' field of the Y – Nb plot (Fig. 7.12), whereas the Glinton data plot just inside the 'within plate' field. A comparison between the Glinton and Great Osgrove Wood data with those for the acidic Charnian volcanics (Tables 7.2 and 7.3) shows a striking contrast. The abundances of K_2O, Rb and Ba as well as Nb and CaO are lower in the Porphyroid, clearly indicating differences in their petrogenesis. There are no geochemical grounds to relate these volcanics and their probable associates from North Creake, Orton and Oxendon Hall with the Charnian volcanics. This conclusion is at variance with the correlations suggested by Kent (1962) and Poole and others (1968) and it is proposed that the acid volcanics from these boreholes could be of Lower Palaeozoic, perhaps Caledonian, age rather than Charnian (see also Evans, 1979).

The Beckermonds Scar microdiorite is more evolved than the Warboys diorite, but the two suites of samples are remarkably similar geochemically (showing similar calc-alkaline characteristics on the AFM diagram (Fig. 7.11b), in spite of their great geographical separation and

petrographic differences. Base-metal contents, excepting Cu, are also similar (Fig. 7.10). The sample of lava from Woo Dale is geochemically very similar to the Beckermonds Scar microdiorite, but a wider range of compositions is present in the more acid tuffs.

7.3 Caledonian plutons

The distribution of Caledonian intrusives in Leicestershire is shown in Fig. 7.1; and the full distribution of exposed and postulated intrusions (Chapter 4) in the East Midlands is shown in Fig. 7.8. North of Leicester the Mountsorrel Granodiorite and its supposed continuation to the Kirby Lane borehole (Fig. 7.8) are the more acid representatives. To the south of Leicester there are intrusives of intermediate composition that have been described by Le Bas (1968, 1972, 1982). Gravity surveys suggest that dioritic to granitic intrusions of presumed Caledonian age form an important part of the basement in the area (Allsop and Arthur, 1983). Dating of the Mountsorrel Granodiorite yielded ages of 423 ± 17 Ma (Cribb, 1975) and 430 ± 7 Ma (Hampton and Taylor, 1983) by Rb–Sr whole rock and 452 Ma by the U–Pb zircon method (Pidgeon and Aftalion, 1978). The latter date is consistent with the U-Pb zircon date for the South Leicestershire Diorites.

The borehole sites from which the concealed Caledonian granites of northern England have been sampled are shown in Fig. 3.1. The extent of the Weardale and Wensleydale Granites has been inferred from their gravity signature Bott (1967). They are younger than the Leicestershire granites, with a date 394 ± 34 Ma (Dunham, 1974, recalculated from Holland and

104

Lambert, 1970) for the Weardale core and a date of 400 ± 10 Ma (Dunham, 1974) for the Wensleydale core.

7.3.1 Petrology and Mineralogy

Mountsorrel Granodiorite

In the Mountsorrel area a range of rock types from gabbro through diorite to granodiorite has been recognised (Le Bas, 1968), which indicates that the pluton should be regarded as an igneous complex. Granodiorite, which has been sampled from the disused Castle Hill quarry and the active Budden Wood quarry, is by far the most extensive lithology. It is a medium-grained, grey, biotite granodiorite with well-formed zoned plagioclases, frequently interstitial K-feldspar and quartz, with biotite and occasional amphibole. There are some fine-grained aplitic veins, intensely kaolinised and haematised alteration zones, and fine-grained dolerite dykes, which are heavily veined with calcite and which might be of much later, perhaps Carboniferous, age. The granodiorite is ex-

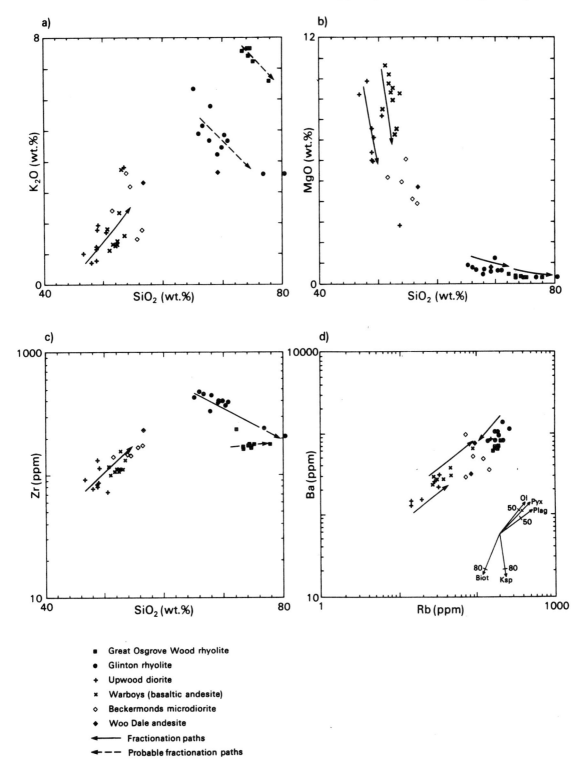

- ■ Great Osgrove Wood rhyolite
- ● Glinton rhyolite
- + Upwood diorite
- × Warboys (basaltic andesite)
- ◇ Beckermonds microdiorite
- ◆ Woo Dale andesite
- ◄——— Fractionation paths
- ◄– – – Probable fractionation paths

Fig. 7.9 Variation diagrams for basement igneous rocks of uncertain age sampled from boreholes. Mineral vectors in Rb–Ba diagram serve to show possible minerals controlling fractionation. Vectors show theoretical effects on melt composition of crystallising single mineral phases (assuming Raleigh fractionation) and are annotated according to proportion (%) of melt remaining. After Tindle and Pearce (1981)

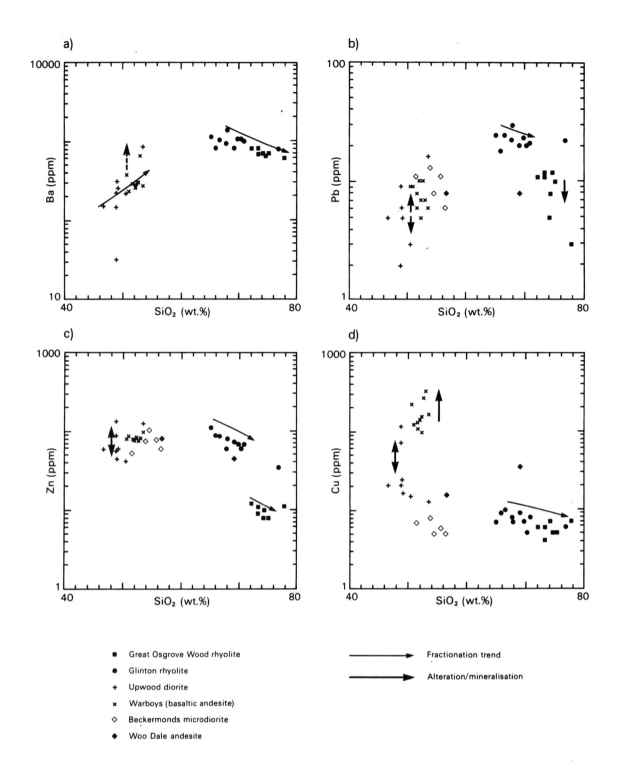

a)

b)

c)

d)

■	Great Osgrove Wood rhyolite
●	Glinton rhyolite
+	Upwood diorite
✕	Warboys (basaltic andesite)
◇	Beckermonds microdiorite
◆	Woo Dale andesite

———————▶ Fractionation trend

━━━━━━━▶ Alteration/mineralisation

Fig. 7.10 Base-metal variation diagrams for igneous rocks of uncertain age sampled from boreholes

tremely heterogeneous, containing many small xenoliths and xenocryst aggregates. Several xenoliths examined have igneous texture and consist dominantly of plagioclase and amphibole. They may represent earlier igneous phases of the Mountsorrel Complex.

The small section of core from the Kirby Lane Borehole consists of medium-grained, highly altered granodiorite. Geophysical evidence suggests that this is part of the same mass as the Mountsorrel Granodiorite (Fig. 7.8 and (Chapter 4). Much of the amphibole and biotite is chloritised, the often euhedral outlines of plagioclase being carbonated, sericitised and haematised. Perthitic K-feldspar and quartz are largely interstitial. There is minor carbonate veining.

South Leicestershire Diorites

There are several exposures of diorite plutons in the Enderby–Croft–Stoney Stanton area SW of Leicester (Fig. 7.1). Many of these are abandoned quarries, now mostly flooded or filled in. Sampling was mainly from the Croft and Enderby quarries. At Croft the medium-grained, pinkish grey diorite is highly altered, well-formed plagioclases are sericitised, biotites are chloritised, and amphiboles are partly replaced by chlorite and epidote. Quartz occupies interstitial sites and was late to crystallise. Radiating prehnite often fills cavities and zeolite veining also occurs. The Enderby diorite is finer-grained, and the alteration is mainly to chlorite and epidote.

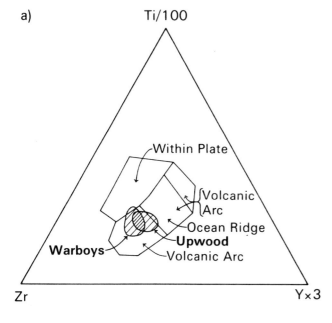

a) Ti/100

Within Plate

Volcanic
Arc

Ocean Ridge

Upwood

Warboys

Volcanic Arc

Zr

Y×3

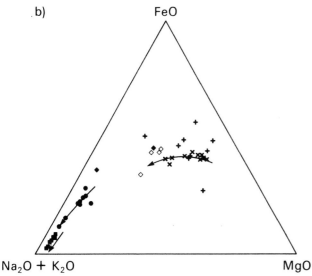

b) FeO

Na₂O + K₂O

MgO

- ■ Great Osgrove Wood (rhyolite)
- ● Glinton (rhyolite)
- + Upwood diorite
- × Warboys basaltic andesite
- ◇ Beckermonds microdiorite
- ◆ Woo Dale andesite
- ← Fractionation trends

Fig. 7.11 Ti–Zr–Y discriminant diagram for basaltic compositions (fields after Pearce and Cann, 1973) and AFM diagram for central and eastern England borehole samples of uncertain age

Weardale Granite

The granite, sampled by approximately 400 m of core in the Rookhope Borehole (Dunham and others, 1965) has a weathered upper surface overlain by Carboniferous sediments. It is a medium- to coarse-grained mica-rich granite with a planar fabric due to the alignment of muscovite and biotite plates and lensoid patches of quartz. Both potash and albitic plagioclase feldspars are white, and quartz is grey, resulting in a white or grey-mottled overall appearance. Apatite is the most abundant accessory mineral, although Ti oxides (after ilmenite, perhaps) are also common, whereas zircon and monazite are rare.

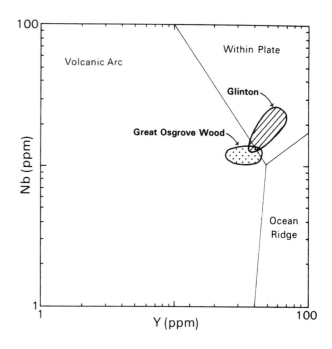

Fig. 7.12 Discriminant diagram for granitic compositions (fields from Pearce and others, 1984) showing affinities of igneous units of uncertain age for central eastern England boreholes

In the upper part of the cored section the fabric is strong, sometimes with 'eyes' developed. Holland and Lambert (1970) argued that the foliation developed before magmatic activity ceased because the unfoliated aplites and pegmatites are intruded along planes sub-parallel to the foliation. However, a preferred orientation of muscovite is present in some aplites. Clearly, the foliation and lensoid fabric, with muscovite replacing biotite and potash feldspar, suggests that significant recrystallisation and/or re-equilibration occurred as a high-temperature subsolidus process. Igneous textures are preserved only in layers dominated by twinned and zoned albitic plagioclase and perthitic potash feldspar.

In the upper part of the core jointing and fracturing are common, and mineralisation includes fluorite, quartz, pyrrhotite, pyrite, galena, tourmaline and molybdenite. Often a yellowish green hue due to sericitisation is developed through zones, tens of centimetres wide, adjacent to mineral veins. At lower levels in the borehole fracturing and the effects of hydrothermal alteration become infrequent, but intrusions of aplite and zones of pegmatite are common throughout.

Wensleydale Granite

The samples analysed are from about 100 m of core from the Raydale Borehole (Dunham, 1974). At the top of the granite a weathered surface is overlain by Carboniferous sediments. The granite is medium-grained, non-porphyritic, unfoliated and pink, with potash feldspar, albitic plagioclase, quartz and chlorite (pseudomorphing biotite). Except for extensive post-magmatic chloritisation, variable sericitisation and the introduction of dolomite into interstitial cavities, the fabric and mineralogy are predominantly related to magmatic processes. There is petrographic evidence that crystallisation of all the major minerals (quartz, K-feldspar, plagioclase, and biotite) was continuous from early to late, each being found both included in and interstitial to the others. In addition, quartz and feldspar sometimes form graphic intergrowths, which suggests a stage of rapid cooling or volatile loss.

The extent of mineralisation in the Wensleydale Granite has not been reported previously. However, a sample of dolomite vein-breccia was found to contain occasional grains of pyrite, chalcopyrite and sphalerite. One fine-grained sample is probably an aplite or microgranite.

7.3.2 GEOCHEMISTRY

Representative mean compositions of the main rock types sampled are given in Table 7.4. Data are also presented for the northern England Wensleydale and Weardale Granites.

Figs. 7.13 and 7.14 demonstrate that there is relatively little compositional variation at the scale of individual quarries in the Leicestershire plutons, excluding aplites and mineralised or altered zones. There are also close similarities between each of the plutons sampled. The diorite at Enderby is the least evolved, but it is, nevertheless, closely related to the slightly more evolved diorite at nearby Croft. The Mountsorrel Granodiorite is more evolved, but clearly related to the South Leicestershire Diorites in terms of such elements as Zr and Y, which are at similar levels in all the intrusions. The Mountsorrel Granodiorite is also closely comparable with the altered granodiorite from the Kirby Lane borehole. Much of the geochemical variation within intrusions reflects primary magmatic processes—for example, the decreasing Fe_2O_3, MgO, CaO and Zn and increasing K_2O and Rb probably result from fractionation of ferromagnesian minerals and plagioclase. The relationships between intrusions may be due to their derivation from a common parental magma, or from different partial melt batches from a similar source region. REE patterns (Fig. 7.15) are similar for all of the rock types, but with a slight overall enrichment from dioritic to granodioritic compositions. The altered Kirby Lane sample, however, is relatively depleted in REE and in Y. Eu anomalies increase in the more evolved rocks, which is compatible with progressive fractionation of a feldspar component. The form of the Rb–Ba trend in Fig. 7.13d is unlike that of suites previously considered and could be explained by biotite fractionation or by biotite remaining in the residue after partial melting.

Many effects of alteration are localized within zones around zeolite- and carbonate-bearing pods and veins at Croft, and kaolinised zones at Mountsorrel. The zeolite/carbonate/sericite alteration at Croft and in the Kirby Lane core results in enrichment of K_2O and depletion of Na_2O and Sr with erratic behaviour of Fe_2O_3, MgO, Al_2O_3, CaO and Rb (e.g. Fig. 7.13 a, b). Kaolinisation enriches K_2O and depletes CaO, Na_2O, Fe_2O_3 and Sr at Mountsorrel. More general mobilisation of the base metals Cu and Pb is reflected in their variable abundances at Mountsorrel and to a lesser extent at Croft (Fig. 7.14 b, d). Mountsorrel aplite compositions reflect extreme fractionation involving K-feldspar, plagioclase and biotite, but are of minor extent and generally do not contain anomalous levels of base metals.

Table 7.4 Geochemistry of Caledonian intrusives

	Weardale Granite Mean(19)	2 × sd	Wensleydale Granite Mean(9)	2 × sd	Mountsorrel Granodiorite Mean(9)	2 × sd	Croft Diorite Mean(8)	2 × sd	Enderby Diorite Mean(4)	2 × sd
SiO_2	70.27	2.00	76.16	1.32	68.87	0.90	62.44	2.90	58.31	2.80
TiO_2	0.18	0.00	0.16	0.16	0.49	0.10	0.57	0.10	0.71	0.10
Al_2O_3	18.03	2.50	12.24	0.60	15.26	0.90	16.49	1.10	15.65	1.20
Fe_2O_3	0.84	0.60	0.64	0.30	2.88	0.60	4.87	0.50	6.47	0.70
MgO	0.58	0.20	0.56	0.25	1.95	0.60	2.98	0.70	5.04	0.60
CaO	0.77	0.60	0.26	0.25	2.37	1.00	4.22	0.60	5.37	0.60
Na_2O	2.90	2.20	2.71	0.46	3.89	0.50	4.33	0.60	3.75	0.60
K_2O	5.89	1.50	5.57	0.58	3.72	0.40	2.18	0.30	1.79	0.40
MnO	0.04	0.00	0.03	0.01	0.08	0.00	0.09	0.00	0.13	0.00
P_2O_5	0.20	0.00	0.07	0.03	0.13	0.00	0.21	0.00	0.23	0.00
Total	99.70		98.40		99.64		98.38		97.45	
As	7	23	0	0	3	6	1	4	3	2
Ba	387	262	0	0	581	300	508	41	468	112
Ce	31	16	0	0	71	4	43	17	41	5
Cl	47	32	0	0	181	74	207	139	133	82
Co	1	5	0	0	7	7	10	3	18	6
Cr	50	122	0	0	22	6	137	255	145	373
Cu	12	21	0	0	11	1	25	6	32	19
Mo	0	0	0	0	1	4	0	0	1	1
Ni	6	3	0	0	9	1	14	2	38	9
Nb	9	3	18	3	10	7	8	2	7	2
Pb	49	132	16	4	17	3	12	5	10	2
Rb	412	141	255	27	121	150	58	14	45	14
S	304	26	0	0	155	38	112	194	62	68
Sc	3	2	0	0	8	1	10	3	14	3
Sr	93	50	20	5	239	54	483	134	514	60
Th	11	4	23	4	13	10	5	4	4	3
V	14	6	0	0	52	6	73	9	107	12
Y	9	5	60	11	25	6	22	2	19	1
Zn	36	21	0	0	35	11	54	70	70	5
Zr	81	15	122	14	166	24	169	12	153	17

The Leicestershire plutons plot on a calc-alkaline trend on an AFM diagram (Fig. 7.16). They plot in the field of 'volcanic arc' granites in both the Y – Nb discrimination diagram of Pearce and others (1984), and the Rb – Hf – Ta diagram of Harris and others (1986). The relation of this group to the known Caledonian subduction zone of northern England/Southern Scotland, which gave rise to Borrowdale Volcanics in the Lake District at about the same time (430 – 450 Ma), is not clear. It is possible, given the poor level of knowledge of pre-Silurian basement in southern and eastern Britain, that an as yet undiscovered zone of subduction was responsible, to which the acid extrusives of uncertain age, so common in East Midlands boreholes, might also be related. Indeed, Soper (1986) has suggested that a northward-dipping Caledonian subduction zone in southern England might have been responsible for development of granites in northern England and the Southern Uplands of Scotland.

The younger granites of northern England are more evolved than the diorite to granodiorite plutons of

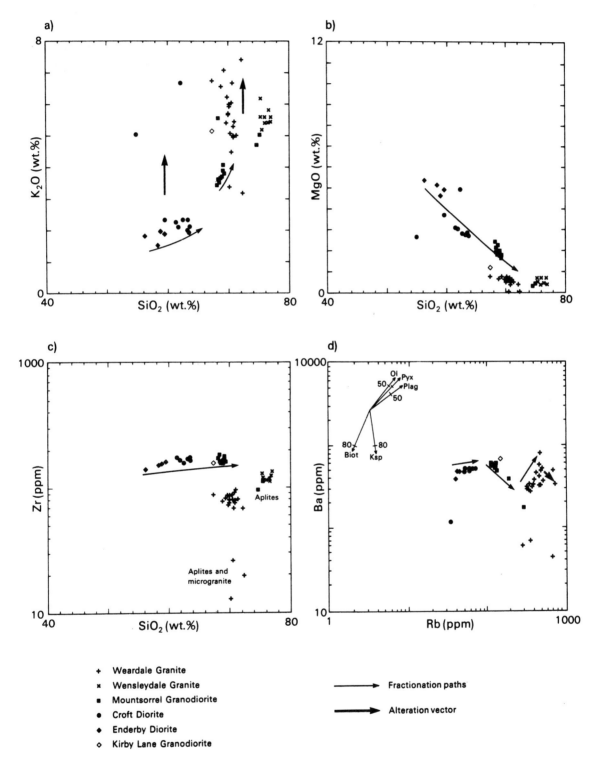

Fig. 7.13 Variation diagrams for Caledonian intrusives of central and northern England. Mineral vectors in Rb – Ba diagram serve to show possible minerals controlling fractionation. Vectors show theoretical effects on melt composition of crystallising single mineral phases (assuming Raleigh fractionation) and are annotated according to proportion (%) of melt remaining. After Tindle and Pearce (1981)

Leicestershire, being richer in SiO_2, K_2O and Rb, but lower in Fe_2O_3, TiO_2, MgO, CaO, Sr, V and Zr (Table 7.4). There are also differences between the Weardale and Wensleydale Granites—not just in alkali and LIL-element compositions (Table 7.4), which might reflect alteration, but also in their relatively immobile high field strength (HFS) element contents (Fig. 7.17). In particular, Zr, Nb, Y and Th are low in Weardale material compared to Wensleydale. REE patterns (Fig. 7.15) also differ; the Weardale pattern slopes steeply and uniformly, with a small negative Eu anomaly, and low content of HREE, whereas Wensleydale has a higher content of REE with a large negative Eu anomaly and relatively high contents of HREE. It is suggested here that magmatic fractionation resulted in greater enrichment of incompatible elements in the Wensleydale Granite than the Weardale Granite. Similar behaviour has been observed in comparing the Cairngorm Granite of

Scotland with the Carnmenellis Granite of SW England (Webb and others, 1985). Primary compositions of Weardale Granite, especially its radioelement content, are considered to have been more uniform with depth than at Wensleydale, where borehole sampling may reflect a relatively enriched high-level portion of the body.

The geochemistry of the Wensleydale Granite core is fairly uniform, and little primary variation is apparent. A complete data set was not obtained owing to the shortage of material for analysis, so the mobility of the full range of base metals cannot be assessed. Of particular note, however, are the high Y contents of a dolomitic vein and of a sericitised/dolomitised sample, suggesting that Y was mobilized and concentrated locally, although it is frequently regarded as an immobile element.

A small degree of magmatic variation is apparent from the geochemistry of the Weardale core, which is sometimes modified by sericitic alteration. There is a high

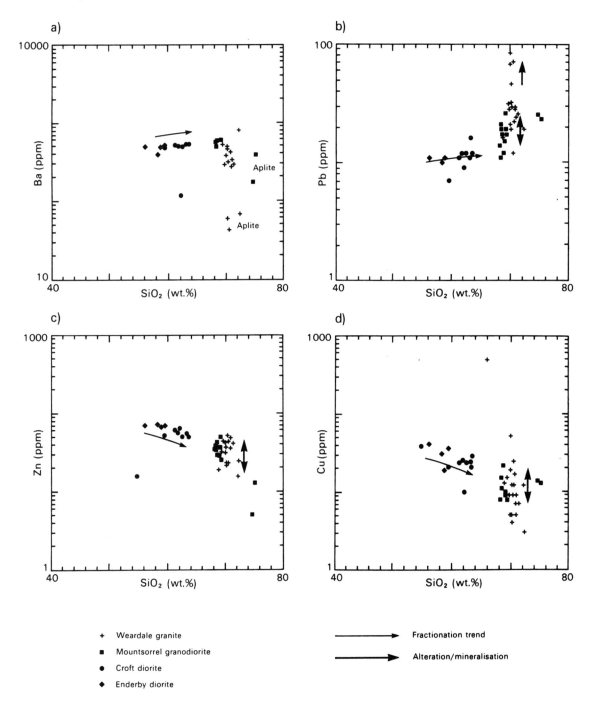

Fig. 7.14 Base-metal variation diagrams for Caledonian intrusives of central and northern England

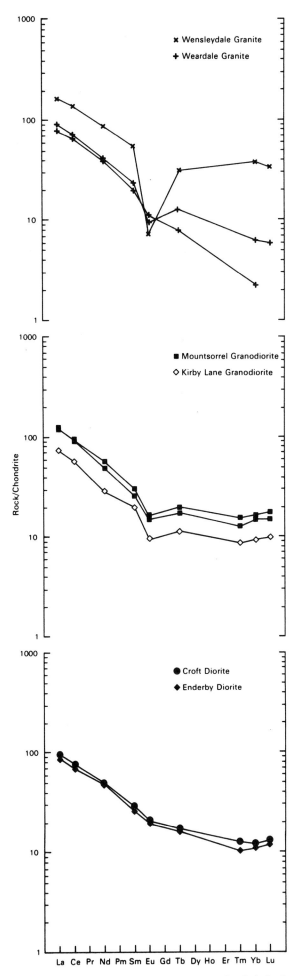

Fig. 7.15 Chondrite-normalised REE plots for Caledonian granites of central and northern England. Normalisation factors of Evensen and others (1978)

degree of scatter in base-metal abundances (Table 7.4), especially Pb and Cu (Fig. 7.14 b and d). Ba concentrations are fairly constant and the low values in aplite and microgranite are considered to be of primary origin. A few mineralized samples have exceptionally high metal values (Pb, for example), so background abundances are much lower than the mean values quoted in Table 7.4. Without statistically representative sampling it is difficult to assess whether there has been net overall loss or gain of these elements as a result of mobilisation and mineralisation events. A suite of U analyses (tabulated in Brown and others, 1987) shows that U abundance is also variable. The presence of only minor secondary U distributions as revealed by fission tracks, together with corroded uraninite grains, suggests that U has been leached from many rocks, probably by hydrothermal fluids. The higher concentrations of U measured, i.e. 10–12 ppm, may represent primary U abundances more closely than the present mean value.

In terms of the tectonic setting diagrams (Fig. 7.16) for granites (Pearce and others, 1984), the Weardale granite is in the 'volcanic arc' field of the Y–Nb diagram, but in the 'syn-collision' field of the Y + Nb–Rb diagram. On the Rb–Hf–Ta diagram of Harris and others (1986) it falls in the 'syn-orogenic' (Group II) field. The Wensleydale Granite features a 'within-plate' component on both the Y–Nb and the Y + Nb–Rb diagram, and lies in the 'post-orogenic' (Group III) granite field of the Rb–Hf–Ta diagram. Although diagrams involving such a mobile component as Rb should be viewed with caution, the general indications are, taking into account multi-element plots (Fig. 7.17), that both granites are subduction-related, but the Wensleydale Granite exhibits more within-plate characteristics. The Weardale batholith lies parallel to the postulated Iapetus suture zone (Fitton and others, 1982) where subduction had been active during the 50–80 Ma prior to its emplacement. Although major tectonism occurred at about the time of intrusion (*ca* 400 Ma), there is no evidence for continental collision with large-scale thrust tectonics, but an arc accretion event between impinging continents is not unreasonable. Extreme LIL enrichment of the Weardale Granite could be a result of water–rock interaction, possibly associated with its intrusion into hydrous crust. The Wensleydale Granite was further away from the site of subduction, and crystallized in a dryer crustal environment, so different fractionation controls may account for its apparent 'within-plate' character.

7.4 Carboniferous basalts

Magmatic activity occurred in many parts of Britain throughout the Carboniferous. In the stable foreland to the north of the foredeep of south west England (Upton, 1982; Floyd, 1982) activity occurred in the Midland Valley of Scotland, the Northumberland Trough, Derbyshire and the English Midlands. Basic volcanism was particularly extensive in the Midland Valley of Scotland during the Viséan and continued in that region into the Permian. Most lavas were subaerial, erupted from small central vent volcanoes along SW–NE- to WSW–ENE- (Caledonoid) trending volcanic lineaments. Extensive volcanism also occurred in the Northumberland Trough during the Dinantian with basalt compositions ranging from tholeiitic to highly alkalic. Further south, volcanism was generally less extensive, and is less well exposed; many occurrences have been discovered as a result of recently drilled boreholes and others may be inferred from geophysical signatures (Chapter 4). Basaltic compositions are again predominant, and they vary from tholeiitic to alkalic both from place to place, and through time at a given locality (Upton, 1982).

Collections of Westphalian basalt samples from borehole sites in the Vale of Belvoir (Fig. 7.19) and Viséan basalts from exposures in Derbyshire (Fig. 7.18) were obtained from Dr S. Kirton of BGS (formerly at Lancaster University) and from Dr R. Macdonald (and K. Gass) of Lancaster University. The samples have been described previously by Kirton (1981, 1984), MacDonald and others (1984) and Gass (1980). The relative ages of the volcanics are well constrained by a detailed biostratigraphy. Maximum ages for intrusive units are provided by the strata that they intrude. Stratigraphic relationships and correlations were given for the Vale of Belvoir succession by Kirton (1981) and, for the Derbyshire succession, by Walters and Ineson (1981).

The Derbyshire rocks are of special significance to the current study because they occur in the sequence of carbonate host rocks of the South Pennine Orefield. It is therefore important to consider possible relationships between the Carboniferous igneous rocks and the mineralisation: in particular, whether they could have provided a source for base metals or mineralising fluids (Chapter 3).

7.4.1 PETROLOGY AND MINERALOGY

Derbyshire basalts

Volcanism in Derbyshire is mainly Viséan (Asbian–Brigantian) in age, and is represented by lava flows, vents, tuffs and ash bands (see Fig. 7.18). K-bentonite

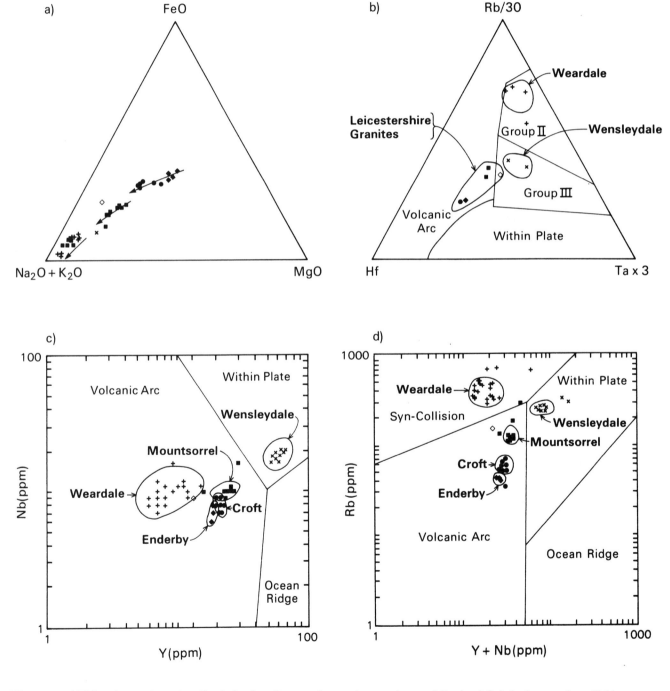

Fig. 7.16 AFM and tectonic setting discrimination diagrams for northern and central England Caledonian granites. Fields taken from (b) Harris and others (1986) and (c) and (d) Pearce and others (1984)

bands, interpreted as altered volcanic ash (e.g. Trewin, 1968; Trewin and Holdsworth, 1972) indicate continuing activity into the Namurian, and radiometric ages on sills suggest that their emplacement may have been as late as Westphalian. Much of the igneous activity was subaerial: the Upper Miller's Dale lavas were extruded onto an erosion surface, and air-fall ash was deposited on karstic surfaces (Walkden, 1977). In other places—for example, at Calton Hill—there are subaqueous basaltic agglomerates, and hyaloclastites are known from boreholes but not at outcrop (Aitkenhead and others, 1985).

Numerous igneous units are exposed at the surface, but few attain more than 50 m in thickness. Greater thicknesses have been proved in boreholes to the east; the Fallgate Borehole, Ashover, penetrated nearly 300 m of volcanics (Ramsbottom and others, 1962) and the Haddonfields Borehole more than 200 m of lava and tuff (Aitkenhead and others, 1985). Walters and Ineson (1981) estimated that some ten lavas extend for at least 10 km^2 and each has a volume of the order of 4 km^3. They

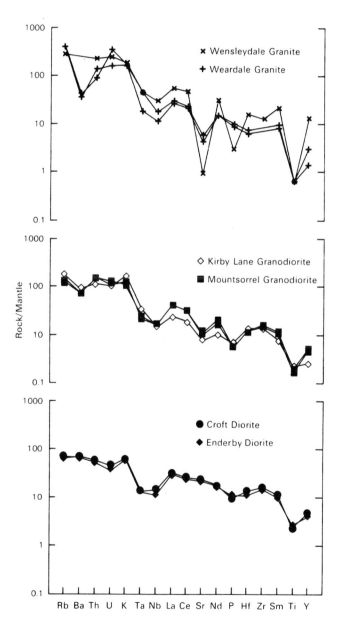

Fig. 7.17 Mantle-normalised plots for Caledonian granites of central and northern England. Normalisation factors of Wood (1979)

concluded also that the surface expression of volcanism in Derbyshire represents only the western margin of an igneous province that extends eastwards under Namurian cover. Geophysical evidence also suggests that thicker sequences of volcanics underlie the eastern margin of the Derbyshire 'Dome' (Chapter 4).

The lavas are generally highly altered and contain amygdales of carbonate, chlorite and chalcedony. The primary minerals olivine, pyroxene and labradoritic plagioclase are replaced by serpentine, chlorite, Fe–Ti oxides, carbonate, albite and epidote. The sills are largely medium-grained olivine dolerites and are usually fresher than the lavas, although the Calton Hill lavas are particularly fresh. The units sampled are listed below and their outcrops are shown in Fig. 7.18.

	Buxton–Tideswell area	Matlock area
Extrusive:	Upper Miller's Dale Lava	Upper Matlock Lava
	Lower Miller's Dale Lava	Lower Matlock Lava
	Calton Hill Lava	
	Cressbrook Dale Lava	
	Cave Dale Lava	
	Shacklow Wood Lava	
	Conksbury Bridge Lava	
Intrusive:	Tideswell Dale Sill	Ible Sill
	Water Swallows Sill	Bonsall Sill
		Peak Forest Sill
		Potluck Sill

Vale of Belvoir basalts

The dominant episode of igneous activity in the Vale of Belvoir occurred in the Westphalian, but there is evidence of intermittent volcanic activity that extends back to the Viséan and Namurian (Falcon and Kent, 1960). A sequence of early Westphalian volcanics up to 200 m in thickness was proved during a recent drilling programme by the NCB. Burgess (1982) and Kirton (1981) suggested a total volume of 45 km^3. Sills were emplaced later in Westphalian times. The main part of the lava sequence is reflected by a magnetic anomaly (Chapter 4). The locations of boreholes sampled by Kirton (1981) and the subject of the present geochemical studies are shown in Fig. 7.19.

The Westphalian volcanics were described by Burgess (1982): they comprise lavas, tuffs, agglomerates and sills. Carbonate- and chlorite-filled amygdales are found in the lavas, which are generally highly altered with no fresh olivine or pyroxene. Sills are often fresh, with only slight or incomplete alteration of olivine and pyroxene. The primary mineralogy of the basalts is olivine, pyroxene, plagioclase (labradorite bytownite) and Fe–Ti oxides. Olivine is replaced largely by chlorite, whereas plagioclase is albitised and replaced by carbonate. The groundmass is replaced by chlorite, smectite and carbonate (Kirton, 1984).

7.4.2 GEOCHEMISTRY

The geochemistry of the Carboniferous basalts from the East Midlands is summarized in Tables 7.5a and 7.5b.

Previous studies (MacDonald and others, 1984; Kirton, 1984; Walters, 1981) have placed considerable emphasis on the effects of alteration on the geochemistry of these basalts. Major elements and alkali trace elements are most likely to have been mobilised, whereas elements such as Y, Zr, Nb, TiO$_2$ and P$_2$O$_5$ are generally regarded as immobile during alteration of basalts (Pearce and

113

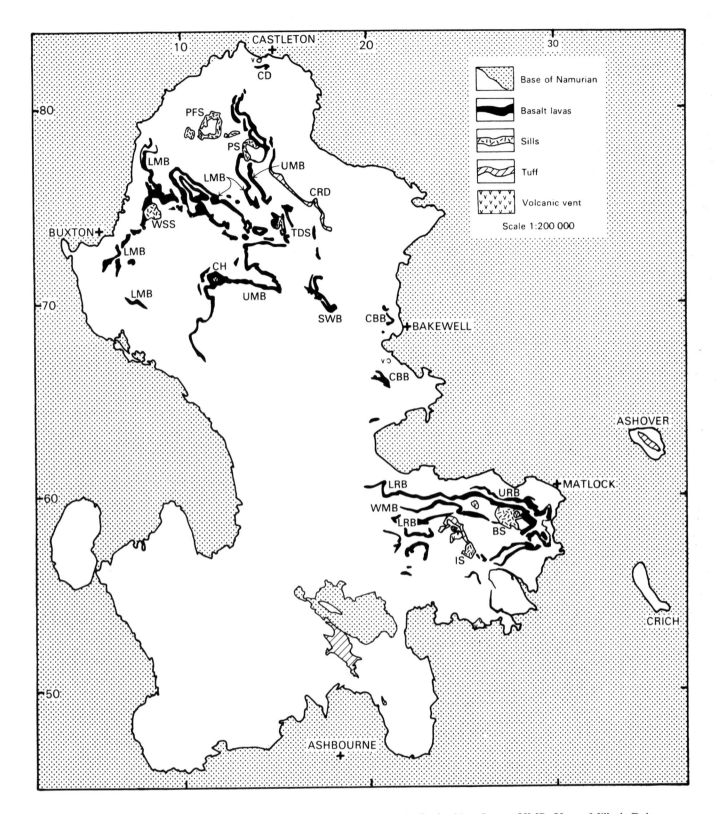

Fig. 7.18 Location of Carboniferous lavas, intrusions and volcanic vents in Derbyshire. Lavas: UMB, Upper Miller's Dale; LMB, Lower Miller's Dale; CD, Cave Dale; CRD, Cressbrook Dale; SWB, Shacklow Wood; CBB, Conksbury Bridge; LOB, Lathkill Lodge; LRB, Lower Matlock; WMB, Winstermoor; URB, Upper Matlock; R, Rowsley boreholes. Sills: PFS, Peak Forest; PS, Potluck; Wss, Water Swallows; TDS, Tideswell; BS, Bonsall; IS, Ible. Volcanic vents: SV, Speedwell; CH, Calton Hill, GM, Grangemill

Cann, 1973; Floyd and Winchester, 1975) and have been used to examine primary variation in such suites (Kirton, 1984; MacDonald and others, 1984).

In both the Viséan of Derbyshire and the Westphalian of the Vale of Belvoir the basalts range from tholeiitic to alkaline in their magmatic affinities, but the tholeiitic suite is dominant. In contrast, the less extensive, basic

magmatism of Westphalian C age in the West Midlands was mainly alkaline. These affinities have been demonstrated by MacDonald and others (1984) and Kirton (1984) by use of immobile element and mineral data. For example, Kirton used the Nb/Y – Zr/P_2O_5 plot for whole rock data and MacDonald and others used Al_2O_3, TiO_2 and SiO_2 compositions of clinopyroxenes to identify

114

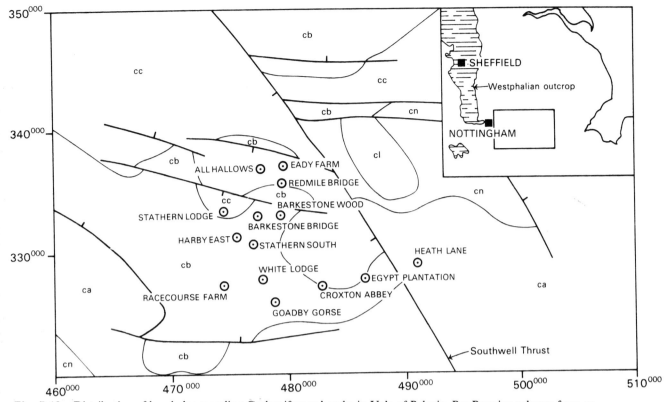

Fig. 7.19 Distribution of boreholes sampling Carboniferous basalts in Vale of Belvoir. Pre-Permian subcrop from an interpretation by N. J. P. Smith (cl Dinantian, cn Namurian, ca Westphalian A, cb Westphalian B, cc Westphalian C)

Table 7.5a Geochemistry of Viséan basalts from Derbyshire

	Upper Millers Dale Lava		Lower Millers Dale Lava		Matlock Lavas		Calton Hill Lava	
	Mean(8)	2 × sd	Mean(8)	2 × sd	Mean(7)	2 × sd	Mean(6)	2 × sd
SiO_2	50.22	7.9	51.01	5.4	47.59	4.8	47.60	2.3
TiO_2	12.47	2.3	12.55	2.3	13.64	2.6	12.19	1.8
Al_2O_3	1.94	0.5	1.93	0.6	1.79	0.6	2.21	0.3
Fe_2O_3	11.76	5.9	12.77	4.1	10.25	6.2	13.34	1.0
MgO	9.07	5.8	7.98	4.8	5.97	4.1	11.20	4.8
CaO	9.72	4.7	8.19	10.9	13.84	8.4	8.97	1.9
Na_2O	2.32	1.1	2.17	0.9	2.04	0.8	2.41	1.1
K_2O	0.47	0.6	1.35	1.8	1.68	1.0	1.26	1.1
MnO	0.12	0.1	0.09	0.1	0.10	0.1	0.19	0.0
P_2O_5	0.27	0.1	0.23	0.1	0.37	0.1	0.50	0.2
Total	98.36		98.27		97.27		99.87	
As	1	2.8	3	4.6	2	6.1	1	4.7
Ba	150	95.9	203	166.4	402	257.0	346	259
Ce	39	20.2	36	19.6	50	12.5	67	24.9
Cl	96	250.3	207	399.0	156	224.0	69	82.5
Co	59	29.2	46	22.8	45	15.7	55	12.1
Cr	349	125.1	323	116.2	260	207.5	363	92.2
Cu	62	10.5	34	36.4	40	34.7	61	36.6
Mo	0	1.4	0	1.0	1	1.9	1	2.7
Ni	476	386.3	288	345.0	157	176.0	442	348.2
Nb	16	11.4	12	2.4	30	6.0	35	16.0
Pb	8	6.1	8	3.9	9	4.4	8	1.0
Rb	9	12.3	19	23.4	29	16.9	24	7.5
S	276	223.8	199	151.1	298	464	159	71.2
Sc	24	6.2	25	5.4	24	2.3	22	3.3
Sr	275	49.4	296	124.0	419	83.6	380	260.0
Th	1	2.6	2	3.3	4	4.0	2	4.1
V	200	42.6	187	36.2	206	53.6	233	45.3
Y	25	4.6	24	7.9	26	3.6	27	3.7
Zn	106	52.8	97	13.2	78	81.9	94	17.9
Zr	120	26.9	113	11.4	141	16.4	166	48.8

* It should be noted that no Cr and Ni data are given for Derbyshire basalts on account of contamination by these elements in many of the samples during sample preparation.

magma types. The Nb–Zr variation diagram (Fig. 7.20) shows a continuum of compositions from tholeiitic (with low Nb and Zr abundances) to alkaline (with high Nb and Zr abundances) in both groups, although the trends are not quite identical. In Derbyshire the alkaline suite is represented by the Calton Hill lavas (Table 7.5a), whereas in the Vale of Belvoir certain units (mean composition given as Egypt II, Table 7.5b) from the Egypt Plantation Borehole are the most alkaline. More alkaline compositions are characterised by high Nb/Y, Zr/Y and P_2O_5/Y, and at Egypt Plantation such samples are rich in Sr, Ba and Ce, but low in Sc and V.

It appears that most of the variations of immobile elements are related to the degree of silica saturation of the magmas and may be regarded as primary magmatic characteristics. No consistent time- or spatially-related compositional variations have been recognised, which suggests that high-level magma chamber fractionation was not important (see also MacDonald and others, 1984). Source compositions and local melting conditions in the mantle are more likely to have controlled the observed variations, particularly the sympathetic behaviour of Zr and Nb (Fig. 7.20). All samples analysed plot as 'within-plate' basalts on the Ti–Zr–Y discriminant diagram (Fig. 7.22a) of Pearce and Cann (1973).

Many elements exhibit variations that are not easily related to the immobile elements: this is particularly true for Rb, Sr and Ba, which are contained dominantly in feldspars. Using Nb to reflect magmatic variation, the plots of Sr (Fig. 7.20c) and Ba (Fig. 7.21a) appear to show

underlying primary magmatic patterns on which enrichment/depletion effects are superimposed.

The degree of alteration of Carboniferous basalts suggests extensive interaction with groundwaters and/or hydrothermal fluids. Many of the major elements, such as Fe_2O_3, MgO, CaO, Na_2O and K_2O, have been mobilised and redistributed (Tables 7.5a and 7.5b). The variability of many base-metal compositions, particularly Cu, suggests that they too were affected to some degree (Fig. 7.21d). Few samples are enriched in Cu, but many are depleted, some to less than 20 ppm, in both the Vale of Belvoir and Derbyshire, compared to an almost uniform primary concentration of 60–80 ppm. Mobilisation of Zn extends the primary range of concentrations of 80–100 ppm, shown by enrichment to 700 ppm and depletion to 30 ppm (Fig. 7.21c). This is more important in the Vale of Belvoir than in Derbyshire. Both Ba and Pb (Fig. 7.21 a and b) appear to have been depleted more in the Vale of Belvoir than in Derbyshire. It could be argued that the Derbyshire samples are fresher, despite the fact that they are from outcrop, but it is more likely that the surface samples were more carefully selected to avoid mineralised and altered zones than was possible with borehole core. Indeed, Walters (1981) has described depletion of many trace elements, including Ba, Cu and Zn, from alteration zones in Derbyshire basalts, along with localised enrichments of Cu, Ni, Zn and Pb in the vicinity of mineralised zones. In volumetric terms the capacity of the basaltic flows and sills to act as sources of base metals appears to be somewhat limited. Although the

Table 7.5b Geochemistry of Westphalian volcanics from Vale of Belvoir

	Barkestone Bridge Mean(6)	2 × sd	Redmile Bridge Mean(8)	2 × sd	Goadby Gorse Mean(8)	2 × sd	Eady Farm Mean(5)	2 × sd	Stathern South Mean(5)	2 × sd	Egypt I Plantn Mean(9)	2 × sd	Egypt II Plantn Mean(4)	2 × sd
SiO_2	45.72	2.1	46.58	6.5	45.08	3.5	45.04	3.6	44.3	4.0	46.36	3.9	45.02	1.6
TiO_2	12.84	1.6	13.04	1.5	12.46	1.0	12.79	1.8	12.5	1.9	13.22	1.5	12.25	1.0
Al_2O_3	1.90	0.5	2.03	0.7	1.85	0.4	1.99	0.5	2.0	0.3	2.21	0.8	1.91	0.3
Fe_2O_3	16.06	3.3	15.87	7.4	14.78	2.1	17.17	5.2	15.6	3.8	15.00	5.6	15.78	0.8
MgO	9.30	7.2	9.19	6.3	11.56	3.0	11.05	7.1	11.4	7.3	9.14	3.4	8.84	6.3
CaO	7.50	4.6	7.04	5.9	8.45	4.8	6.38	8.2	8.5	10.4	6.65	3.3	7.76	0.7
Na_2O	2.18	0.8	2.41	1.7	2.33	0.4	1.96	1.0	1.9	0.5	2.53	1.1	2.75	1.9
K_2O	0.74	0.4	1.03	0.6	0.79	0.2	0.89	0.9	0.9	0.8	1.16	1.1	1.47	0.4
MnO	0.19	0.1	0.17	0.2	0.23	0.2	0.23	0.2	0.2	0.3	0.17	0.1	0.20	0.1
P_2O_5	0.28	0.1	0.33	0.2	0.36	0.1	0.32	0.1	0.3	0.1	0.43	0.2	1.04	0.1
Total	96.71		97.69		97.89		97.82		97.47		96.87		97.02	
As	3	4.2	1	3.3	0	1.4	2	5.5	2	5.5	0	2.0	2	3.8
Ba	126	63.2	157	119.0	126	25.7	137	116.5	96	42.3	177	118.0	467	387.2
Ce	41	29.5	51	16.3	35	21.1	35	28.9	39	24.1	56	439.9	117	10.7
Cl	197	260.6	245	446.7	164	203.3	333	149.8	209	327.5	772	768.0	200	144.4
Co	50	13.3	53	16.2	52	10.9	49	20.1	62	30.6	49	15.5	49	4.8
Cr	245	27.6	260	67.2	261	59.0	287	74.2	235	56.5	223	113.6	202	108.4
Cu	53	42.7	48	49.6	53	64.1	76	103.3	64	46.0	38	29.8	44	17.2
Mo	2	1.7	1	1.7	1	1.5	0	0.9	0	0.0	1	1.5	5	3.5
Ni	300	143.0	332	308.1	349	292.0	257	22.9	281	90.4	226	157.0	318	198.1
Nb	15	4.8	19	10.4	18	7.1	18	3.6	17	8.0	25	10.7	52	3.8
Pb	8	3.7	7	2.9	4	4.1	4	1.8	4	1.8	6	3.8	10	3.0
Rb	10	5.3	14	9.1	13	3.4	14	15.0	13	5.8	17	16.2	16	5.7
S	387	466.8	247	197.1	420	719.1	314	308.2	386	704.3	256	252.0	228	135.0
Sc	24	8.6	23	10.2	21	11.7	24	4.0	24	9.5	23	9.7	12	1.9
Sr	315	132.7	329	143.7	367	91.7	293	179.2	310	84.8	387	83.0	1126	184.5
Th	3	4.3	2	5.2	4	4.8	3	5.0	1	3.3	2	3.3	34	3.0
V	188	43.6	190	37.7	190	54.1	192	27.7	200	46.6	206	53.8	152	15.6
Y	23	6.4	22	6.9	22	6.5	22	6.2	24	10.7	26	9.3	22	2.0
Zn	115	30.5	100	112.2	142	193.3	112	113.4	71	62.8	167	227.9	154	47.8
Zr	121	26.9	135	37.6	136	27.7	132	27.0	129	51.1	172	58.7	252	31.2

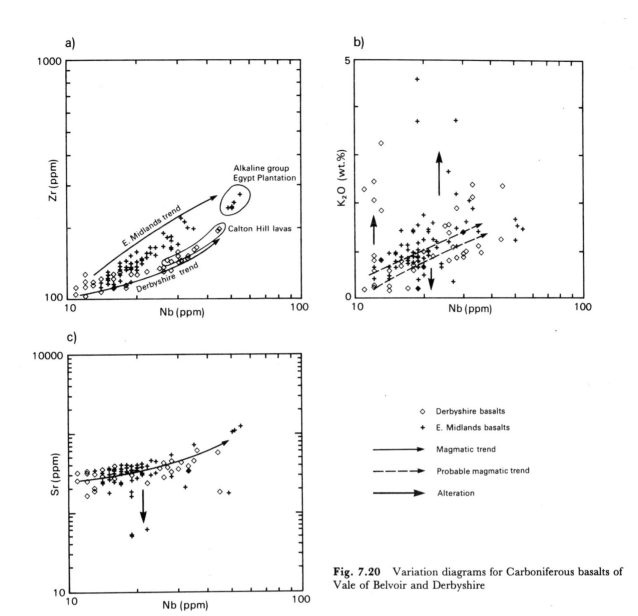

Fig. 7.20 Variation diagrams for Carboniferous basalts of Vale of Belvoir and Derbyshire

Table 7.6 Heat production in Caledonian and Precambrian intrusives

	Mean heat production $\mu W\ m^{-3}$	Number of samples	Mean uranium ppm	Mean thorium ppm	Mean potassium %
Weardale	2.7	3	6.8	10.1	4.1
Granite	3.4	16*	9.6	10.8	4.0
	4.4	6+	12.9	12.3	4.0
	3.7★				
Wensleydale	3.4	6=	6.7	21.1	4.8
Granite	3.3	19*	6.5	19.6	4.2
Mountsorrel	2.1	2	3.2	14.8	2.6
Granodiorite	2.5	8*	3.8	19.2	3.1
Kirby Lane	2.0	1	2.8	12.9	4.1
Granodiorite	1.7	1	1.4	14.0	4.1
Croft diorite	0.8	1	1.2	5.7	1.5
Enderby diorite	0.7	1	1.0	5.2	1.2
Northern Diorite	0.7	1	1.1	5.4	0.8
Southern Diorite	1.4	1	2.2	8.6	2.1

* Unpublished BGS data (1981)
+ Hennessy 1981
= BGS neutron activation data 1985
★ Preferred value (Lee and others, 1984)

117

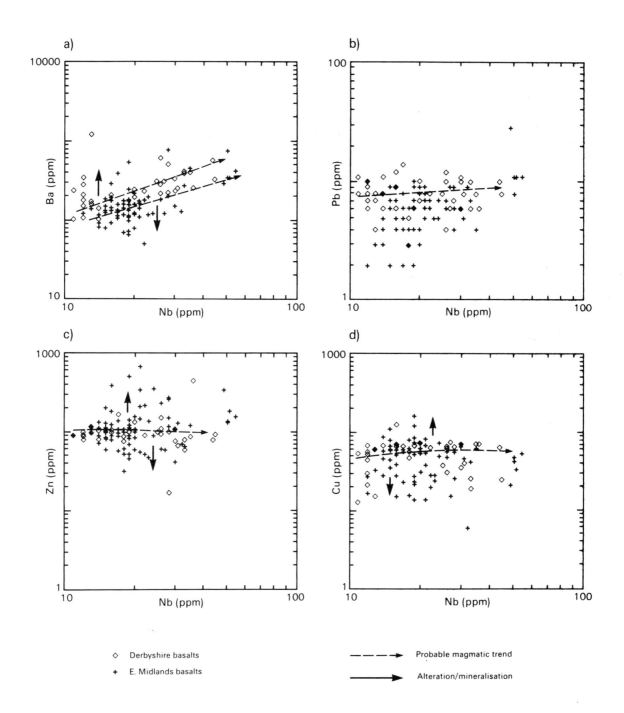

Fig. 7.21   Base-element variation diagrams for Carboniferous basalts of Vale of Belvoir and Derbyshire

main igneous centres typically extend for 10–20 km, their maximum thickness is around 200 m in the Vale of Belvoir, and nearly 300 m in Derbyshire. As the larger centres are known only from boreholes and geophysical data (Chapter 4), it is possible that other large centres are present at depths not yet explored. The source potential of the basalts for Ba, Pb, Zn and Cu appears greater in the Vale of Belvoir than in Derbyshire.

7.5   Heat production of granitic basement rocks

Zones of crust with above-average heat production, such as highly radiothermal granites, give rise to elevated geotherms and may provide the heat source to drive convective fluid flow and develop low-temperature hydrothermal mineralisation (Fehn and others, 1978). Such a model was proposed by Brown and others (1980) to account for syn- and post-Carboniferous mineralisation

associated with the Caledonian granites of northern England.

The Precambrian and Caledonian dioritic and granodioritic intrusives of Leicestershire are poorly radiothermal (Table 7.6) and are unlikely to develop any more heat than average crust. They do not form significant radiothermal anomalies in relation to their country rocks, and therefore would not be expected to develop postemplacement mineralisation.

The Wensleydale Granite has a greater heat production than 'average crust' (Table 7.6). The material sampled, however, probably represents some of the more evolved granite of the intrusion (section 7.3), and its measured heat production is likely to be an overestimate of the mean bulk heat production for the intrusion (as proposed in the case of the Cairngorm Granite, Scotland, by Webb and others (1985)). This is reflected in the only slightly enhanced surface heat flow of 65 μW m^{-2} (England and others, 1980), and probably suggests that the pluton had

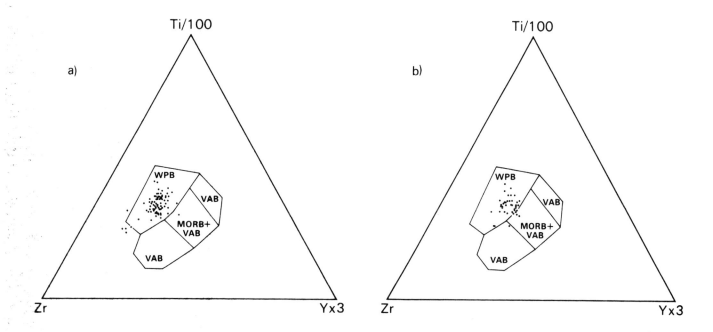

Fig. 7.22 Ti–Zr–Y discriminant diagrams for basaltic compositions (fields of Pearce and Cann, 1973) for Vale of Belvoir basalts (a) and Derbyshire basalts (b). WPB—within-plate basalts; VAB—volcanic arc basalt; MORB—mid-ocean ridge basalt

only a limited capacity for raising geotherms and causing convection of hydrothermal fluids.

The Weardale Granite, although apparently only slightly more radiothermal than the Wensleydale Granite (Table 7.6), was probably depleted in U (section 7.3.2) during hydrothermal alteration. On the basis of an estimated primary U content of 10–12 ppm, the primary heat production value would have been nearer 4.5 μW m^{-3}. This may still be true of parts of the batholith where hydrothermal activity was less intense. Geochemical evidence (section 7.3.2) suggests that the primary distribution of radioelements was probably more uniform to great depth (as proposed for the Carnmenellis Granite, SW England, by Webb and others (1985)). The Weardale

Granite, also more extensive volumetrically than the Wensleydale Granite (Chapter 4), is therefore more radiothermal in bulk terms, as borne out by the enhanced surface heat flow of 95 μW m^{-2} measured at the Rookhope borehole (England and others, 1980) and the marked heat flow anomaly over the batholith on the UK heat flow map (British Geological Survey, 1984). These properties suggest that the Weardale Granite had the greater potential for raising geotherms, driving convective hydrothermal systems and developing post-magmatic mineralisation. This is in line with the close relationship between mineralisation and the geophysically inferred granite cupolas at Weardale but not at Wensleydale (as described in Chapter 3 and also by Brown and others (1987)).

Table 7.7 Mobility of elements in the igneous basement

	Depleted elements	Enriched/Depleted	Enriched elements	Process of Mobilization
Charnian Porphyroid	Mg, Ca, Pb	Na, Ba, Sr		CH, E, S, CB, H
Northern Diorite	Mg, Ca	Na, Sr, Cu, *Pb*	Fe	CH, E, S, CB, H, M
Southern Diorite	Fe, Mg, Ca, Na, Co, V	Sr, *Cu*, Zn, *Pb*	K, Rb, Ba	CH, E, S, CB, H, M
Upwood B/H		Fe, Ca, Zn, Pb, Co	Ca	Z, CH, S, CB, SI
Warboys B/H	Ca	Co	Na, Cu	CH, S, CB
Glinton B/H	(V)	Na, Ca	K?	S, CB, SI
Gt. Osgrove Wood B/H	Ca	Na, Sr	K?	S, CB, SI
Beckermonds B/H				CH, CB, S
Enderby Diorite				CH, E, S
Croft Diorite	Na, Sr, Th, Cu, Co	Ca, Pb	K	Z, CB, S
Kirby Lane B/H	Ca, Na, Sr, Zn, Pb, Co, Y		K, Co	CH, S, CB
Mountsorrel Granodiorite	Fe, Ca, Na, Ba, Sr	Cu, Zn, Pb, Co	K	CH, S, K
Weardale Granite	Sr	Fe, K, *Cu*, Zn, Y	Ca, Rb, Ba, Pb, Co, Cr	CH, S, M
Wensleydale Granite			Y	CH, CB, H, S, M
Carboniferous basalt				CH, CB

Italics = general rather than local mobilisation

Key to mobilisation processes:

CB, Carbonatisation; CH, Chloritisation; E, Epidotisation; H, Haematisation; K, Kaolinisation; M, Mineralisation; S, Sericitisation; SI, Silicification; Z, Zeolitisation.

7.6 Mobilisation of elements in the igneous basement

Patterns of elemental redistribution, superimposed on primary magmatic variations, illustrated by variation diagrams in sections 7.1 to 7.4, provide evidence for the recognition of enrichments and depletions in basement igneous units. The results obtained from a wide range of diagrams are summarised in Table 7.7.

Enrichments may reflect the localised occurrence of sulphides, oxides, carbonates and sulphates in mineralised zones. This results in erratic, often extreme, enhancement of whole-rock abundances. Enrichments may also reflect deposition from mineralising fluids of components that are incorporated in alteration products, as is often the case when pervasive sericitisation, kaolinisation or epidotisation occurs. These effects are usually less extreme, their magnitude being related to the extent of alteration. Metal enrichments therefore provide an insight into the ore-forming potential of hydrothermal fluids and enable the metals introduced to be identified. Metal depletions reflect the decomposition, often selective, of primary minerals and the ability of fluids to remove metals that are not stable in the mineral structure of alteration products. Those metals become part of the hydrothermal fluid and the affected host rocks are therefore potential sources of mineralisation.

Several different mobilisation processes may occur, separately, and on different scales in hydrothermal systems. Each may give rise to enrichment or depletion in a suite of rocks. When individual processes are of local extent, such as kaolinisation in the Mountsorrel granodiorite, it is possible to characterise specific effects on the whole-rock geochemistry (Table 7.7). More commonly, several processes are involved in alteration and detailed petrographic and mineralogical (microprobe) analysis would be required to identify the sequence of alteration effects and to characterise the chemistry of each process. The data from the present study provide only general information about elemental mobilities and an indication of the scale of mobilisation; more information could be obtained through combined microprobe, fluid inclusion and stable isotope studies.

Most of the effects of remobilisation of elements recognised in Table 7.7 are local: only in a few cases (italics) are they more extensive. The occurrence of local base-metal enrichments and extensive depletions of metals may indicate potential sources of mineralisation and the presence of fluid pathways that could constrain mineral deposit models of the region.

Elemental mobilisation (Table 7.7) is most evident in the more basic extrusives, such as the Carboniferous basalts and the Upwood agglomerate, and in the intermediate to acid intrusives, such as the Charnwood Southern Diorite and the Weardale Granite. Most commonly, mobilisation affects the alkali- and alkaline-earth elements and base metals. Deposition of base metals through mineralisation has occurred in the Charnwood Southern Diorite (Ba), the Weardale Granite (Pb) and the Vale of Belvoir Carboniferous basalts (Zn). Evidence of extensive fluid transport of base metals is provided by the Charnwood Diorites (Cu, Pb), the Weardale Granite (Cu), the Vale of Belvoir basalts (Cu, Ba, Zn) and the Derbyshire basalts (Cu). The Charnwood Diorites (Pb) and the Mountsorrel and Weardale Granites could be sources for Cu and Pb, and the Carboniferous basalts, especially in the Vale of Belvoir, sources for Cu, Pb, Ba and Zn.

7.7 Tectonic evolution in the Palaeozoic based on igneous rocks

The main problem in interpreting the late Precambrian and Lower Palaeozoic tectonics of central England is that evidence is confined to small inliers projecting through Upper Palaeozoic and Mesozoic rocks. Although boreholes across central and eastern England sample basement rocks, these are of uncertain age, and at present any temporal framework for these rocks, as implied by Table 7.1, is provisional. There remain large areas where basement is inaccessible.

In England and Wales there are no rocks known to be older than *ca* 700 Ma, and Sr isotopic evidence suggests that no basement is older than *ca* 900 Ma (Thorpe and others, 1984). This is broadly in agreement with Rb–Sr isotope evidence of Hampton and Taylor (1983).

The oldest rocks of the area are the Charnian of Leicestershire, consisting of a varied sequence of sediments with minor interbedded volcanics believed to be 680–700 Ma old (Thorpe, 1982). The Charnian andesites and dacites have calc-alkaline affinities and are believed to have formed in an accretionary volcanic arc environment (see section 7.1.1; Le Bas, 1982; and Thorpe and others, 1984). This sequence was intruded at about 550–600 Ma by a suite of altered diorites that probably predates the early Cambrian sediments of the Nuneaton area. These magmas are also typical of an arc environment, but their higher LIL contents suggest a greater crustal involvement in their formation, as might be expected in an arc environment evolving with time. Thorpe and others (1984) suggested that a SE-dipping subduction zone was active in late Precambrian time, its axis being represented by the Mona Complex of Anglesey. There is no firm evidence of late Precambrian basement between the Midlands and the Channel Isles, but it is probable that the whole area of southern Britain was an accretionary arc environment (Pharaoh and others, 1988b). It is possible that the basic diorite and basaltic agglomerate of Warboys and Upwood are related to the diorite intrusions at Charnwood because of their geochemical similarities, but in the absence of reliable dating this is speculative. The other igneous borehole material from the East Midlands comprises rhyolitic lavas and pyroclastics that are geochemically distinct from the Charnian volcanics. They could be related to the younger Charnwood diorites, or to the Caledonian intrusive suite of Leicestershire. There is, as yet, no firm geochemical evidence to support either option.

The Caledonian intrusives of central England are also of volcanic arc type, but they are almost 300 km south of the Iapetus suture, and might instead be related to some form of subduction zone active in southern Britain (cf.Soper, 1986). Indeed, Siluro-Devonian calc-alkaline volcanics are known in parts of southern England and South Wales. The Caledonian granites of northern England are easier to reconcile with arc magmatism, although Iapetus had probably closed by the time of their emplacement (Brown and others, 1985) A south-easterly dipping subduction zone was probably active through much of Ordovician time, leading to the calc-alkaline volcanics of the Lake District and the rhyolitic volcanics of North Wales (Fitton and others, 1982). The combined subduction and within-plate character of the Wensleydale Granite may reflect its separation in time and/or space from the active zone of subduction and its derivation from a source affected by less hydrous melting than is normally associated with volcanic arc granites. The LIL-enriched, subduction-related character of the Weardale Granite could reflect the earlier subduction of a volcanic arc with sediments just prior to closure of Iapetus. The magma was probably formed largely by hydrous melting near the base of tectonically thickened crust, the LIL-enriched fluid component coming from the subducted slab.

Following Caledonian times there was no destruction of crust due to lateral movements in central and northern England. During the Carboniferous, Britain was well within the Laurasian continent, the nearest plate boundary being to the south of the British Isles. Francis (1978) recognised that volcanism was not associated with hotspot development or subduction but with deep-seated faults at the hinge lines between stable blocks and subsiding basins. Upton (1982) extended this scheme, suggesting that sites of Carboniferous 'within-plate' volcanism relate to lines of weakness in the underlying basement that were reactivated by tensional stresses developed in the continental crust due to impactions along the Hercynian zone of plate convergence to the south of Britain. A model of lithospheric stretching in an inhomogeous crust (containing major granitic plutons) has been proposed by Leeder (1982) to explain Carboniferous basin formation and rifting in northern England and Scotland.

In northern England the low-density Weardale and Wensleydale batholiths formed buoyant cores to the fault-bounded Alston and Askrigg blocks. In Derbyshire, where intrusive bodies and volcanic vents line up along WNW–ESE anticlinal axes, and not along the SW–NE Caledonide trend, the volume of volcanics is much less than in the Scottish Midland Valley or in the Northumberland Trough, where reactivated Caledonian weaknesses are considered to have allowed melts easier access to the surface (MacDonald and others, 1984). The Derbyshire lavas were regarded by MacDonald and others (1984) as products of localised partial melting of the underlying upper mantle that came to the surface along fractures in attenuated crust and in later Carboniferous time along the flanks of the South Pennine sag. In the Vale of Belvoir the magmas are believed to have ascended deep basement faults at the margin of the Widmerpool Gulf (Kirton, 1981). Leeder (1982) has suggested that a change in the predominant style of igneous activity from volcanism to sill emplacement is reflected by a change in tectonic regime from fault-bounded rifting in the Dinantian to a general crustal sagging in the Namurian. No corresponding geochemical changes of significance have been observed. A consequence, however, of extensive sill emplacement is that magmatic heat is trapped, which leads to an enhancement of geothermal gradients in the upper crust.

121

8 METALLOGENIC MODELS

T. B. Colman, D. G. Jones, J. A. Plant and K. Smith

8.1 Introduction

In the previous sections published information on the geochemistry, isotope systematics, mineralogy and metallogeny of the Pennine orefields has been reviewed and new information on the geophysics, tectonic history and the geochemistry of igneous rocks and shales has been presented. Here we attempt to use these different sources of evidence to propose a detailed metallogenic model that accounts for Pennine-style mineralisation in relation to the structural and stratigraphic evolution of the region. Our starting points are (1) the suggestion of Dunham that the Pennine mineralisation represents a sub-type of the Mississippi Valley type (MVT) genus of ore deposits and (2) the basinal brine models, originally proposed by White (1958), for such deposits.

According to this theory, warm brines from sedimentary basins carry dissolved metals from depth to the basin margins, where ore deposits are formed. The processes involved are controversial: some authors—for example, Dozy (1970), Noble (1963) and Jackson and Beales (1967)—favoured a basin dewatering, or stratafugic, process whereby brines are generated as a result of sediment compaction during basin evolution. Others—for example, Cathles and Smith (1983)—suggested that episodic dewatering occurs as a result of excess pore pressures, the overpressuring driving sudden bursts of deep brines towards the basin margins. An alternative model, proposed by Garven and Freeze (1985) and Garven (1984), holds that gravity-driven groundwater flow, due to topographic differences across the basin, carries warm fluids from deeply buried strata into shallow sediments. The application of each of these models for the Pennines is considered below, together with more detailed evidence on the source of fluids, metals, heat and the processes involved in ore deposition.

There is mounting evidence that Tournaisian–Chadian rocks, of similar age to those which host Irish-style deposits, occur at depth in Britain. Hence models for Irish-style deposits are also briefly considered. Criteria suitable for exploration for Pennine- and Irish-style deposits buried beneath later cover in the study region are proposed and combinations of exploration criteria are used to evaluate the prospectivity of the study region for the two types of carbonate-hosted ore deposits in Chapter 9.

8.2 Mississippi Valley type mineralisation: general characteristics and models

Mississippi Valley type ores constitute a major class of deposits, known from many parts of the world, which form in anorogenic settings distinct from those in which volcanogenic ore deposits are found. As in the case of petroleum and natural gas, such deposits are now viewed as being part of the normal evolution of sedimentary basins.

MVT deposits are characterised generally by the following.
(1) A mainly Phanerozoic age of formation with many deposits in rocks of Cambro-Ordovician and Carboniferous age.
(2) Epigenetic deposition in orogenically inactive settings in platform carbonate, or rarely sandstone, host rocks with normal geothermal gradients of 20–30°C/km.
(3) A close spatial and genetic relationship with shale-filled basins.
(4) A lack of association with contemporaneous igneous activity.
(5) The common association of Pb–Zn–Ba–F in ore deposits of simple mineralogy, mainly comprising galena, sphalerite and dolomite with minor pyrite, marcasite, and chalcopyrite and, locally, calcite, fluorite, quartz, baryte, K-feldspar and K-mica.
(6) Formation at low temperatures (mainly in the 90–100°C range) from high-salinity fluids.
(7) An association with hydrocarbons, especially petroleum.
(8) Low Ag content of galena.

Sangster (1983) and Anderson and MacQueen (1982) stressed the diversity of individual deposits, which vary in their mineralogy, deposit type, host rock lithology and chronology within the overall scheme.

8.2.1 GALENA- AND/OR FLUORITE-DOMINANT OREFIELDS

One of the most important types of MVT deposits for the purpose of this study are the galena, and/or fluorite, dominant deposits in North America, which, as noted by Dunham (1983), are closely similar to the Pennine mineralisation. The main Pb-dominant area in North America is the South-East Missouri orefield, containing the Old Lead Belt and the Viburnum Trend, which has produced more than 20 Mt Pb with minor Zn, in mainly stratiform bedding replacement orebodies. In contrast, the main Tri-State district has produced 11.5 Mt Zn and 2.65 Mt Pb (Laznicka, 1985). Other areas with largely fracture-controlled veins, and only minor stratiform orebodies, may be fluorite-dominant (for example, the Central Kentucky and Illinois–Kentucky Orefields) or baryte-dominant (for example, the Sweetwater (Tennessee) Orefield). The latter is unusual in that all the deposits comprise residual baryte, with minor fluorite, in soils overlying thin, uneconomic, primary veins (Laurence, 1960). The host rocks are Ordovician or Viséan carbonates and there appears to be a spatial relationship with alkali volcanism, which is absent in the other MVT areas (Heyl, 1969). The Central Kentucky field has a crudely concentric zonation with inner fluorite/calcite/sphalerite, centre fluorite/barite/calcite and outer barite/galena/sphalerite + calcite + fluorite zones that are centred over the Lexington Dome on the Cincinnati Arch (Jolly and Heyl, 1964). This zonal pattern is similar to that of the northern Pennines (Dunham, 1948; Chapter 3).

8.2.2 MISSISSIPPI VALLEY TYPE MINERALISATION MODELS

The general model applied to MVT deposits is that originally proposed by Jackson and Beales (1967) for the Devonian Pine Point deposit of the Canadian North-West Territories. The formation of this deposit is suggested to have resulted from early diagenetic dewatering of basin facies shales with lateral migration into karstified limestone on adjacent emergent blocks. Dewatering occurs initially during the compaction of the basin sediments by lithostatic pressure. Complex phase transformations may occur as the clay-dominated sediments lose their pore water and as the temperature and lithostatic pressure rise with depth of burial. The fluids, charged with metals in solution, move up-dip towards the basin margins and deposition of the metals and other ore-forming elements occurs following incursion of the fluids into permeable areas of adjacent platform carbonates—for example, those of biostromal character, in back- or fore-reef settings and areas of karstification. This general model for MVT mineralisation and the modified model of episodic dewatering emphasise compaction as the principal process involved in the mobilisation of ore fluids. However, fluids capable of transporting metals may also move in sedimentary basins in response to such factors as thermal gradients and variations in topography, gravity and deformation (Hanor, 1979; Anderson and Mac-Queen, 1982). Bethke (1986) has prepared mathematical models to investigate the formation of MVT deposits in the Upper Mississippi Valley mineral district. The results suggest that the orefields formed at the basin margin during a period of regional gravity-driven groundwater flow, across the Illinois Basin, initiated by the uplift of the opposed basin margin (the Pascola Arch). According to these models, fluids displaced from the deep basin by compaction-driven flow are shown to move too slowly to avoid conductive cooling before reaching the orefields.

The problem of understanding the chemistry of Mississippi Valley ore deposits by use of simple experimental Cl–S systems, in which the phase assemblage varies in response to pH and fO_2, has been discussed by Barton (1981). Particular difficulties include the problem of transporting sulphur and the ore metals in solutions at the temperatures and pH indicated by fluid inclusion studies of MVT ores. Hence some authors prefer mixing models of ore deposition, whereby the metals and sulphur are derived from separate fluids, whereas others prefer a non-mixing model. Several authors, e.g. Barnes (1979), Giordano and Barnes (1981) and Giordano (1985), considered that some of the difficulties in understanding the chemistry of the ore deposits can be overcome if organic complexes are involved in the transport and deposition of base-metal sulphides, in addition to chloride complexes. The general association of ore deposits and hydrocarbons, including petroleum, in the MVT orefields supports this suggestion.

8.3 Preferred metallogenic model for the Pennine Orefields

8.3.1 FLUID SOURCES

Overall there is strong evidence that one component of the ore-forming system was formational water derived from shales. Oxygen isotope data (enrichment in ^{18}O), the sulphide sulphur isotopic data, δD values obtained on fluorites and the composition of fluid inclusions (which are in the Na–Ca–Cl system) are in the range of those from the main stage of Mississippi Valley mineralisation which are of undisputed oilfield brine affinity (Taylor, 1974). Shales occur within the study region in the Lower Palaeozoic basement and in late Dinantian–Namurian basins. Largely undeformed Middle Cambrian–Tremadoc shales, with high trace-element contents, occur on the Midlands Microcraton to the south-west of the study area. Although the ore-forming elements appear to have an organic association, and therefore constitute a good potential fluid and metal source, they are separated from the South Pennine Orefield by a zone of deformed and in-

durated Lower Palaeozoic shales with only average, or below average, contents of the main ore-forming elements. These rocks were deformed by late Caledonian events a substantial time (+ 100 Ma) before mineralisation took place and it is therefore highly unlikely that they could have provided significant quantities of fluids and metals.

Several lines of evidence suggest that the orefields were generated from shales in the late Dinantian–Namurian basins. These shales are enriched in the trace elements As, Cu, Mo, Ni, P_2O_5, Pb, S, V and Zn compared to average shale. Moreover, the trace-element suite, with enrichment in P_2O_5, and high U/Th ratios, suggests that the ore-forming metals may be largely associated with the organic and/or carbonate fraction of the shale from which they could have been readily mobilised (Chapter 6). They also have high contents of Rb and U, which accounts for the enrichment of the ores in radiogenic Sr and Pb isotopes and the *J*-type model ages of the ore assemblage. Moreover, calculated heat production values for the Viséan–Namurian shales of 3.6 μW m^{-3} (Chapter 6) approximate to those of the buried high heat production granites of Weardale and Wensleydale (Chapter 7). Together with the thermal blanket effect of the flat-lying, low-conductivity shales their high heat prodution would have promoted the maturation of hydrocarbons and ore fluids. The late Dinantian–Namurian shales also occur adjacent to the orefields and structural and stratigraphical analysis (Chapter 5) suggests that the evolution of the basins was such that ore fluids of appropriate temperatures could have been generated in late Westphalian–Lower Permian times.

Evolution of Edale Gulf

The model for derivation of the fluids has been examined in detail for the Edale Gulf, a sub-basin that is adjacent to the South Pennine Orefield (Fig. 5.8), by use of a thermal history derived from basin decompaction procedures constrained by vitrinite reflectance measurements (Fig. 8.1). A similar exercise could equally have been performed with use of the Widmerpool Gulf. The structural and stratigraphic evolution of the Edale Gulf is summarised in Fig. 5.6. The basin reached its maximum depth in Westphalian times, prior to Variscan deformation. Following uplift at the end of the Carboniferous, erosion of the area is likely to have continued intermittently throughout the Permian, Mesozoic and Tertiary. The post-Carboniferous sedimentary cover of the area was either thin or entirely absent.

A realistic estimate of the actual depth of a Carboniferous basin such as the Edale Gulf in late Westphalian times can be obtained by decompacting the sedimentary pile. In this method the youngest Carboniferous rocks are assigned densities appropriate to recently deposited sediments and the remainder of the sequence is restored to the thickness that it would have had if it had only been compacted by the overlying Carboniferous sediments. A simple decompaction of the Edale Gulf section in Fig. 8.1 indicates that the restored Silesian sequence may have been up to 450 m thicker before compaction. The base of the Namurian in the basin would then have been at a depth of about 2.9 km. In order to generate ore fluids that had temperatures equivalent to the highest fluid inclusion temperatures observed in the South Pennine Orefield (170°C, Chapter 3) at the base of the Namurian in the Edale Gulf, the geothermal gradient at the time of deepest burial would

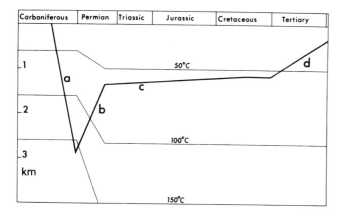

Fig. 8.1 Burial history of basal Namurian shales in Edale Gulf. Isotherms have been calculated assuming an elevated geothermal gradient during the Carbonifeous which declined to present-day values in Lower Permian times

a Total Silesian isopach for Edale Gulf was decompacted using standard lithological density–depth curves to provide more realistic estimate of actual depth of Silesian basin in area in late Westphalian times

b Silesian compaction was enhanced during Variscan deformation. Uplift and erosion reduced depth to basal Namurian shales in area

c Evidence of Permian and Mesozoic sedimentation in Pennines is largely lacking. It is assumed that area remained elevated throughout most of this interval and that Pemian and Mesozoic sediments were thin or absent

d Pennine area was further uplifted during Tertiary as a fairly uniform eastern tilt was imparted to Eastern England Shelf

For discussion see text

have to have been about 50°C/km. This figure is close to the Westphalian (pre-Whin Sill) geothermal gradient of 54°C/km that has been proposed by Suggate (1981) for the Alston Block. The development of a similar elevated gradient in the East Midlands may be partly related to the presence of active volcanism in the area throughout the Carboniferous (Chapters 2, 3 and 7) and to high heat production, low-conductivity shales in the basin (Chapter 6).

The inversion of the Pennine area during Variscan deformation and the subsequent removal of most of the Westphalian cover ensured that high temperatures could not be attained subsequently within the Namurian shale succession without the assumption of unrealistically high geothermal gradients (exceeding 100°C/km). In fact, a post-Carboniferous rise in the geothermal gradient much above the present-day value of 25–30°C/km can be eliminated because of the effect it would have had on the Permian and Mesozoic rocks that are preserved elsewhere in the East Midlands (Kirby and others, 1987). These reconstructions of basin evolution indicate that high-temperature fluids (in excess of 150°C) were likely to exist in the deeper parts of the East Midlands basins only for a short period of time at the end of the Westphalian. Cooler mineralising fluids would have been available from shallower parts of the basin or higher parts of the Namurian sequence for a longer period, but the main mineralising event was probably curtailed over the whole basin by dewatering of the Namurian shales during Variscan deformation (see below). The lower temperature limit of the South Pennine fluid inclusions (74°C, Chapter 3) may help to define an eastward boundary beyond which the basal Namurian shales would not have

been buried sufficiently deeply to generate mineralising fluids.

In a 50°C/km geothermal gradient this limit would coincide with approximately 1 km of total Silesian cover (which is equivalent to about 800 m of compacted sediments), but fluid migration from the deeper parts of the basin towards the margins, together with the development of mineralising fluids at stratigraphic levels below the basal Namurian, would complicate the simple distribution of mineralisation. If the assumptions made here are valid, however, the intensity of carbonate-hosted mineralisation can be inferred to decrease regionally eastwards across the East Midlands.

Composition of ore-forming fluids

Sheppard and Langley (1984) have argued on the basis of $\delta^{18}O$ and D data and the concentration of major elements in modern Na–Ca–Cl brines (which contain 8360–197 000 mg/l of total dissolved solids) from collieries in north east England that ore fluids could be generated rapidly by dissolution, exchange and/or filtration reactions between meteoric water, evaporites and clays. They also suggest that such fluids may have been available for flushing out to produce mineralisation on more than one occasion during the formation of the Pennine orefields. Some of the brines have high Br:Cl ratios consistent with reactions with organic matter (hydrocarbons), they are moderately acid (pH < 6.1) and they may be enriched in Pb, Zn and Ba or sulphate depending on the particular unit in the Westphalian from which they are derived. The fluids studied by Sheppard and Langley vary considerably, however, at the local (mine) scale, whereas the fluid inclusion data of Rogers (1977), Shepherd and others (1982) and Atkinson (1983) indicate the relative homogeneity of the ore fluids in the individual Pennine orefields. The REE data on fluorites also suggest fluid homogenisation throughout the source region of each orefield with infinite reservoir conditions of fluid generation. Hence a model that involves dewatering following equilibration of fluids within sedimentary basins is preferred to one that involves recharge and interaction of meteoric water with sediments.

There remains the problem of Pb dominance, the quantities of fluorite in the orefields and the association with tectonism. Oilfield brines encountered at 2.4–4 km depth (comparable with that proposed for the generation of the Pennine ore fluids) in Central Mississippi carry up to 367 mg/l Zn and 182 mg/l Pb (Carpenter and others, 1974) and the modern colliery brines from northern England described above (Sheppard and Langley, 1984) are also enriched in Zn. The enrichment of Zn relative to Pb in these fluids is consistent with the 5:1 ratio of Zn:Pb in the crust generally and in the basement rocks and Carboniferous shales of the East Midlands (Chapter 6). It is also consistent with the higher solubility of Zn than Pb in chloride brines generally (Barrett and Anderson, 1982). The high ratio of Pb:Zn throughout the Pennine orefields (Chapter 3) has been attributed by Atkinson (1983) to a fine control of pH during ore deposition whereby Zn was retained in solution following deposition of Pb. The slight enrichment of Zn in shales overlying the blocks (Chapter 6) could be interpreted as evidence of this process. However, the associated trace elements suggest that it is more likely that the Zn reflects the organic- and/or illite-rich nature of the condensed sequences overlying blocks. It is considered generally unlikely that Zn would remain in a system in which excess H^+ ions would have been readily consumed by reaction with limestone, as indicated by the presence of an outer calcite zone in the orefields (see below). Moreover, mineral textures suggest an approach to equilibrium conditions (Atkinson, 1983), whereas only relatively small amounts of sphalerite are found throughout the range of structural settings of mineralisation in the Pennines, including against permeability barriers such as volcanic horizons and shales. The evidence thus suggests a predominance of $Pb > Zn$ in the ore fluid at or near to source that is difficult to account for by simple dissolution models. Atkinson (1983) also suggested that F-bearing aluminosilicates such as biotite would have been unstable in the ore-forming fluids at 100°C, providing a potential source of F; muscovite, K feldspar, plagioclase (and biotite) could also have supplied Pb and Ba. The REE patterns in fluorites from the North Pennines could be derived by simple dissolution of feldspar (see Chapter 3; Fig. 3.4). In the South Pennine Orefield, however, the REE pattern in fluorites, which show only moderate light REE enrichment with a small negative Eu anomaly, are inconsistent with simple dissolution of either feldspar or biotite and instead appear to reflect equilibration with clay minerals.

The source of fluoride for the mineralisation is important in developing a genetic model for the Pennine orefields. Fluorine transport in aqueous solutions is controlled mainly by the solubility of fluorite (Ellis and Mahon, 1964). Surface and near-surface sources can be discounted since the low solubility of fluorite provides an equilibrium control on dissolved fluoride activity: fluoride contents rarely exceed 1.6 mg/l for air temperatures up to 26°C (Nordstrom and Jenne, 1977). Higher concentrations of fluoride (up to 10 – 15 mg/l) occur in deep brines, however, following the removal of Ca by precipitation or base exchange (Kraynov and others, 1969; Browne and Ellis, 1970). In the study region the rock types most likely to have high fluoride contents are the organic phosphatic rich black shales. In Scotland fluoride in water anomalies are associated with the suite of Devonian high heat production granites and black phosphatic shales are generally enriched in F with contents up to 7600 ppm (Fleischer and Robinson, 1963). The F may be in clays or held in detrital silicates, apatite or fluorite and released by desorption or dissolution processes. In the South Pennine Orefield the REE contents of fluorite are consistent with ore fluids derived from a shale source and in the North Pennines with dissolution of feldspar of the HHP Weardale Granite, although other evidence suggests that these fluids, their metals and F were derived mainly by shale dewatering. The alkaline-tholeiitic basaltic lavas are not considered important as a source of F or Pb since the ore assemblage (low Cu, Co and Ni) is generally inconsistent with extensive leaching of basaltic rocks.

Possible explanations for Pb-dominant assemblages in MVT deposits generally include (a) the preferential complexing of Pb by organic solvents, (b) stabilisation of Zn-bearing phases in the source rocks and (c) control of ore fluids by diagenetic phase transformations rather than by simple solubility in chloride brines.

According to Hanor (1979), one of the dominant diagenetic reactions that occurs in the argillaceous sediments of the Gulf Coast is the dehydration of montmorillonite, or mixed-layer montmorillonite – illite, clays by the coupled removal of interlayer water and the conversion of montmorillonite to illite. Sources of potassium required for the conversion of montmorillonite to illite may include subsurface water, feldspars or detrital micas. Because Ba and Pb can both substitute for K in feldspars and micas, diagenetic destruction of those mineral phases

during the production of illite provides a potential source of these elements for formation waters. The coupled dehydration and alteration of montmorillonite occurs rapidly when sediments reach certain threshold temperatures during progressive burial. One stage of water expulsion appears to occur at a temperature of approximately 90°C, and a second phase of dehydration at about 120°C. Geochemical data from Viséan – Namurian shales (Chapter 6) indicate that Pb may have been associated with the organic fraction of the sediment, whereas Zn correlated with illite content, and therefore Pb may have been more readily available for leaching into the orefluids.

In the Pennine orefields and those of Central Kentucky and Illinois – Kentucky in North America there appears to be an association between Pb dominance in the sulphide assemblage, a high fluorite content in the orefields generally and the fracture control of orebodies. We tentatively suggest that fracturing released highly evolved basinal fluids in which the activity of major species was controlled by diagenetic phase transformations (see above) rather than by solubility relations. Therefore regional tectonism may be one of the most important factors in the genesis of Pb – F-dominant MVT deposits generally.

8.3.2 FLUID TRANSPORT

The stratigraphic and structural evolution of the Pennine region, described in Chapter 5, indicates that the Pennine axis was a N – S zone of uplift at intervals from the Lower Permian Variscan tectonic event until the present. Hence it is difficult to apply a gravity model of fluid movement for the Pennine orefields since the ore fluids would be driven away from the orefields towards the eastern seaboard of Britain. Moreover, the scale of the Illinois Basin, for which a numerical model of fluid movement was developed by Bethke (1986), is markedly different from that of the Pennines, where fluids probably travelled distances of the order of only a few tens of kilometres and geochemical evidence, discussed above, is also inconsistent with a simple basin recharge – discharge model of ore deposition. This model is therefore not considered further. The early diagenetic basin dewatering model generally applied to MVT deposits cannot be applied directly to the Pennine orefields, however, since the time interval between the development of the shale-filled basins (Dinantian – Namurian) and the main episode of mineralisation (Lower Permian) was too great to represent penecontemporaneous dewatering (Chapter 3 and above). In the South Pennines the reconstruction of the structural-stratigraphical evolution of the area (Chapter 5) also indicates that mineralisation occurred when the host carbonate rocks were at depths of the order of 2 km.

In the Pennine orefields and the fluoritic orefields in the Mississippi Valley there appears to an association between Pb dominance in the sulphide assemblage, a high fluorite content in the orefields generally (Central Kentucky and Illinois – Kentucky) and the fracture control of the orebodies, porosity and permeability being related to tectonism rather than karstification or near-surface groundwater movement.

We tentatively suggest that the high content of Pb and F in Pennine mineralisation reflects a control of ore fluid compositions by diagenetic phase transformations that may, in turn, reflect overpressuring and/or the heat production and thermal blanketing of shales in the sedimentary basins that are thought to be the source of the ore fluids. In the Pennines release of these evolved basinal fluids appears to have depended on seismic pumping and it is suggested that regional tectonism may be one of the most important factors in the genesis of Pb–F-dominant MVT deposits generally. Tectonism during the Lower Permian would not only explain the disposition of the mineralisation in regional vein systems but is also consistent with time constraints whereby fluids of appropriate temperature were available in the sedimentary basins for only a short time after the Westphalian. The amount of metals in the orefields can also readily be accounted for in this way. If the volume of the Viséan–Namurian sediments in Edale Gulf is calculated, assuming a conservative thickness of 2 km, it is necessary to extract only very small percentages of the Pb, Zn, Ba and F present to provide all the recorded output from the South Pennine Orefield (2%, 0.1%, 0.03% and 0.1%, respectively; Chapter 6). It does not, therefore, seem necessary to derive fluids from more distant sources such as the North Sea Basin, as has been proposed by Ford (1976) and Ineson and Ford (1982).

8.3.3 Deposition of Mineralisation

Evidence for a mixing or non-mixing model of sulphide ore formation in the Pennine orefields requires consideration. The formational fluids pumped from the shale basins, which are described above, are considered to have been evolved fluids the chemistry of which was controlled by diagenetic phase changes. They could have provided Pb, Zn, Ba, Cl and F, but the origin of the sulphur is unclear. Transport of base metals and sulphide species in the same solution requires temperatures in excess of 200°C or exceptionally low pH values that are outside the ranges indicated for the Pennine orefields by fluid inclusion studies (see Chapter 3). In the Pennines thermal and other evidence also precludes bacterial reduction of SO_4^{2-} as a source of reduced sulphur species. The two most important potential sources of sulphide and sulphate for the Pennine mineralisation include (1) sulphate carried in the brines, with non-bacterial reduction of the SO_4^{2-} by organic matter or (2) derivation of H_2S by reduction of a separate source of preexisting sulphate such as evaporites or formational fluids in the limestone. The supply of sulphur, by reducing SO_4^{2-} in the brines, is difficult to reconcile with the quantity and distribution of baryte, since the solubility of baryte is so small in SO_4^{2-}-bearing solution (Barnes, 1979). Moreover, in the Pennine orefields, sulphide-rich zones are generally surrounded by oxidised SO_4^{2-}-rich zones consistent with a low SO_4^{2-} but Ba^{2+}-bearing brine encountering a region in its path rich in H_2S. Sulphur isotopic determinations on the sulphides and sulphates of the Pennine orefields (Chapter 3) provide additional constraints.

Hence the sulphide isotopic systems of the Pennines are typical of deep formation waters, whereas those of sulphate are consistent with a limestone formational fluid. The evidence thus favours a mixing model of ore deposition, with the deposits formed by one of the following mechanisms.

(a) The interaction of metalliferous brine with a zone enriched in H_2S in the limestones in which the formational fluid was mainly SO_4^{2-}. Thermal degradation of organic sulphur compounds in the more deeply buried parts of the carbonate sequences could release H_2S (Hunt, 1979), SO_4^{2-} fluids predominating at shallower levels. The extent to which the distribution of sulphide and

sulphate in the Pennines reflects such a gradient or is related to the ingress of H_2S from lower in the sequence as a result of tectonism associated with the mineralising events is not known.

(b) The interaction of metalliferous brines containing hydrocarbons with a sulphate formational fluid in the limestone to produce a sulphide zone surrounded by a sulphate zone. This would require extensive re-equilibration of the sulphur isotope systems of the sulphides with the basinal fluids.

A mixing model of ore deposition is also favoured by the distribution and composition of fluorite (and baryte). Although the ore solutions may be in equilibrium with carbonate during transport, deposition of sulphides from chloride complexes releases acid according to the reaction

$$MeCl_2 + H_2S = MeS + 2HCl$$
$$\text{metal chloride} \qquad \text{metal sulphide}$$

The acid could not escape from the system without dissolving carbonate; the consumption of H^+ ions produced by sulphide precipitation involving neutralisation reactions of metal chloride brines of the type

$$MeCl_2 + CaCO_3 + H_2S = MeS + CaCl_2 + H_2CO_3$$

Hence, since neutralisation reactions are important in promoting sulphide precipitation, carbonate is the most likely host rock for deposition of MVT and Pennine deposits. The release of calcium can precipitate fluorite, as in the case of geothermal waters in the Western USA (Nordstrom and Jenne, 1977), and the chemistry of fluid inclusions in fluorites from the Pennines is most readily explained by the mixing of fluids with similar NaCl but different $CaCl_2$ contents consistent with limestone formational fluids. The REE patterns and $^{87}Sr/^{86}Sr$ ratios of fluorites are also consistent with reaction between the ore-forming brines and a limestone formational fluid. Deposition of baryte and fluorite by mixing rather than as a result of a rapid decrease in solubility because of a temperature drop is also more consistent with the mineralogical, geochemical and isotopic evidence.

8.3.4 Role of Granites and Carboniferous Basic Volcanics

In the case of the Alston Block a buried radiothermal granite appears to have played an important role in localising mineralisation. The effect on the geothermal gradient may also have been slightly enhanced by the thermal blanket of Upper Silesian sediments. In this field ore formation may have involved a degree of convection through the buried granite, although the ore-forming fluids are nevertheless thought to represent brines derived initially from adjacent Carboniferous basins (Dunham and Wilson, 1985). The Weardale and Wensleydale Granites are shown to be highly evolved geochemically with heat productions of 3.7 and 3.4 $\mu W\ m^3$, respectively, although in the case of the Wensleydale intrusion this value may decline in depth as a result of fractionation (Chapter 7). Their chemistry, isotopic age dates and geophysical characteristics indicate that they are members of the post-orogenic, metalliferous suite of large volume granites emplaced throughout the Caledonides in Lower Devonian times (Brown and others, 1987). The geochemistry of the Weardale Granite is consistent with that of an evolved calc-alkaline granite; the Wensleydale intrusion, which is considerably more fractionated than that of

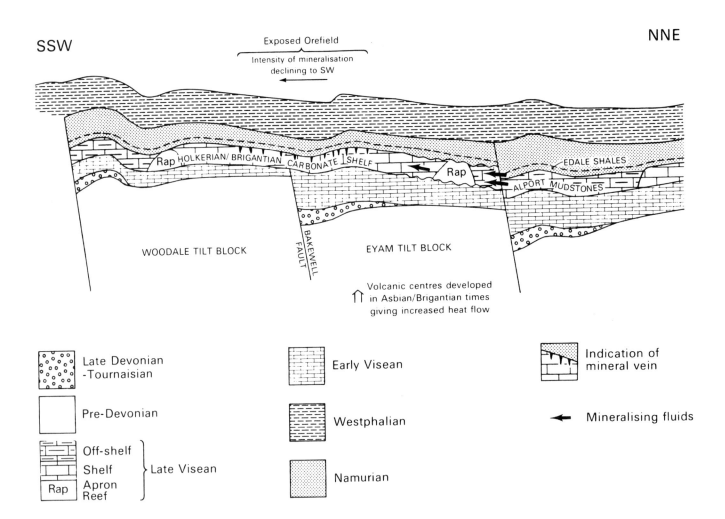

SSW

Exposed Orefield

Intensity of mineralisation
declining to SW

NNE

Rap HOLKERIAN/ BRIGANTIAN CARBONATE SHELF

Rap

EDALE SHALES

ALPORT MUDSTONES

WOODALE TILT BLOCK

BAKEWELL FAULT

EYAM TILT BLOCK

Volcanic centres developed
in Asbian/Brigantian times
giving increased heat flow

Late Devonian
-Tournaisian

Pre-Devonian

Off-shelf
Shelf
Apron
Reef
} Late Visean

Rap

Early Visean

Westphalian

Namurian

Indication of
mineral vein

Mineralising fluids

Fig. 8.2 Proposed model for Pennine-style mineralisation by seismic pumping of fluids from overpressured Viséan–Namurian shales into Dinantian carbonates using growth faults

Weardale, is similar in its chemistry to that of the evolved sub-alkaline Sn–U granites of the East Grampians of Scotland. Petrographical and geochemical evidence, such as the dispersion of the LIL elements, and the variability of the Th/U, Rb/Sr and K/Rb ratios, suggests that high-low temperature alteration and re-equilibration occurred in the Weardale Granite penecontemporaneously with its emplacement (Brown and others, 1987; O'Brien and others, 1985). Regeneration of the Weardale hydrothermal system, which occurred as a result of tectonism and the higher than normal (but declining) geothermal gradient in Lower Permian times being reflected in the overlying orefield of the Alston Block. In the case of the Wensleydale Granite, which is relatively unaltered, it seems likely that it formed only a generalised hot spot and its main role in mineralisation may have been to buoy up the Askrigg Block.

Differences in such features as the REE patterns in fluorites, the presence of REE-bearing mineral phases in veins overlying the buried granite cupolas of the Weardale intrusion, the K/Na ratios of the fluids and the different temperatures of ore deposition between the orefields can be related to the extent to which reaction occurred between buried high heat production granites and the migrating ore-forming brines.

In the case of the Carboniferous igneous rocks it has been suggested that these may have provided a source of F (Dunham, 1983). This is unlikely, however, since the ore assemblage (low Cu, Co, Ni, etc.) is generally incompatible with extensive leaching of basic rocks. The main role

of the volcanics appears to have been as permeability barriers to the ore fluids; they may also have provided a supply of bases for acid neutralisation reactions, promoting ore deposition.

Hence, in common with MVT deposits generally, there appears to be no association between ore deposition and contemporaneous igneous activity, although the presence of igneous rocks in the zone of fluid flow and fluid/rock interaction has modified the MVT signature of the Alston Block (Brown and others, 1987).

8.3.5 Overall Pennine–style Model

The model developed here for Pennine-style mineralisation involves seismic pumping of Viséan–Namurian basinal brines into limestones (Fig. 8.2). The brines were of oilfield affinity, but their chemistry, with Pb > Zn/Ba and with high content of F, was probably controlled by phase transformations during burial and overpressuring that culminated in end-Westphalian times. Hence the deposition of metal sulphides, baryte and fluorite appears to have occurred following pumping of Pb-, Ba-, F-, Cl- and (Zn-) enriched fluids containing hydrocarbons into limestones where they mixed and reacted with SO_4^{2-} and $CaCO_3$; it is not clear whether an additional source of reduced sulphur (model (*a*) above) was involved or whether sulphur was generated locally by the reduction of SO_4^{--} by hydrocarbons (model (*b*) above). In either case the role of hydrocarbons in reducing sulphate to sulphide

128

was probably critical. In the South Pennine Orefield basic volcanics, in addition to acting as physical barriers that ponded up the ore fluids, could also have supplied bases for acid neutralisation reactions promoting ore deposition. In the North Pennines HHP granites locally reacted with, and focused, metalliferous brines.

Mineralisation is thought to have occurred during Variscan tectonism in Lower Permian times with catastrophic dewatering of shale basins such as the Edale Gulf into regional ENE fracture systems generated in response to dextral transcurrent movement that reactivated NNW-trending early Carboniferous growth faults. Such a model is consistent with detailed basin compaction studies that show that fluids of appropriate composition and temperature were only available for a short period from end-Westphalian to Lower Permian times. It is also consistent with new geochemical data obtained from Viséan–Namurian basinal shales in the region. Calculations of the Time and Temperature Index (TTI) for basal Namurian shales show that most of the Pennine area lay within the oil-generative window in late Westphalian and Lower Permian times. In the deepest parts of the basin Namurian shales had already begun to generate gas during this interval (Kirby and others, 1987).

The Pennine mineralisation cannot be explained in terms of a simple basin dewatering model. It has been suggested, however, that considerable subsidence occurred in Namurian–late Westphalian times as a major crustal sag developed in the Central Pennines following decay of the thermal anomaly in the mantle (Leeder, 1982). This is reflected in the Edale Gulf by the steep downward deepening of the basin during late Namurian–late Westphalian times (Fig. 8.1). Simple MVT mineralisation could have occurred during this period, the formation of economic ore deposits such as those of the Mississippi Valley depending on factors such as the availability of open structures for mineralisation.

8.4 Irish-style mineralisation

Irish-style deposits differ from Pennine style deposits in many respects, despite their occurrence in comparable but older Dinantian host rocks.

Table 8.1

	PENNINE	IRISH
Style	Epigenetic veins	Syn-diagenetic stratiform
Host rock	Late Viséan carbonates	Tournaisian carbonates
Mineralisation	F > Pb > Ba > Fe > Zn	Fe > Zn > Pb and Ba
Salinity, wt% NaCl	18 – 25	10 – 15
Ore fluids	Homogeneous	Variable
Fluid temperature	70–150°C (higher in the NPO)	100–250°C
Pb isotopes	J-type	Normal
Igneous rocks	Not contemporaneous with mineralisation	Penecontemporaneous
Geothermal gradient	High (decreasing)	High (increasing)

8.4.1 GENERAL CHARACTERISTICS OF IRISH–STYLE MINERALISATION

The stratabound nature of the major Irish Carboniferous hosted base-metal deposits and isotopic age evidence support a synsedimentary or syndiagenetic mode of ore emplacement during the Tournaisian. Pb isotopic model age dating (Boast and others, 1981) is consistent with the stratigraphic dating of the deposits, although interpretation of the Pb isotope age dates for Navan and Silvermines is difficult. In addition to the different age and syngenetic, rather than epigenetic, style of the mineralisation, the Irish deposits differ from those of the Pennines in their Zn > Pb > Ba with little F assemblages. They also have lower salinities (in the range 10–15 Wt%) and higher temperatures, which may have exceeded 250°C, are indicated by fluid inclusion studies (Samson and Russell, 1983). Unlike the Pennines, which are characterised by relative homogeneity of fluids, temporal changes from neutral hypersaline to acid less saline solutions are indicated in the case of Irish deposits. Also, in contrast to the Pennine orefields, Pb isotope values vary from mine to mine, reflecting changes in the Caledonian basement—for example, across the Iapetus suture (H. Mills, personal communication). There is also variation at the local scale; in Navan mine, for example, changes in fluid chemistry are accompanied by systematic changes in Pb isotopic compositions (H. Mills, personal communication).

The Irish deposits are associated with tensional tectonics over reactivated deep faults of inherited Caledonian trend. Navan and Silvermines have been suggested to be located directly over the line of the Caledonian Iapetus suture (Phillips and others, 1976) and there is considerable structural and stratigraphic evidence for a direct association between faulting and mineralisation. Hence the Navan deposit has been related to the release of pulses of hydrothermal fluids through contemporaneous ENE–NE-trending tensional fractures into shallow water carbonate sedimentary/diagenetic environments from Tournaisian to Arundian times to produce a modified sedimentary-exhalative deposit (Andrew and Ashton, 1985). Although alkali basaltic volcanism in Viséan times followed shortly after formation of the Irish deposits, especially in the Limerick area to the south of the Iapetus suture (Upton, 1982), there is no direct association between volcanism and mineralisation.

8.4.2 MODELS FOR IRISH–STYLE MINERALISATION

As was discussed earlier, Irish-style deposits have been suggested by Russell (1978) and Boyce and others (1983) to result from warm, moderately saline, brines that vented onto the contemporaneous sea-floor, the ore fluids being derived from Caledonian basement by deeply excavating hydrothermal convective systems. Fig. 8.3 shows a conjectural application of this model to the South Pennine Orefield. Deep circulation of fluids is required by this model because a normal geothermal gradient of 30°C/km is assumed. Pb isotope systematics also tend to support interaction of fluids with Caledonian rocks, post-Caledonian cover formations having been considered to be relatively thin, with negative gravity anomalies mainly interpreted as buried granites.

In contrast, Brown and Williams (1985) preferred a basin dewatering model and presented geophysical evidence that deep linear Upper Palaeozoic basins, with thick sequences of sediments, as yet unproven by drilling,

SSW NNE

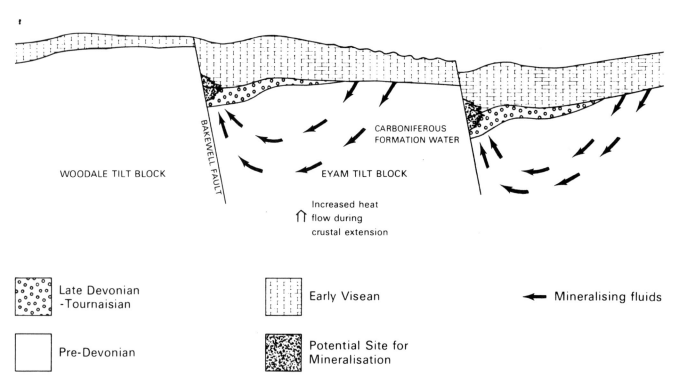

Fig. 8.3 Irish-style mineralisation model of Russell (1978) involving 'deeply excavating hydrothermal cells' applied to South Pennines

may exist in some parts of the Irish Midlands. In the light of this evidence models that argue mainly on geothermal grounds that hydrothermal fluids must have circulated through Lower Palaeozoic basement must be re-evaluated. In the case of the Navan deposit, for example, a negative gravity anomaly to the south-east of the mine,

which was previously interpreted as a granite (Murphy, 1952), has been reinterpreted as a deep sedimentary basin, part of which was intersected by the Trim No. 1 borehole, which proved at least 1700 m of Carboniferous sediments (Sheridan, 1972). If this basin were sufficiently deep, it is possible that the temperatures of the Navan

SSW NNE

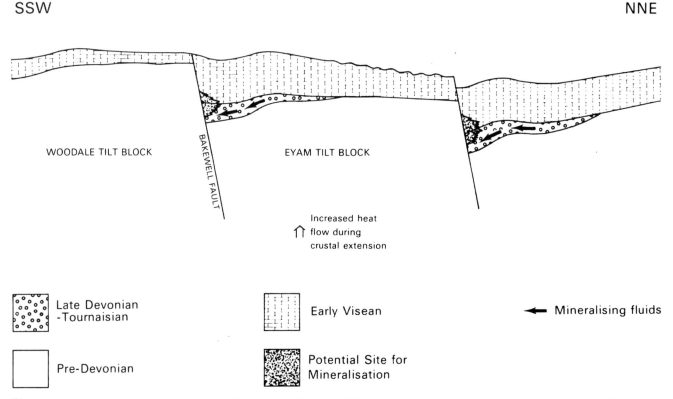

Fig. 8.4 Irish-style mineralisation model of Brown and Williams (1985) involving early Dinantian basin dewatering applied to South Pennines

mineralising fluids may have been attained within the basin infill, particularly if the geothermal gradient were steepened at the time. The Brown and Williams (1985) model is shown in Fig. 8.4 in a conjectural South Pennine Orefield setting.

The Irish deposits were formed at exhalative centres on the sea-floor following the transgression of the Tournaisian sea onto the margin of the Old Red Sandstone continent, in a tectonic regime of regional thinning and extension of the crust (Anderton and others, 1979; Leeder, 1982). According to the McKenzie model of sedimentary basin formation by lithospheric stretching (McKenzie, 1978), this early, rifting stage of basin evolution is characterised by isothermal bunching, high heat flow and fracturing at the rifted shoulders of the basin. This is followed by alkaline or tholeiitic basaltic magmatism, depending on the degree of stretching (Dewey, 1982). The probable presence of a relict zone of high heat flow during the early Dinantian, coincident with the line of the Iapetus suture across Central Ireland, has also been inferred by Leeder (1982).

The presence of steep geothermal gradients in the Irish Tournaisian would have promoted the circulation of high-temperature fluids through the coarse basal clastics within the basins and particularly near extensional fault zones (permeability may increase from 0.05 mD to 300 mD where fault zones intersect sediment piles at 1 km depth (Barnes, 1979)). The lower salinity and hence relatively high buoyancy of the fluids involved in Irish-style deposition may also have assisted fluid circulation. The ore assemblage with $Zn > Pb$ and minor Ba and F is consistent with ore fluid compositions controlled mainly by dissolution–precipitation reactions with rocks of approximately intermediate calc-alkaline bulk composition. The ore fluid and Pb isotope chemistry of the deposits could thus be explained by interaction of the fluids with Caledonian basement. Alternatively, they could reflect interaction with clastic sediments derived from basement. The chemistry of Old Red Sandstone basins, for example, clearly follows that of the Caledonian mountain hinterland from which they were derived (Plant and others, 1986). Hence water–rock interaction in such sediments could account for the Pb isotope systematics of the mineralisation. The extent of hydrocarbon anomalies associated with Irish deposits (Carter and Cazalet, 1984) also supports a predominantly basinal source of fluids.

8.4.3 PREFERRED IRISH-STYLE MODEL

Overall, a model involving intrabasinal fluid flow, as a result of the heating, rapid subsidence and compaction of immature sedimentary infills in tectonically controlled basins, accounts for many of the geochemical, stratigraphic and tectonic features of the Irish deposits. Mineralising fluids migrated from the basin towards the surface along the main basin, controlling (listric) faults during the Tournaisian, and mineralised the thin Tournaisian successions at the basin margins. Continued movement on the main growth faults caused the submarine erosion of some of the orebodies by the process of scarp retreat in mid-Dinantian times (Boyce and others, 1983). Rapid subsidence of the basins followed ore deposition with differential uplift of adjacent blocks. This is indicated by the dolomitic and chert breccias that overlie the ore deposits at Silvermines and by boulder conglomerates deposited by debris flows over the ore zone at Navan—a feature that was important in the preserva-

tion of the deposits (Andrew and Ashton, 1985).

Such a model is consistent with the extensional tectonic regime and high geothermal gradient characteristic of the Irish Tournaisian, but may not adequately explain all the features of the other known Irish-style deposits. Such ore-forming systems differ from those involved in the MVT deposits, and from geothermal fields—for example, Cheleken—that are characterised by normal geothermal gradients of 20–30°C/km (Barnes, 1979). Instead, they have affinities with such geothermal systems as those of the Red Sea deeps that are associated with zones of high heat flow and tectonism in areas of crustal thinning and extension.

The results of the present study suggest that such mineralisation may occur at depth in England, although uplift and erosion to bring the deposits to the level at which they can be discovered by conventional geochemical or geophysical prospecting methods have not generally occurred.

8.5 Structural-stratigraphic events leading to formation of carbonate-hosted deposits

The following events in the structural and stratigraphic evolution of the region are considered to be significant in the formation of Pennine- and Irish-style mineralisation.

8.5.1 LATE PROTEROZOIC–SILURIAN

Heterogeneities developed in the basement during Precambrian–Caledonian orogenesis. In Ireland, Wales, the Lake District and Western England the lineaments and structural grain of the basement have a NE–SW Caledonian trend, whereas in Eastern England NW–SE-trending basement structures have been inferred.

8.5.2 LOWER DEVONIAN

Low-density high heat production Caledonian granites such as those of the North Pennines were emplaced. There is evidence that the extent to which the granites re-equilibriated in the crust with metamorphic, formational or meteoric water was a function of their crustal setting. Hence the Weardale Granite, which was emplaced into hydrous Caledonian basement like that exposed in Wales, Ireland and the Lake District, was hydrothermally altered, whereas the Wensleydale Granite, which was emplaced into a NW–SE-trending magnetic basement ridge, was not. In Ireland the emplacement of the Leinster, Galway and Donegal granites was controlled by major Caledonian trending lineaments (Leake, 1978).

8.5.3 LATE DEVONIAN–TOURNAISIAN

Basement tilt blocks and half graben were initiated associated with listric faulting. The listric faults frequently follow the line of ancient basement lineaments and in places dextral displacement has been demonstated (Arthurton, 1984; Coller, 1984). Low-density granites appear to have played a role locally in buoying up crust that formed basement highs with only thin sedimentation on them throughout the Lower–Middle Carboniferous. In Ireland the early onset of basin formation was related to the proximity of the Variscides foredeep, and there was

131

widespread development of Waulsortian facies sedimentation.

The extension and thinning of the crust in Tournaisian times was followed in Ireland by the emplacement of minor amounts of alkaline basic volcanics. The Irish ore deposits formed at this time in response to the high geothermal gradient and high rate of crustal extension. In Ireland unconformities developed at basin margins above the Tournaisian sequence as a result of intra-basinal submarine scarps that are possibly linked to the development of later Dinantian basins.

The rapid subsidence of the basins following ore deposition, with differential uplift of adjacent blocks, as indicated by the dolomitic and chert breccias that overlie the ore deposits at Silvermines and by boulder conglomerates deposited by debris flows over the ore zone at Navan, was important in their preservation (Andrew and Ashton, 1985).

8.5.4 CHADIAN–ARUNDIAN

Carbonate sedimentation over a wide area of the Pennine basin was associated with the continuation of active growth faulting (such as that inferred in Derbyshire) controlling sedimentation with progressive onlap onto the culminations of tilt blocks. Waulsortian-type facies developed in the deeper parts of the basins.

8.5.5 HOLKERIAN–EARLY NAMURIAN

Some of the early Dinantian growth faults, such as the Bakewell and Bonsall faults, ceased to move, possibly because they were sealed by contemporaneous (Asbian–Brigantian) volcanism; the volcanic centres appear to lie along these structures. A late Dinantian carbonate platform developed as a drape structure over the early sealed faults, and thus the apron reefs, which mark the boundary between shelf and basin facies, have no direct relationship with the major basement faults. In the early Namurian sedimentation in the basins continued to be dominated by shales. Continued growth on the remaining Dinantian faults is inferred to have produced erosional scarps on the newly stabilised blocks analogous to those formed in the Arundian in Ireland. Upper Dinantian basinal facies sediments accumulated in the active half graben and overlapped onto the eroded scarps at the block margins. The intensity of mineralisation in the South Pennine Orefield also increases towards the late Dinantian basin margin that is inferred to exist beneath the Edale Gulf. In the shale basin the chemical conditions of deposition resulted in the accumulation of large amounts of U and other trace elements, such as Pb, Zn and Mo, mainly in the organic/carbonate fraction. Heat production related to the shale composition affected diagenesis, maturation and fluid migration in the basins. A heat production of $3.4\ \mu W\ m^{-3}$ has been calculated for the Viséan shale-dominated basins (Chapter 6). Late Dinantian–early Namurian growth faulting effectively juxtaposed the potential source of fluids (shale basins) with carbonate host rocks.

Similar conditions are believed to have existed in Ireland where only minor late Dinantian and early Namurian areas have been preserved from the effects of later erosion.

8.5.6 LATE NAMURIAN–WESTPHALIAN C

Early Namurian shales were buried and overpressured beneath late Namurian and Westphalian fluvial deltaic sediments. The metalliferous brines (and hydrocarbons) evolved in the overpressured basin. The centre of the Pennine basin is inferred to have reached its maximum depth in late Westphalian C times. The high heat production of the shales and their low conductivity may have contributed to the maturation of the fluids.

8.5.7 END WESTPHALIAN–EARLY PERMIAN

Catastrophic dewatering of shale basins occurred as a result of an E–W-directed stress system associated with the ending of the Variscan orogeny over NW Europe. The deformation reactivated basement faults, small-scale transcurrent movement taking place in a dextral sense in Caledonide-trending basement and in a sinistral sense in NW–SE-trending basement. Inversion of the Pennine basin took place, the maximum uplift coinciding with areas of thickest Silesian deposition. Overpressured fluids moved by seismic pumping processes into fractures propagated in brittle limestone and sandstone sequences. Mineral deposition occurred following the incursion of metalliferous brines carrying hydrocarbons into limestones.

The enhanced geothermal gradient over the buried radiothermal Weardale granite locally focused fluids so that there is a spatial association between mineral zonation and the granite subcrop.

8.5.8 UPPER PERMIAN–RECENT

Intermittent uplift and erosion along the Pennine axis continued from Permian to Recent times and exposed the North and South Pennine orefields.

8.6 Ore formation in relation to Carboniferous basin evolution

Overall, the formation of Irish-style deposits was associated with the onset of crustal rifting, extension and conditions of high heat flow in the Lower Dinantian that transformed the margins of the newly consolidated Caledonian continent into deep marine basins. In contrast, MVT deposits generally form during periods of declining or normal geothermal gradients in relation to regional basin subsidence. In terms of the McKenzie model of basin formation (McKenzie, 1978) Irish-style deposits are thus suggested to characterise the waxing phase of the cycle as hot asthenosphere rises beneath a thinned, rifted continental crust, whereas MVT deposits characterise the waning phase, which is associated with subsidence (sagging) and isostatic adjustment of the crust, with uplift of basin margins, following the decline of the thermal anomaly in the mantle. These conditions have been suggested to have prevailed in the study region in the late Namurian (Leeder, 1982). The presence of alkaline-tholeiitic magmas located along some of the basin margins was also related to this regime of crustal extension. In the case of the Pennines, end Westphalian–lower Permian mineralisation associated with Variscan tectonism produced a modified tectonically-driven type of MVT ore deposit in which Pb and F are dominant. Ore fluids are suggested to have been evolved diagenetic fluids from basins overpressured during the extensive Westphalian sedimentation. The geothermal gradient is also thought to have been moderately elevated. Different sources of evidence suggest that the main episode of mineralisation was in end Westphalian – early Permian times. Structural models of Pennine basin evolution preclude major episodes of mineralisation related to basin dewatering after the Variscan deformation, although isotopic and other evidence has been used to suggest that mineralisation occurred episodically until at least Jurassic times (Ineson and Mitchell, 1972; Walters, 1981). These events probably represent minor lower-temperature mineralisation or redistribution/re-equilibration of the ore/gangue mineral assemblage.

9 EXPLORATION CRITERIA

T. B. Colman, J. D. Cornwell, D. G. Jones, J. A. Plant, K. Smith and A. S. D. Walker

As was discussed in Chapters 4 and 6, direct geophysical and geochemical methods are unlikely to be of value for the location of buried Pennine- or Irish-style mineralisation, although lithogeochemical criteria may increase the target size for drilling operations. Studies using remotely sensed data on the exposed Pennine orefields by Hunting Geology and Geophysics were of only limited value for identifying further exploration targets and it is unlikely that methods based on remotely sensed data would help to locate buried ore deposits in the East Midlands since most of the structures that controls ore deposition have not affected the basal Permian (or later) cover rocks. As was suggested in Chapter 8, however, a number of stratigraphic and structural events can be shown to have played a role in the formation of Pennine- and Irish-style ore deposits and several indirect geophysical and geological exploration criteria can be derived to identify such features.

Basin dewatering models suggest a close analogy between the formation of MVT deposits and the origin of hydrocarbons in sedimentary basins. As a result, many of the exploration criteria for buried MVT deposits are directly analogous to those used in the search for hydrocarbons. Hydrocarbons are also difficult to detect by direct methods at the surface and, in an unknown area, the main exploration effort must be directed towards the study of the stratigraphic and structural evolution of the basin, with particular emphasis on the location of suitable source rocks and host rocks or reservoirs.

Several regional exploration criteria can be derived from the metallogenic models developed in Chapter 8. The principal criteria for the location of buried Pennine-style mineralisation in the study region include (1) the presence of thick Viséan and Namurian sedimentary basin sequences (the main source of mineralising fluids), (2) platform carbonates (the main host rock) and their relationship with buried granites, (3) structural information and the migration of ore fluids, (4) basement lineaments, (5) geochemistry and (6) potential sources of high heat flow.

9.1 Viséan and Namurian basins

In the East Midlands the most important stratigraphical and structural criterion is the regional disposition of late Dinantian platform carbonates (the main host rock) and late Dinantian–early Namurian shale basins (the main source of mineralising fluids). The margins of the late Dinantian shale basins at outcrop can be mapped as a facies change between the platform carbonates and the basinal shales, where the transition is commonly marked by patchily preserved apron reefs. At depth borehole information can be used together with gravity and seismic data to map the distribution of the basins, which may be identified as Bouguer anomaly lows because of the significant density contrast between shales and the underlying, or laterally equivalent, Viséan limestones. The detailed internal structure of the basins can be interpreted from seismic reflection surveys. In some circumstances, when the known thickness of Namurian rocks is insufficient to explain the amplitude of the observed gravity anomaly, the presence of accumulations of low-density sediments in the underlying Viséan sequence may be suspected.

Without additional evidence, however, it is difficult to distinguish between the gravity effect of shale sequences above the Viséan carbonates from those due to arenaceous sequences below it. Plate 12 shows the main Dinantian–Namurian basin developments superimposed on the stripped Bouguer anomaly map; the presence of a coincidental gravity low provides evidence that the known thickness of Namurian sediments is insufficient to explain the amplitude of the anomaly and therefore suggests the possible existence of Viséan basins. The potential of the basins for providing a source of metalliferous brines must be considered in relation to their subsequent burial history.

9.2 Platform carbonates and their relationship to buried granites

Although the top of the Viséan carbonates, which form the dominant host rock in the study area, can be readily distinguished using gravity and seismic data, it is difficult to distinguish them from the underlying Lower Palaeozoic, or even Precambrian, basement. The density of limestone is slightly lower or overlaps with those of most basement rocks and the gravity response is difficult to evaluate unless younger rocks are absent. There is similar uncertainty with the interpretation of seismic reflection data (Chapter 5), although recent seismic refraction studies of the Askrigg Block (Banks and Gurbuz, 1984) suggest that Carboniferous limestones have slightly lower velocities and higher Poisson's ratios than the adjoining Lower Palaeozoic sediments.

On a more local scale the interpretation of the regional gravity data provides a technique for rapidly evaluating areas in terms of extensions of the limestone surface beneath thinner developments of Namurian, or younger, cover rocks. Local Bouguer anomaly highs around the margins of the Viséan limestone outcrop in the Derbyshire 'Dome' coincide in places with concealed anticlinal structures or structural highs in the limestone surface: for example, the Wirksworth anticline and Crich inlier. This suggests that gravity data could be used to predict the existence of similar, previously unrecorded, structures elsewhere.

The southern margin of the Derbyshire 'Dome' is characterised by the appearance of Bouguer anomaly ridges trending southwards into an area of Permo-Triassic cover rocks. One coincides with a structural high in Viséan rocks and forms the western margin of the thick Namurian sediments preserved in the Widmerpool Gulf, and the suggestion that the other is due to a concealed ridge of limestone is supported by the existence of a small inlier of limestone at Snelston that also hosts minor copper–lead mineralisation (Chapter 3).

Platform carbonates are invariably developed in those parts of the Pennines which are situated above Caledonian granites. This relationship may be used to indicate the presence of a suitable carbonate host rock, but cannot be used to infer the presence of mineralisation. The position of the Wensleydale Granite beneath the Askrigg Block, for example, does not directly indicate which parts of the overlying carbonate platform are mineralised. Similarly, the absence of a granite beneath a structural block, as is the case in Derbyshire, does not preclude the development of mineralisation. The major contribution of the Caledonian granites is probably their structural role in controlling parts of the pre-Carboniferous basement that nucleated as stable blocks. Hence where the main Carboniferous basin faults appear to have originated as re-

Several intrusions of low-density (granitic) rocks have been postulated to explain the existence of Bouguer anomaly lows (Chapter 4; Plates 10 and 12) and these can be grouped as follows:

(a) discordant (NE–SW elongated) granites – Market Weighton and Hornsea

(b) concordant (NW–SE elongated) granites – Newark and Rotherham

(c) granites in the Wash area – these show some similarities with both (a) and (b).

This grouping probably reflects differences in ages and/or composition of the intrusions, but all of them occur in areas of somewhat elevated basement. Large-volume, low-density granites, characterised by distinct Bouguer anomaly lows, are also likely to contain a higher proportion of radiogenic minerals. The significance of buried granites in this context is discussed below.

Within the study region the Market Weighton Block shows the greatest resemblance, at least geophysically, to those of the North Pennines, although there is little geological evidence of Carboniferous shelf sediments. The concealed Newark and Rotherham granites have had a different effect on block development. The occurrence of the low-density mass along the shallow southern margin of the Gainsborough Trough could indicate the presence of a granite or tilted block (cf. Smith and others, 1985).

In the North Pennines the faulted margins of the Askrigg and Alston Blocks are well defined in places by gravity anomalies due to the juxtaposition of high-density Lower Palaeozoic sediments and thick basinal accumulations of lower-density Namurian and Westphalian rocks. The anomalies take the form of linear gradient zones or linear highs. Similar features in the study region can possibly be interpreted in the same way—for example, the Askern–Spital lineament (Plate 10). In the East Midlands basement highs may be coincident with magnetic anomalies— for example, at Nocton.

9.3 Structural information and migration of ore fluids

Structural data, such as the measurement of fault, vein and joint trends, together with vein widths and lengths in the exposed Pennine orefields, can be used to predict the likely shape, size and trend of orebodies in the subsurface. Regional analysis suggests that many of the mineralised structures originated as a result of pervasive east–west-directed stress during Variscan deformation in end-Carboniferous to early Permian times.

Fluid migration directions from the shale-filled basins were also partly controlled by the structure of the adjacent carbonate platforms. At the time of mineralisation the Derbyshire carbonate platform was a buried anticlinal drape, which culminated structurally above the concealed sites of the early Dinantian Bakewell and Bonsall growth faults (Fig 5.6g). Mineralising fluids migrated towards these structural culminations beneath an impermeable cover of Namurian shales. Under deep burial minor mineralisation might also be expected to occur by the downward dewatering of thin Namurian block successions into the underlying fractured platform carbonates.

Geological and geophysical data indicate that the late Dinantian–early Namurian shale basins are asymmetric graben or half graben (Chapter 5). In the modified dewatering model proposed (Chapter 8) the more intense mineralisation is predicted to occur in the carbonate plat-form that is adjacent to the thickest part of the shale sequence, alongside the main bounding fault of the basin; the intensity of mineralisation in the platform carbonates is expected to decrease with distance from the shale basins. This is in contrast to predictions based on the Jackson and Beales (1967) model for basin dewatering whereby the main mineralisation would be predicted to occur up-dip of the deepest part of the basin, beyond its outer margin.

The use of derivatives of the potential field data (the second derivative of the gravity field (e.g. Plates 4b, 8b), tends to emphasise the shorter-wavelength anomalies that are more likely to be due to the near-surface extensions of faults that affect Carboniferous rocks.

9.4 Basement lineaments

In the study region many basement lineaments can be recognised from the regional geophysical evidence and these have been described in Chapter 4 (see Fig 4.8; Plates 8, 10, 12). Some appear to coincide with known geological features, including major faults affecting the Carboniferous cover sequence—for example, the southern margin of the Askern–Spital high, suggesting that reactivation of basement faults may have played a role in controlling the disposition of Carboniferous basins. Magnetic anomalies interpreted as being due to Carboniferous igneous activity appear to lie along the western end of the Grantham lineament. The significance of some major lineaments, including the eastern section of the Grantham lineament, is not understood. These are identified in Fig 4.8 on the basis of the existence of elongated magnetic anomalies that are interpreted as magnetic intrusions.

9.5 Geochemistry

The geochemistry of the potential source rocks can be related to the known geochemical characteristics of the mineralisation. The data outlined in Chapters 6 and 7, taken in conjunction with other evidence, support the derivation of fluids and ore-forming elements from Viséan–Namurian basins, such as the Edale and Widmerpool Gulfs.

9.6 Heat flow

In the preliminary studies of present-day heat flow in Britain (Richardson and Oxburgh, 1978) a belt of high values (exceeding $60\,mW\,m^-$) crosses the study area with a NW–SE trend. The coincidence of this belt elsewhere with indications of metalliferous mineralisation suggested that the two might be linked, possibly through the presence of high heat production granites (Brown and others, 1980). More recent data from the study area (Downing and Gray, 1986) confirm that the areas of high heat flow are more restricted in size and are confined mainly to the belt around the Eakring anticline. Although the Eakring heat flow high has been ascribed to ascending groundwater in Carboniferous limestones, it occurs close to, and runs parallel with, the postulated Newark granite. The relationship of the heat flow to the Namurian basins and to the regional gravity and magnetic data is shown in Plate 9.

Additional constraints on the thermal evolution of sedimentary basins, which can be identified by use of geophysical and other evidence, include the presence of concealed high heat production granites, contemporaneous volcanic provinces and major fracture systems. Possible relationships between these features, as represented by evidence on the stripped Bouguer anomaly map, the pseudo-gravity map and the distribution of geophysical lineaments, with the main Namurian basin development are illustrated by the combined images in Plates 7 and 12. Vitrinite reflectance data and other indications of thermal maturity, including the concentration of radioelements in the basinal shales, provide further constraints.

9.7 Discussion

Interpretations of the structural and stratigraphic history of the area can be used together with thermal maturity indicators, such as vitrinite reflectance data, to reconstruct the geothermal history of the shale basins and carbonate host rocks. The application of sedimentary decompaction models will achieve a more realistic estimate of basin depth variation with time. The geothermal histories of various parts of the basin can be compared with fluid inclusion data from the exposed orefields, which give some indication of the expected range of temperatures in the mineralising fluids. Calculations of the amount of fluid produced by the compaction of the source rocks can be linked with observations on the known metal contents of formational fluids to estimate the volume of potential ore deposits within a basin. The application of this method in basins other than the Pennine basin, however, has shown that it is not a major constraint on the development of ore deposits. Fluids and metals are generally available in sufficient abundance to produce mineralisation as a result of basin dewatering if other conditions, such as temperature and host rock permeability, are fulfilled.

In the case of Irish-style deposits, as in the Pennines, the regional disposition of Carboniferous basins and blocks is one of the major controlling factors in the location of deposits, although listric fault zones, and the areas immediately adjacent to them, have more direct exploration significance than in the case of Pennine-style mineralisation. The Tynagh Fault zone in Ireland is strongly mineralised, for example. In the Pennines ore deposition appears to be directly related to reaction between metalliferous brines and formational fluids in carbonates in extensive vein systems. In Ireland, on the other hand, deposition followed exhalation of fluids onto the sea-floor or into unconsolidated sea-floor sediments and resulted from mixing of ore fluids with sea water or near-surface brines so that carbonates may have played a less direct role. In the Irish-style deposits reactivation of Caledonian basement faults and features such as the Iapetus suture, as listric Carboniferous faults, suggest that the identification of basement lineaments may be more directly relevant to exploration than in the Pennines.

Although the main emphasis in this report is on carbonate-hosted mineralisation, it is interesting to note that the geophysical data provide evidence for environments that might contain other styles of mineralisation. These include intrusions of acid, intermediate and basic composition and major faults. Additional criteria are necessary in order to eliminate some of the potential targets, but the interpretation of combined geophysical data sets can be useful; for example, the disparity in the sizes of the magnetic anomaly and high-density source at Nocton could indicate serpentinisation of a basic intrusion.

9.8 Prospectivity

It is apparent from the regional exploration criteria that sets and subsets of exploration criteria can be readily combined to indicate prospectivity for Pennine- and Irish-style ore deposits. In the case of Irish-style mineralisation the following criteria are probably the most powerful subset – Dinantian carbonate basins along Caledonoid (and probably also NW – SE) trending lineaments in regions of alkaline-tholeiitic basic volcanism. The resulting prospectivity map is shown in Plate 12d. The areas that are considered prospective for these ore deposits in the study region include the Castleton and Alport areas and, in England generally, the Craven Basin (particularly the southern margin of the Askrigg Block) and the northern margin of the Northumberland Basin. These areas not only contain rocks of comparable age and facies to the Irish-style mineralisation but are also underlain by basement faults of Caledonide, rather than NW – SE, trend (although the significance of lineament direction is not known). In the case of Pennine-style mineralisation the most important features include basement highs separated by listric faults from shale basins that became overpressured and inverted. The faults were later reactivated during uplift and compression and formed pathways for fluids. Buried radiothermal granites are also important. Prospectivity maps for Pennine-style mineralisation are shown in Plates 12a – c. They are based on a combination of gravity lows, considered indicative of buried granites; basement lineaments identified by gravity and aeromagnetic data, for which there is additional seismic/borehole evidence of listric movement in Carboniferous times; and Asbian/Brigantian carbonate shelf facies limestones juxtaposed by such faults against deep Viséan – Namurian shale basins. The prospectivity map correctly identifies the eastern margin of the South Pennine Orefield and the value of combinations of criteria can be shown by the identification of buried granites in The Wash region that lack appropriate Carboniferous cover and hence are not considered prospective for Irish-style or Pennine-style carbonate-hosted deposits.

10 CONCLUSIONS AND RECOMMENDATIONS

Overall, it is clear that the formation of Irish-style deposits was associated with the onset of crustal rifting, extension and conditions of high heat flow in the Lower Dinantian that transformed the margins of the newly consolidated Caledonian continent into deep marine basins. New evidence suggests that rocks of comparable facies, affected by contemporaneous listric faulting, in a similar structural regime, occur at depth in the study region. In contrast, MVT deposits generally form during declining or normal geothermal gradients in relation to regional basin subsidence. In terms of the McKenzie model of basin formation (McKenzie, 1978) Irish-style deposits are thus suggested to characterise the waxing phase of the cycle as hot asthenosphere rises beneath a thinned continental crust, whereas MVT deposits characterise the waning phase, which is associated with subsidence and isostatic adjustment of the crust following the decline of the thermal anomaly in the mantle. In the Pennines and East Midlands of England such deposits could have formed during Namurian–Westphalian times, when a major crustal sag is inferred to have developed in the Central Pennines, but the formation of large replacement ore deposits comparable with those of the Mississippi Valley would depend on such factors as the availability of open structures for mineralisation and hence the depth of burial of likely host rocks.

In the case of the Pennine orefields mineralisation is associated with Lower Permian (Variscan) tectonism, which produced a modified, tectonically-driven type of MVT ore deposit, in which Pb and F are dominant. Different sources of evidence suggest that the main episode of mineralisation was in end-Westphalian to early Permian times and there is a strong genetic link between the formation of the Pennine orefields and the oilfields of the region. Thermal models of Pennine basin evolution preclude major episodes of mineralisation, related to basin dewatering, after the Variscan deformation, and do not therefore support isotopic and other evidence that indicates episodic mineralisation until at least Jurassic times (e.g. Ineson and Mitchell, 1972; Ineson and Ford, 1982). These events probably represent minor lower-temperature mineralisation or redistribution/re-equilibration of the ore and gangue mineral assemblage.

The identification of exploration criteria based on metallogenic models provides a valuable new method of exploration and resource evaluation for Irish-style and Pennine-style mineralisation at the regional to district scale. The approach is particularly suitable for use with image analysis systems that can be used for interactive studies in real time. Such systems provide a powerful means of modelling ore deposits by screening and searching multidisiplinary data sets and also enable sets and subsets of exploration criteria developed from the models to be employed in regional analysis. Further research and development is required, however. In particular, the methods developed should be tested and refined in other areas, particularly areas of greater structural complexity. Other types of systematic data sets, such as high-resolution remotely sensed data from Seasat, SPOT, thermal imagery and systematic regional geochemistry, should be evaluated for modelling, exploration and resource evaluation for metalliferous minerals.

REFERENCES

AITKENHEAD, N. 1977. The Institute of Geological Sciences Borehole at Duffield, Derbyshire. *Bulletin of the Geological Survey of Great Britain* no. 59, 1–38.

— and CHISHOLM, J. I. 1982. A standard nomenclature for the Dinantian formations of the Peak District of Derbyshire and Staffordshire. *Report of the Institute of Geological Sciences* no. 82/8, 17p.

— — and STEVENSON, I. P. 1985. Geology of the country around Buxton, Leek and Bakewell. *Memoir of the British Geological Survey*, Sheet 111, 168p.

ALLEN, J. R. L. 1960. The Mam Tor Sandstones: a 'turbidite' facies of the Namurian deltas of Derbyshire. *Journal of Sedimentary Petrology*, **30**, 193–208.

— 1968. The Cambrian and Ordovician systems. In *The geology of the East Midlands* SYLVESTER-BRADLEY, P. C. and FORD, T. D. eds. (Leicester: Leicester University Press), 20–40.

ALLISON, I. and RUSSELL, M. J. 1985. Anhydrite discovered in the Fucoid Beds of northwest Scotland. *Journal of Sedimentary Petrology*, **55**, 917–918.

ALLSOP, J. M. 1983. Geophysical appraisal of two gravity minima in the Wash district. *Report of the Institute of Geological Sciences* no. 83/1, 28–31.

— 1984. Geophysical appraisal of a Carboniferous basin in northeast Norfolk, England. *Proceedings of the Geologists' Association*, **95**, 175–180.

— and ARTHUR, M. J. 1983. A possible extension of the South Leicestershire Diorite complex. *Report of the Institute of Geological Sciences* no. 83/10, 25–30.

ANDERSON, G. M. and MACQUEEN, R. W. 1982. Ore deposits models – 6. Mississippi Valley-type lead–zinc deposits. *Geoscience Canada*, **9**, 108–117.

ANDERSSON, A., DAHLMAN, B. and GEE, D. G. 1982. Kerogen and uranium resources in the Cambrian alum shales of the Billingen–Falbygden and Narke areas, Sweden. *Geologiska Föreningens i Stockholm Förhandlingar*, **104**, 197–209.

ANDERTON, R., BRIDGES, P. H., LEEDER, M. R. and SELLWOOD, B. W. 1979. *A dynamic stratigraphy of the British Isles* (London: George Allen & Unwin), 312p.

ANDREW, C. J. and ASHTON, J. H. 1985. Regional setting, geology and metal distribution patterns of Navan orebody, Ireland. *Transactions of the Institution of Mining and Metallurgy* (Section B), **94**, B66–93.

ARMANDS, G. 1972. Geochemical studies of uranium, molybdenum and vanadium in a Swedish alum shale. *Stockholm University Contributions in Geology*, **27**, 1–148.

ARTHAUD, F. and MATTE, P. 1977. Late Palaeozoic strike-slip faulting in southern Europe and northern Africa: result of right-lateral shear zone between the Appalachians and the Urals. *Bulletin of the Geological Society of America*, **88**, 1305–1320.

ARTHURTON, R. S. 1983. The Skipton Rock Fault—an Hercynian wrench fault associated with the Skipton Anticline, NW England. *Geological Journal*, **18**, 105–114.

— 1984. The Ribblesdale fold belt, NW England—a Dinantian–early Namurian dextral shear zone. In *Variscan tectonics of the North Atlantic region* HUTTON, D. H. W. and SANDERSON, D. J. eds. *Special Publication of the Geological Society of London*, no. 14, 131–138.

ATKINSON, P. A fluid inclusion and geochemical investigation of the fluorite deposits of the Southern Pennine Orefield. Unpublished PhD thesis, University of Leicester.

— Moore, J. McM. and EVANS, A. M. 1982. The Pennine orefields of England with special reference to recent structural and fluid inclusion investigations. *Bulletin du Bureau de Recherches Géologiques et Minières, section II*, no. 2, 149–156.

BADHAM, J. P. N. 1982. Strike-slip orogens—an explanation for the Hercynides. *Journal of the Geological Society of London*, **139**, 493–504.

BALCON, J. 1981. Quelques idées sur les mineralisation plombo-zincifères dans les formations carbonatées en Belgique [Some thoughts on the lead–zinc mineralisation in the carbonate formations in Belgium]. *Bulletin de la Société Belge de Géologie*, **90**, no. 1, 9–61.

BALL, T. K., NICHOLSON, R. A. and PEACHEY, D. 1985. Gas geochemistry as an aid to detection of buried mineral deposits. *Transactions of the Institution of Mining and Metallurgy* (Section B), **94**, B181–188.

BAMFORD, D., FABER, S., JACOB, B., KAMINSKI, W., NUNN, K., PRODEHL, C., FUCHS, K., KING, R. and WILLMORE, P. 1976. A lithospheric seismic profile in Britain – 1. Preliminary results. *Geophysical Journal of the Royal Astronomical Society*, **44**, 145–160.

— NUNN, K., PRODEHL, C. and JACOB, B. 1977. LISPB-III Upper crustal structure of Northern Britain. *Journal of the Geological Society of London*, **133**, 481–488.

— — — — 1978. LISPB-IV Crustal structure of Northern Britain. *Geophysical Journal of the Royal Astronomical Society*, **54**, 43–60.

BANKS, R. J. and GURBUZ, C. 1984. An investigation of the crustal structure of northwest England using quarry blasts as seismic sources. *Proceedings of the Yorkshire Geological Society*, **45**, 19–25.

BARANOV. V. 1957. A new method for interpretation of aeromagnetic maps: pseudo-gravimetric anomalies. *Geophysics*, **22**, 359–383.

BARNES, H. L. ed. 1979. *Geochemistry of hydrothermal ore deposits, 2nd edn* (New York: John Wiley & Sons), 798p.

BARRETT, T. J. and ANDERSON, G. M. 1982. The solubility of sphalerite and galena in NaCl brines. *Economic Geology*, **77**, 1923–1933.

BARTON, P. B. 1981. Physical-chemical conditions of ore deposition. In *Chemistry and geochemistry of solutions at high temperatures and pressures* RICKARD, D. T. and WICKMAN, F. E. eds. (Oxford: Pergamon Press), 509–526.

— 1986. US activities in mineral deposit modelling. In *Mineral deposit modelling. IUGS commission on storage, automatic processing and retrieval of geological data. Report of meeting, Paris, 1984* VOKES, F. M. (Rapporteur), 3–10.

BATESON, J. H., EVANS, A. D. and JOHNSON, C. C. 1984. Investigation of magnetic anomalies and potentially mineralized structures in Whin Sill, Northumberland, England. *Transactions of the Institution of Mining and Metallurgy* (Section B), **93**, B71–77.

— and JOHNSON, C. C. 1984. A regional geochemical soil investigation of the Carboniferous Limestone areas south of Kendal (south Cumbria and north Lancashire). *Mineral Reconnaissance Programme Report, British Geological Survey*, no. 71. 44p.

BEALES, F. W. and JACKSON, S. A. 1966. Precipitation of lead–zinc ores in carbonate reservoirs as illustrated by Pine Point orefield, Canada. *Transactions of the Institution of Mining and Metallurgy* (Section B), **75**, B278–285.

BERRIDGE, N. G. 1982. Petrography of the Pre-Carboniferous rocks of the Beckermonds Scar borehole in the context of the magnetic anomaly at the site. *Proceedings of the Yorkshire Geological Society*, **44**, 89–98.

BESLY, B. M. and TURNER, P. 1983. Origin of red beds in a moist tropical climate (Etruria Formation, Upper Carboniferous, UK). In *Residual deposits: surface related weathering, processes and materials* WILSON, R. C. L. ed. *Special Publication of the Geological Society of London,* no. 11, 131–147.

BETHKE, C. M. 1986. Hydrologic constraints on the genesis of the Upper Mississippi Valley mineral district from Illinois Basin brines. *Economic Geology*, **81**, 233–249.

BHAN, S. K. and HEDGE, V. S. 1985. Targeting areas for mineral exploration—a case study from Orissa, India. *International Journal of Remote Sensing*, **6**, 473–480.

BHATTACHARYYA, B. K. 1972. Design of spatial filters and their application to high resolution aeromagnetic data. *Geophysics*, **37**, 68–91.

BJØRLYKKE, K. 1974. Geochemical and mineralogical influence of Ordovician island arcs on epicontinental clastic sedimentation. A study of Lower Palaeozoic sedimentation in the Oslo Region, Norway. *Sedimentology*, **21**, 251–272.

— and GRIFFIN, W. L. 1973. Barium feldspars in Ordovician sediments, Oslo Region, Norway. *Journal of Sedimentary Petrology*, **43**, 461–465.

BLAIR, D. G.1975. Structural styles in North Sea oil and gas fields. In *Petroleum and the Continental Shelf of North-West Europe. Volume 1. Geology* WOODLAND, A. W. ed. (Barking, Essex: Applied Science Publishers), 327–338.

BLUCK, B. J. 1978. Geology of a continental margin 1: the Ballantrae Complex. In *Crustal evolution in north-western Britain and adjacent regions* BOWES, D. R. and LEAKE, B. E. eds. *Geological Journal Special Issue* no. 10, (Liverpool: Seel House Press), 151–162.

BOAST, A. M., SWAINBANK, I. G., COLEMAN, M. L. and HALLS, C. 1981. Lead isotope variation in the Tynagh, Silvermines and Navan base-metal deposits, Ireland. *Transactions of the Institution of Mining and Metallurgy* (Section B), **90**, B115–119.

BONHAM, L. C. 1980. Migration of hydrocarbons in compacting basins. In *Problems of petroleum migration* ROBERTS, W. H. and CORDELL, R. J. eds. *American Association of Petroleum Geologists, Studies in Geology* no. 10, Tulsa, Oklahoma, 69–88.

BOTT, M. H. P. 1961. Geological interpretation of magnetic anomalies over the Askrigg Block. *Quarterly Journal of the Geological Society of London*, **117**, 481–493.

— 1962. A simple criterion for interpreting negative gravity anomalies. *Geophysics*, **27**, 376–381.

— 1967. Geophysical investigations of the northern Pennine basement rocks. *Proceedings of the Yorkshire Geological Society*, **36**, 39–168.

— and MASSON-SMITH, D. 1957a. The geological interpretation of a gravity survey of the Alston Block

and the Durham coalfield. *Quarterly Journal of the Geological Society of London*, **113**, 93–117.

— — 1957b. Interpretation of a vertical field magnetic survey in north-east England. *Quarterly Journal of the Geological Society of London*, **113**, 119–136.

— ROBINSON, J. and KOHNSTAMM, M. A. 1978. Granite beneath Market Weighton, east Yorkshire. *Journal of the Geological Society of London*, **135**, 535–543.

— SWINBURN, P. M. and LONG, R. E. 1984. Deep structure and origin of the Northumberland and Stainmore Troughs. *Proceedings of the Yorkshire Geological Society*, **44**, 479–495.

BOWIE, S. H. U., DAWSON, J., GALLAGHER, M. J. and OSTLE, D. 1966. Potassium-rich sediments in the Cambrian of northwest Scotland. *Transactions of the Institution of Mining and Metallurgy* (Section B), **75**, B125–145.

— PARKER, A. and RAYNOR, E. J. 1979. Uranium reconnaissance survey of British Lower Palaeozoic shales: trace elements related to clay mineralogy. *Transactions of the Institution of Mining and Metallurgy* (Section B), **88**, B61–64.

BOYCE, A. J., ANDERTON, R. and RUSSELL, M. J. 1983. Rapid subsidence and early Carboniferous base-metal mineralization in Ireland. *Transactions of the Institution of Mining and Metallurgy* (Section B), **92**, B55–66.

BRADLEY, D. C. 1982. Subsidence in late Palaeozoic basins in the Northern Appalachians. *Tectonics*, **1**, 107–123.

BREWER, J. A. and SMYTHE, D. K. 1984. MOIST and the continuity of crustal reflector geometry along the Caledonian–Appalachian orogen. *Journal of the Geological Society of London*, **141**, 105–120.

BRITISH GEOLOGICAL SURVEY. 1984. Settle. 1:50 000 geological map series. Sheet No. 60.

BROMEHEAD, C. E. N., EDWARDS, W., WRAY, D. A. and STEPHENS, J. V. 1933. The Geology of the country around Holmfirth and Glossop. *Memoir of the Geological Survey of England and Wales*. Sheet 86, 209p.

BROOME, J. and TESKEY, D. 1985. Presentation of magnetic anomaly map data by stereo projection of magnetic shadowgrams. *Canadian Journal of Earth Sciences*, **22**, 311–314.

BROWN, A. G. ed. 1979. *Prospecting in areas of glaciated terrain, 1979.* Excursion Handbook (Dublin: Irish Association for Economic Geology), 92p.

BROWN, C. and WILLIAMS, B. 1985. A gravity and magnetic interpretation of the structure of the Irish Midlands and its relation to ore genesis. *Journal of the Geological Society of London*, **142**, 1059–1075.

BROWN, G. C., CASSIDY, J., OXBURGH, E. R., PLANT, J., SABINE, P. A. and WATSON, J. V. 1980. Basement heat flow and metalliferous mineralisation in England and Wales. *Nature, London*, **288**, 657–659.

BROWN, G. C., FRANCIS, E. H., KENNAN, P. and STILLMAN, C. J. 1985. Caledonian igneous rocks of Britain and Ireland. In *The nature and timing of orogenic activity in the Caledonian rocks of the British Isles* HARRIS, A. L. ed. *Memoir of the Geological Society of London* no. 9, 1–15.

— IXER, R. A., PLANT, J. A. and WEBB, P. C. 1987. Geochemistry of granites beneath the North Pennines and their role in mineralization. *Transactions of the*

Institution of Mining and Metallurgy (Section B), **96**, B65–76.

BROWN, M. J. and OGILVY, R. D. 1982. Geophysical and geochemical investigations over the Long Rake, Haddon Fields, Derbyshire. *Mineral Reconnaissance Programme Report, Institute of Geological Sciences*, no. 56, 48p.

BROWNE, P. R. L. and ELLIS, A. J. 1970. The Ohaki–Broadlands hydrothermal area, New Zealand: mineralogy and related geochemistry. *American Journal of Science*, **269**, 97–131.

BULLARD, E. C. and NIBLETT, E. R. 1951. Terrestrial heat flow in England. *Royal Astronomical Society, Geophysical Supplement*, **6**, 222–238.

BURGESS, I. C. 1982. The stratigraphical distribution of Westphalian volcanic rocks in the area east and south of Nottingham, England. *Proceedings of the Yorkshire Geological Society*, **44**, 29–44.

BURGESS, I. C. and HOLLIDAY, D. W. 1979. Geology of the Country around Brough-under-Stainmore. *Memoir of the Geological Survey of Great Britain*, Sheet 31, 132p.

BUSBY, J. P. 1985. Interactive 2.5D gravity and magnetic modelling on the ICL PERQ 2. *Report of the Regional Geophysics Research Group, British Geological Survey*, No. RGRG 86/9.

BUTCHER, N. J. D. 1976. Aspects of the structural control of fluorite mineralisation in the South Pennine orefield with notes on the mining potential. Unpublished PhD thesis, University of Leicester.

BUTLER, D. E. 1981. Marine faunas from concealed Devonian rocks of southern England and their reflection of the Frasnian transgression. *Geological Magazine*, **118**, 679–697.

BUTTERLEY, A. D. and MITCHELL, G. H. 1945. Driving of two drifts by the Desford Coal Co Ltd at Merry Lees, Leicestershire. *Transactions of the Institution of Mining Engineers*, **104**, 703–713.

CALVER, M. A. 1969. Westphalian of Britain. *Comptes rendus, 6ième Congrès International de stratigraphie et de Géologie du Carbonifère*, (Sheffield, 1967), **1**, 133–254.

CARLON, C. J. 1979. *The Alderley Edge mines* (Altrincham: J. Sherrat and Son), 144p.

CARPENTER, A. B., TROUT, M. L. and PICKETT, E. E. 1974. Preliminary report on the origin and chemical evolution of lead- and zinc-rich oil field brines in Central Mississippi. *Economic Geology*, **69**, 1191–1206.

CARTER, J. S. and CAZALET, P. C. D. 1984. Hydrocarbon gases in rocks as pathfinders for mineral exploration. In *Prospecting in areas of glaciated terrain, 1984* (London: Institution of Mining and Metallurgy), 11–20.

CATHLES, L. M. 1977. An analysis of the cooling of intrusives by ground-water convection which includes boiling. *Economic Geology*, **72**, 804–826.

— 1981. Fluid flow and genesis of hydrothermal ore deposits. *Economic Geology: seventy fifth anniversary volume, 1905–1980*, 424–457.

— and SMITH, A. T. 1983. Thermal constraints on the formation of Mississippi Valley-type lead–zinc deposits and their implications for episodic basin dewatering and deposit genesis. *Economic Geology*, **78**, 983–1002.

CAUET, S., WEIS, D. and HERBOSCH, A. 1982. Genetic study of Belgian lead zinc mineralizations in carbonate environments through lead isotopic geochemistry. *Bulletin du Bureau de Recherches Géologiques et Minières, Section II*, no. 4, 329–341.

CAVE, R. 1965. The Nod Glas sediments of Caradoc age in North Wales. *Geological Journal*, **4**, 279–298.

CHADWICK, R. A., KENOLTY, N. and WHITTAKER, A. 1983. Crustal structure beneath southern England from deep seismic reflection profiles. *Journal of the Geological Society of London*, **140**, 893–911.

CHISHOLM, J. I. 1977. Growth faulting and sandstone deposition in the Namurian of the Stanton Syncline, Derbyshire. *Proceedings of the Yorkshire Geological Society*, **41**, 305–323.

— CHARSLEY, T. J. and AITKENHEAD, N. In press. Geology of the country around Ashbourne. *Memoir of the British Geological Survey*, Sheet 124.

CHROSTON, P. N., ALLSOP, J. M. and CORNWELL, J. D. 1988. New seismic refraction evidence on the origin of the Bouguer anomaly low near Hunstanton, Norfolk. *Proceedings of the Yorkshire Geological Society*, **46**, 311–319.

CHROSTON, P. and SOLA, M. A. 1982. Deep boreholes, seismic refraction lines and the interpretation of gravity anomalies in Norfolk. *Journal of the Geological Society of London*, **139**, 255–264.

CHURCH, W. R. and GAYER, R. A. 1973. The Ballantrae ophiolite. *Geological Magazine*, **110**, 497–510.

CLARKE, F. W. 1924. The data of geochemistry. *Bulletin of the United States Geological Survey* 770. 841p.

COATS, J. S., SMITH, C. G., FORTEY, N. J., GALLAGHER, M. J., MAY, F. and McCOURT, W. J. 1980. Strata-bound barium–zinc mineralization in Dalradian schist near Aberfeldy, Scotland. *Transactions of the Institution of Mining and Metallurgy*, (Section B), **89**, B110–122.

COCKS, L. R. M. and FORTEY, R. A. 1982. Faunal evidence for oceanic separation in the Palaeozoic of Britain. *Journal of the Geological Society of London*, **139**, 465–478.

COLLER, D. W. 1984. Variscan structures in the Upper Palaeozoic rocks of west central Ireland. In *Variscan tectonics of the Northern Atlantic region* HUTTON, D. H. W. and SANDERSON, D. J. eds. *Special Publication of the Geological Society of London* no. 14, 185–194.

COLLINSON, J. D., JONES, C. M. and WILSON, A. A. 1977. The Marsdenian (Namurian R_2) succession west of Blackburn: implications for the evolution of Pennine delta systems. *Geological Journal*, **12**, 59–76.

CONDIT, C. D. and CHAVEZ, P. S. JR. 1979. Basic concepts of computerised digital image processing for geologists. *Bulletin of the United States Geological Survey* 1462, 16p.

COOK, A. H. and THIRLAWAY, H. I. S. 1955. The geological results of measurements of gravity in the Welsh Borders. *Quarterly Journal of the Geological Society of London*, **111**, 47–70.

COOMER, P. C. and FORD, T. D. 1975. Lead and sulphur isotope ratios of some galena specimens from the South Pennines and North Midlands. *Mercian Geologist*, **5**, 291–304.

COPE, F. W. 1973. Woo Dale Borehole near Buxton, Derbyshire. *Nature, London*, **243**, 29–30.

— 1979. The age of the volcanic rocks in the Woo Dale Borehole, Derbyshire. *Geological Magazine*, **116**, 319–320.

CREANEY, S. 1980. Petrographic texture and vitrinite reflectance variation on the Alston Block, North-East England. *Proceedings of the Yorkshire Geological Society*, **42**, 553–580.

CRIBB, J. 1976. Application of the generalised linear inverse to the inversion of static potential data. *Geophysics*, **41**, 1365–1369.

CRIBB, S. J. 1975. Rubidium-strontium ages and strontium isotope ratios from the igneous rocks of Leicestershire. *Journal of the Geological Society of London*, **131**, 203–212.

CRITCHLEY, M. F. 1979. A geological outline of the Ecton copper mines, Staffordshire. *Bulletin of the Peak District Mines Historical Society*, **7**, 177–191.

— 1984. Variscan tectonics of the Alston Block, northern England. In *Variscan tectonics of the North Atlantic region* HUTTON, D. H. W. and SANDERSON, D. J. eds. *Special Publication of the Geological Society of London* no. 14, 139–146.

DAVIDSON, C. F. 1966. Some genetic relationships between ore deposits and evaporites. *Transactions of the Institution of Mining and Metallurgy* (Section B), **75**, B216–225.

DAVIS, J. C. and McCULLAGH, M. J. 1975. *Display and analysis of spatial data* (New York: John Wiley & Sons), 392p.

DAY, G. A., COOPER, B. A., ANDERSON, C., BURGERS, W. F. J., RØNNEVIK, H. C. and SCHÖNEICH, H. 1981. Regional seismic structure maps of the North Sea. In *Petroleum geology of the continental shelf of north-west Europe* ILLING, L. V. and HOBSON, G. D. eds. (London: Institute of Petroleum), 76–84.

DEJONGHE, L., RYE, R. O. and CAUET, S. 1982. Sulfur isotopes of barite and lead isotopes of galena from the stratiform deposit in Frasnian carbonate and shale host rocks of Chaudfontaine (Province of Liège, Belgium). *Annales de Société Géologique de Belge*, **105**, 97–103.

DEWEY, H. and EASTWOOD, T. 1925. Copper ores of the Midlands, Wales, the Lake District and the Isle of Man. *Special Report on the Mineral Resources of Great Britain (Memoir of the Geological Survey of Great Britain)* no. 30, 87p.

DEWEY, J. F. 1982. Plate tectonics and the evolution of the British Isles. *Journal of the Geological Society of London*, **139**, 371–412.

— and SHACKLETON, R. M. 1984. A model for the evolution of the Grampian tract in the early Caledonides and Appalachian. *Nature, London*, **312**, 115–121.

DONATO, J. A. and TULLY, M. C. 1982. A proposed granite batholith along the western flank of the North Sea Viking Graben. *Geophysical Journal of the Royal Astronomical Society*, **69**, 187–195.

DOWNING, R. A. and GRAY, D. A. eds. 1986. *Geothermal energy—the potential in the United Kingdom* (London: HMSO for British Geological Survey), 187p.

— and HOWITT, F. 1969. Saline groundwaters in the Carboniferous rocks of the English East Midlands in relation to the geology. *Quarterly Journal of Engineering Geology*, **1**, 241–269.

DOZY, J. J. 1970. A geological model for the genesis of the lead–zinc ores of the Mississippi Valley, USA. *Transactions of the Institution of Mining and Metallurgy*, (Section B), **79**, B163–B170.

DUNHAM, K. C. 1934. Genesis of the North Pennine ore deposits. *Quarterly Journal of the Geological Society of London*, **90**, 689–720.

— 1948. Geology of the Northern Pennine Orefield: **1**, Tyne to Stainmore. *Memoir of the Geological Survey of Great Britain*, 357p.

— 1952. Age relations of the epigenetic mineral deposits of Britain. *Transactions of the Geological Society of Glasgow*, **21**, 395–429.

— 1952. Fluorspar. *Special Reports on the Mineral Resources of Great Britain (Memoir of the Geological Survey of Great Britain)* no. 4, 143p.

— 1973. A recent deep borehole near Eyam, Derbyshire. *Nature, London*, **241**, 84–85.

— 1974. Granite beneath the Pennines in North Yorkshire. *Proceedings of the Yorkshire Geological Society*, **40**, 191–194.

— 1983. Ore genesis in the English Pennines: A fluoritic subtype. In *International conference on Mississippi valley type lead–zinc deposits. Proceedings volume*: KISVARSANYI, G., GRANT, S. K., PRATT, W. P. and KOENIG, J. W. eds (Rolla, Missouri: University of Missouri-Rolla Press), 86–112.

— 1987. Contributed remarks on G. C. Brown and others. 1987. Geochemistry of granites beneath the north Pennines and their role in orefield mineralisation. *Transactions of the Institution of Mining and Metallurgy* (Section B), **97**, B229–230.

— DUNHAM, A. C., HODGE, B. L. and JOHNSON, G. A. L. 1965. Granite beneath Viséan sediments with mineralization at Rookhope, Northern Pennines. *Quarterly Journal of the Geological Society of London*, **121**, 383–417.

— FITCH, F. J., INESON, P. R., MILLER, J. A. and MITCHELL, J. G. 1968. The geochronological significance of argon-40/argon-39 age determinations on White Whin from the northern Pennine orefields. *Proceedings of the Royal Society*, London, Series A, **307**, 251–266.

— and WILSON, A. A. 1985. Geology of the Northern Pennine Orefield: volume 2, Stainmore to Craven. *Economic Memoir, British Geological Survey*, 247p.

DUNNING, F. W. ed. 1985. *Geological structure of Great Britain, Ireland and surrounding seas* (London: Geological Society), Mapchart.

EARP, J. R., MAGRAW, D., POOLE, E. G., LAND, D. H. and WHITEMAN, A. J. 1961. Geology of the country around Clitheroe and Nelson. *Memoir of the Geological Survey of Great Britain* Sheet 68, 346p.

ECKSTRAND, O. R. ed. 1984. Canadian mineral deposit types: a geological synopsis. *Geological Survey of Canada, Economic Geology Report*, no. 36, 86p.

EDEN, R. A., STEVENSON, I. P. and EDWARDS, W. 1957. Geology of the country around Sheffield. *Memoir of the Geological Survey of Great Britain*, Sheet 100, 238p.

EDMUNDS, W. M. 1971. Hydrogeochemistry of groundwater in the Derbyshire Dome with special reference to trace constituents. *Report of the Institute of Geological Sciences* no. 71/7, 52p.

EDWARDS, W. 1951. The concealed coalfield of Yorkshire and Nottinghamshire. 3rd Edition. *Memoir of the Geological Survey of Great Britain*, 274p.

— 1967. Geology of the country around Ollerton. *Memoir of the Institute of Geological Sciences*, Sheet 113, 297p.

— WRAY, D. A. and MITCHELL, G. H. 1940. Geology of the country around Wakefield. *Memoir of the Geological Survey of Great Britain*, Sheet 78, 215p.

ELKINS, T. A. 1951. The second derivative method of gravity interpretation. *Geophysics*, **16**, 29–50.

ELLIOTT, R. E. 1968. Facies, sedimentation successions and cyclothems in productive coal measures in the East Midlands, Great Britain. *Mercian Geologist*, **2**, 351–372.

— 1969. Deltaic processes and episodes; the interpretation of productive coal measures occurring in the East Midlands, Great Britain. *Mercian Geologist*, **3**, 111–135.

ELLIS, A. J. and MAHON, W. A. J. 1964. Natural hydrothermal systems and experimental hot water/rock interactions. *Geochimica et Cosmochimica Acta*, **28**, 1323–1357.

EMBLIN, R. 1978. A Pennine model for the diagenetic origin of base metal ore deposits in Britain. *Bulletin of the Peak District Mines Historical Society*, **7**, 5–20.

ENGLAND, P. C., OXBURGH, E. R. and RICHARDSON, S. W. 1980. Heat refraction and heat production in and around granite plutons in north-east England. *Geophysical Journal of the Royal Astronomical Society*, **62**, 439–455.

EVANS, A. D. and CORNWELL, J. D. 1981. An airborne geophysical survey of the Whin Sill between Haltwhistle and Scots Gap, south Northumberland. *Mineral Reconnaissance Programme Report, Institute of Geological Sciences* no. 47, 31p.

— PATRICK, D. J., WADGE, A. J. and HUDSON, J. M. 1983. Geophysical investigation in Swaledale, North Yorkshire. *Mineral Reconnaissance Programme Report, Institute of Geological Sciences*, no. 65, 25p.

EVANS, A. M. 1979. The East Midlands aulacogen of Caledonian age. *Mercian Geologist*, **7**, 31–42.

— FORD, T. D. and ALLEN, J. R. L. 1968. Precambrian rocks. In *The geology of the East Midlands* SYLVESTER-BRADLEY, P. C. and FORD, T. D. eds (Leicester: Leicester University Press), 1–19.

— and MAROOF, S. I. 1976. Basement controls on mineralisation in the British Isles. *Mining Magazine, London*, **134**, 401–411.

EVENSEN, N. M., HAMILTON, P. J. and O'NIONS, R. K. 1978. Rare-earth abundances in chondritic meteorites. *Geochimica et Cosmochimica Acta*, **42**, 1119–1212.

FAIRBANK, B. D., OPENSHAW, R. E., SOUTHER, J. G. and STANDER, J. J. 1981. Medger Creek geothermal project; an exploration case history. *Geological Survey of Canada, Geothermal Resource Council Bulletin*, 10:6, 3–7.

FAIRBRIDGE, R. W. ed. 1972. *The encyclopedia of geochemistry and environmental sciences.* (*Encyclopedia of earth sciences series,* **4A**) (New York: Van Nostrand Reinhold).

FALCON, N. L. and KENT, P. E. 1960. Geological results of petroleum exploration in Britain 1945–1957. *Memoir of the Geological Society of London* no. 2, 56p.

FEHN, U., CATHLES, L. M. and HOLLAND, H. D. 1978. Hydrothermal convection and uranium deposits in abnormally radioactive plutons. *Economic Geology*, **73**, 1556–1566.

FETTES, D. J., GRAHAM, C. M., HARTE, B. and PLANT, J. A. 1986. Lineaments and basement domains: an alternative view of Dalradian evolution. *Journal of the Geological Society of London*, **143**, 453–464.

FIRMAN, R. J. 1977. Derbyshire wrenches and ores— a study of the rakes' progress by secondary faulting. *Mercian Geologist*, **6**, 81–96.

FITCH, F. J. and MILLER, J. A. 1967. The age of the Whin Sill. *Geological Journal*, **5**, 233–250.

— — and WILLIAMS, S. C. 1970. Isotopic ages of British Carboniferous rocks. *Comptes rendus, 6ième Congrès International de Stratigraphie et de Géologie du Carbonifère*, (Sheffield, 1967), **2**, 771–789.

FITTON, J. G., THIRLWALL, M. F. and HUGHES, D. J. 1982. Volcanism in the Caledonian orogenic belt of Britain. In *Andesites* THORPE, R. S. ed. (New York: John Wiley & Sons), 611–636.

FLEET, A. J. 1984. Aqueous and sedimentary geochemistry of the rare earth elements. In *Rare earth element geochemistry* HENDERSON, P. ed. (Amsterdam: Elsevier), 343–373. (*Developments in Geochemistry* no. 2.)

FLEISCHER, M. and ROBINSON, W. O. 1963. Some problems of the geochemistry of fluorine. In *Studies in analytical geochemistry* SHAW, D. M. ed. *Royal Society of Canada, Special Publication* no. 6, 58–75.

FLOYD, P. A. 1982. Introduction: Geological setting of Upper Palaeozoic magmatism. In *Igneous rocks of the British Isles* SUTHERLAND, D. S. ed. (Chichester: John Wiley & Sons), 217–225.

— and WINCHESTER, J. A. 1975. Magma type and tectonic setting discrimination using immobile elements. *Earth and Planetary Science Letters*, **27**, 211–218.

FORD, T. D. 1976. The ores of the South Pennines and Mendip Hills, England—a comparative study. In *Handbook of strata-bound and stratiform ore deposits, II. Regional Studies and specific deposits, volume 5, Regional studies* WOLF, K. H. ed. (Amsterdam: Elsevier), 161–195.

FRANCIS, E. H. 1978. Igneous activity in a fractured craton: Carboniferous volcanism in northern Britain. In *Crustal evolution in north-west Britain and adjacent regions* BOWES, D. R. and LEAKE, B. E. eds. *Geological Journal Special Issue*, no. 10, (Liverpool: Seel House Press), 279–296.

— 1982. Magma and sediment-I. Emplacement mechanism of late Carboniferous tholeiite sills in northern Britain. *Journal of the Geological Society of London*, **139**, 1–20.

FROEHLICH, P. N., KLINKHAMMER, G. P., BENDER, M. L., LEUDTKE, N. A., HEATH, G. R., CULLEN, D., DAUPHIN, P., HAMMOND, D., HARTMAN, B. and MAYNARD, V. 1979. Early oxidation of organic matter in pelagic sediments of the eastern equatorial Atlantic: suboxic diagenesis. *Geochimica et Cosmochimica Acta*, **43**, 1075–1090.

FROST, D. V. and SMART, J. G. O. 1979. Geology of the country north of Derby. *Memoir of the Institute of Geological Sciences*, Sheet 125, 199p.

FROST, R. T. C., FITCH, F. J. and MILLER, J. A. 1981. The age and nature of the crystalline basement of the North Sea Basin. In *Petroleum Geology of the Continental Shelf of NW Europe*. ILLING, L. V. and HOBSON, G. D. eds. (London: Heyden), 43–57.

GARVEN, G. 1984. The role of regional fluid flow in the genesis of the Pine Point deposit, Western Canada Sedimentary Basin. *Economic Geology*, **80**, 307–324.

— and FREEZE, R. A. 1985. Theoretical analysis of the role of groundwater flow in the genesis of strata-bound ore deposits. 1. Mathematical and numerical model. 2. Quantitative results. *American Journal of Science*, **284**, 1085–1124 and 1125–1174.

GASS, K. N. 1980. A field, petrographic and geochemical study of the Miller's Dale Carboniferous volcanic rocks, North Derbyshire. Unpublished BSc dissertation, University of Lancaster.

GAWTHORPE, R. L. 1986. Sedimentation during carbonate ramp-to-slope evolution in a tectonically active area: Bowland Basin (Dinantian), northern England. *Sedimentology*, **33**, 185–206.

GEOLOGICAL SURVEY OF GREAT BRITAIN. 1965. *Aeromagnetic map of Great Britain Sheet 2. 1:625 000* (Southampton, Ordnance Survey).

GEORGE, T. N. 1974. Lower Carboniferous rocks in Wales. In *The Upper Palaeozoic and post Palaeozoic rocks of Wales* OWEN, T. R. ed. (Cardiff: University of Wales Press), 85–115.

— Johnson, G. A. L., Mitchell, M., Prentice, J. E., Ramsbottom, W. H. C., Sevastopulo, G. D. and WILSON, R. B. 1976. A correlation of Dinantian rocks in the British Isles. *Special Report of the Geological Society of London* no. 7, 87p.

GIBBS, R. J. 1977. Clay mineral segregation in the marine environment. *Journal of Sedimentary Petrology*, **47**, 237–243.

GIFFARD, H. P. W. 1923. The recent search for oil in Great Britain. *Transactions of the Institution of Mining Engineers*, **65**, 221–250.

GILLIGAN, A. 1920. The petrography of the Millstone Grit of Yorkshire. *Quarterly Journal of the Geological Society of London*, **75**, 251–294.

GIORDANO, T. H. 1985. A preliminary evaluation of organic ligands and metal-organic complexing in Mississippi Valley-type ore solutions. *Economic Geology*, **80**, 96–106.

— and BARNES, H. L. 1981. Lead transport in Mississippi Valley-type ore solutions. *Economic Geology*, **76**, 2200–2211.

GREEN, P. M. 1984. Digital image processing of integrated geochemical and geological information. *Journal of the Geological Society of London*, **141**, 941–949.

GUINNESS, E. A., ARVIDSON, R. E., LEFF, C. E., EDWARDS, M. H. and BINDSCHADLER, D. L. 1983. Digital image processing applied to analysis of geophysical and geochemical data for Southern Missouri. *Economic Geology*, **78**, 654–663.

GUION, P. D. 1978. Sedimentation of interseam strata and some relationships with coal seams in the East Midlands Coalfield. Unpublished PhD thesis, City of London Polytechnic.

GULSON, B. L. 1986. *Lead isotopes in mineral exploration* (Amsterdam: Elsevier), 345p. (*Developments in economic geology* 23.)

GUTTERIDGE, P. 1987 Dinantian sedimentation and the basement structure of the Derbyshire Dome. *Geological Journal*, **22**, 25–41.

HALL, W. E. and HEYL, A. V. 1968. Distribution of minor elements in ore and host rock, Illinois–Kentucky fluorite district and Upper Mississippi Valley zinc–lead district. *Economic Geology*, **63**, 655–670.

HAMPTON, C. M. and TAYLOR, P. N. 1983. The age and nature of the basement of southern Britain: evidence from Sr and Pb isotopes in granites. *Journal of the Geological Society of London*, **140**, 499–509.

HANOR, J. S. 1979. The sedimentary genesis of hydrothermal fluids. In *Geochemistry of hydrothermal ore deposits* BARNES, H. L. ed. (New York: John Wiley & Sons), 137–172.

HARLAND, W. B., COX, A. V., LLEWELLYN, P. G., PICKTON, C. A. G., SMITH, A. G. and WALTERS, R. 1982. *A geologic time scale* (Cambridge: Cambridge University Press), 131p.

HARRIS, N. B. W., PEARCE, J. A. and TINDLE, A. G. 1986. Geochemical characteristics of collision zone magmatism. In: *Collision tectonics* SHACKLETON, R. M. RIES, A. C. and COWARD, M. P. eds *Special Publication of the Geological Society of London* no. 19, 67–81.

HART, P. E., DUDA, R. O. and EINAUDI, M. I. 1978. PROSPECTOR—a computer based consultation system for mineral exploration. *Journal of the International Association of Mathematical Geologists*, **10**, 589–610.

HASKIN, L. A., WILDMAN, T. R., FREY, F. A., COLLINS, K. A., KEEDY, C. R. and HASKIN, M. A. 1966. Rare earths in sediments. *Journal of Geophysical Research*, **71**, 6091–6105.

— HASKIN, M. A., FREY, F. A. and WILDEMAN, T. R. 1968. Relative and absolute terrestrial abundances of the rare earths. In *Origin and distribution of the elements* AHRENS, L. H. ed. (Oxford: Pergamon Press), 889–912.

HASZELDINE, R. S. 1984. Carboniferous North Atlantic palaeogeography: stratigraphic evidence for rifting, not megashear or subduction. *Geological Magazine*, **121**, 443–463.

HENNESSY, J. 1981. A classification of British Caledonian granites based on uranium and thorium contents. *Mineralogical Magazine*, **44**, 449–454.

HEYL, A. V. 1969. Some aspects of genesis of zinc–lead–barite–fluorite deposits in the Mississippi Valley, USA. *Transactions of the Institution of Mining and Metallurgy* (Section B), **78**, B148–160.

— 1973. Some fluorite–barite deposits in the Mississippi Valley in relation to major structures and zonation. In *A symposium on the geology of fluorspar* HUTCHESON, D. W. ed. *Kentucky Geological Survey, special publication* 22, 55–57.

HOLLAND, J. G. and LAMBERT, R. St J. 1970. Weardale Granite. *Transactions of the Natural History Society of Northumberland*, **41**, 103–123.

HOLMES, I., CHAMBERS, A. D., IXER, R. A., TURNER, P. and VAUGHAN, D. J. 1983. Diagenetic processes and the mineralization in the Triassic of Central England. *Mineralium Deposita*, **18**, 365–377.

HOWITT, F. and BRUNSTROM, R. G. W. 1966. The continuation of the East Midlands Coal Measures into Lincolnshire. *Proceedings of the Yorkshire Geological Society*, **35**, 549–564.

HUDSON, R. G. S. 1931. The pre-Namurian knoll topography of Derbyshire and Yorkshire. *Transactions of the Leeds Geological Association*, **5**, 49–64.

— and COTTON, G. 1945a. The Lower Carboniferous in a boring at Alport, Derbyshire. *Proceedings of the Yorkshire Geological Society*, **25**, 254–330.

— — 1945b. The Carboniferous rocks of the Edale anticline, Derbyshire. *Quarterly Journal of the Geological Society of London*, **101**, 1–36.

HUNT, J. M. 1979. *Petroleum geochemistry and geology* (San Francisco: Freeman), 169–172.

HUNTING GEOLOGY AND GEOPHYSICS LTD. 1983. Computer correlation of geological, geochemistry and geophysical prospecting data with enhanced satellite imagery. Report of CREST programme of the European Economic Community.

INESON, P. R. 1970. Trace-element aureoles in limestone wallrocks adjacent to fissure veins in the Eyam area of the Derbyshire orefield. *Transactions of the Institution of Mining and Metallurgy*, (Section B), **79**, B238–245.

— and FORD, T. D. 1982. The South Pennine Orefield: its genetic theories and eastward extension. *Mercian Geologist*, **8**, 285–304.

— and MITCHELL, J. G. 1972. Isotopic age determination on clay minerals from lavas and tuffs of the Derbyshire orefield. *Geological Magazine*, **109**, 501–512.

INGHAM, J. K. 1978. Geology of a continental margin 2: middle and late Ordovician transgression, Girvan. In *Crustal evolution in northwestern Britain and adjacent regions* BOWES, D. R. and LEAKE, B. E. eds *Geological Journal Special Issue*, no. 10. (Liverpool: Seel House Press), 163–176.

INSTITUTE OF GEOLOGICAL SCIENCES. 1965. Aeromagnetic map of Great Britain. Sheet 2. 1:625 000. (Southampton: Ordnance Survey).

INSTITUTE OF GEOLOGICAL SCIENCES. Summary of Progress. 1966. *Annual report for 1965, Part I* (London: HMSO), 114p.

INSTITUTE OF GEOLOGICAL SCIENCES. 1977a. Humber-Trent (provisional edition). Sheet 53°N–02°W. *1:250 000 Bouguer Gravity Anomaly map series*, (London: Institute of Geological Sciences.)

INSTITUTE OF GEOLOGICAL SCIENCES. 1977b. Humber-Trent, (provisional edition). Sheet 53°N–02°W. *1:250 000 Aeromagnetic anomaly map series*. (London: Institute of Geological Sciences.)

INSTITUTE OF GEOLOGICAL SCIENCES. 1978. IGS boreholes, 1977. *Report of the Institute of Geological Sciences 78/21*, 2.

INSTITUTE OF GEOLOGICAL SCIENCES. 1979. IGS boreholes, 1978. *Report of the Institute of Geological Sciences 79/12*, 1–6.

INSTITUTE OF GEOLOGICAL SCIENCES. 1982. Coalville, 1:50 000 Map Series. Sheet 155.

IXER, R. A. and TOWNLEY, R. 1979. The sulphide mineralogy and paragenesis of the South Pennine Orefield, England. *Mercian Geologist*, **7**, 51–63.

JACKSON, S. A. and BEALES, F. W. 1967. An aspect of sedimentary basin evolution: the concentration of Mississippi Valley-type ores during the late stages of diagenesis. *Bulletin of Canadian Petroleum Geology*, **15**, 393–433.

JACOB, A. W. B., KAMINSKI, W., MURPHY, T., PHILLIPS, W. E. A. and PRODEHL, C. 1985. A crustal model for a NE/SW profile through Ireland. *Tectonophysics*, **113**, 75–103.

JOLLY, J. L. and HEYL, A. V. 1964. Mineral paragenesis and zoning in the Central Kentucky mineral district. *Economic Geology*, **59**, 596–624.

JONES, C. M. 1980. Deltaic sedimentation in the Roaches Grit and associated sediments (Namurian R_{2b}) in the south-west Pennines. *Proceedings of the Yorkshire Geological Society*, **43**, 39–67.

KAY, R. W., SUN, S–S. and LEE HU, C–N. 1978. Pb and Sr isotopes in volcanic rocks from the Aleutian Islands and Pribilof Islands, Alaska. *Geochimica et Cosmochimica Acta*, **42**, 263–273.

KENT, P. E. 1947. A deep boring at North Creake, Norfolk. *Geological Magazine*, **84**, 2–18.

— 1962. A borehole to basement rocks at Glinton, near Peterborough, Northamptonshire. *Proceedings of the Geological Society of London*, 1595, 40–42.

— 1966. The structure of the concealed Carboniferous rocks of north-eastern England. *Proceedings of the Yorkshire Geological Society*, **35**, 323–352.

— 1967. A contour map of the sub-Carboniferous surface in the north-east Midlands. *Proceedings of the Yorkshire Geological Society*, **36**, 127–133.

— 1974. Structural history. In *The geology and mineral resources of Yorkshire* RAYNER, D. H. and HEMINGWAY, J. E. eds (Leeds: Yorkshire Geological Society), 13–28.

— 1980. British Regional Geology: Eastern England from the Tees to the Wash (2nd Edition) (London: HMSO for Institute of Geological Sciences).

— 1985. UK onshore oil exploration, 1930–1964. *Marine and Petroleum Geology*, **2**, 56–64.

KING, R. J. 1968. Mineralisation; F. Permo-Triassic mineralisation. In *The geology of the East Midlands* SYLVESTER-BRADLEY, P. C. and FORD, T. D. eds. (Leicester: Leicester University Press), 123–132.

KIRBY, G. A., SMITH, K., SMITH, N. J. P. and SWALLOW, P. W. 1987. Oil and gas generation in eastern England. In *Proceedings of the 3rd conference on the Petroleum Geology of North West Europe, London, 1986* (London: Graham and Trotman.)

KIRTON, S. R. 1981. Petrogenesis and tectonic relationships of Carboniferous lavas of the English Midlands. Unpublished PhD thesis, University of Lancaster, 264p.

— 1984. Carboniferous volcanicity in England with special reference to the Westphalian of the E. and W. Midlands. *Journal of the Geological Society of London*, **141**, 161–170.

KNOWLES, B. 1964. The radioactive contents of the Coal Measures sediments in the Yorkshire-Derbyshire coalfield. *Proceedings of the Yorkshire Geological Society*, **34**, 413–450.

KRAUSKOPF, K. B. 1979. *Introduction to geochemistry 2nd edn* (New York: McGraw-Hill.)

KRAYNOV, S. R., MER'KOV, A. N., PETROVA, N. E., BATURINSKAYA, I. V. and ZHARIKOVA, V. M. 1969. Highly alkaline (pH12) fluosilicate waters in the deeper zones of the Lovozero massif. *Geochemistry International*, **6**, 635–640.

LAURENCE, R. A. 1960. Geologic problems in the Sweetwater barite district, Tennessee. *American Journal of Science*, **258-A**, 170–179.

LAZNICKA, P. 1985. *Empirical metallogeny volume 1: Part A* (Amsterdam: Elsevier). (*Developments in Economic Geology* 19), 1002p.

LEAKE, B. E. 1978. Granite emplacement: the granites of Ireland and their origin. In *Crustal evolution in north-western Britain and adjacent regions* BOWES, D. R. and LEAKE, B. E. eds. *Geological Journal Special Issue*, no. 10, (Liverpool: Seel House Press) 221–248.

LE BAS, M. J. 1968. Caledonian igneous rocks. In *The geology of the East Midlands* SYLVESTER-BRADLEY, P. C. and FORD, T. D. eds (Leicester: University of Leicester Press), 41–58.

— 1972. Caledonian igneous rocks beneath central and eastern England. *Proceedings of the Yorkshire Geological Society*, **39**, 71–86.

— 1981. The igneous basement of southern Britain with particular reference to the geochemistry of the pre-Devonian rocks of Leicestershire. *Transactions of the Leicestershire Literary and Philosophical Society*, **75**, 41–57.

— 1982. Geological evidence from Leicestershire on the crust of southern Britain. *Transactions of the Leicestershire Literary and Philosophical Society*, **76**, 54–67.

LEE, M. K. 1986. Hot dry rock. In *Geothermal energy—the potential in the United Kingdom* DOWNING R. A. and GRAY, D. A. eds (London: HMSO for British Geological Survey), 21–41.

— WHEILDON, J., WEBB, P. C., BROWN, G. C., ROLLIN, K. E., CROOK, C. N., SMITH, I. F., KING, G. and THOMAS-BETTS, A. 1984. Hot Dry Rock prospects in Caledonian granites. *Investigation of the Geothermal Potential of the UK.* (Keyworth: British Geological Survey.)

LEEDER, M. R. 1976. Sedimentary facies and the origins of basin subsidence along the northern margin of the supposed Hercynian Ocean. In *Sedimentary basins of continental margins and cratons* BOTT, M. H. P. ed. *Tectonophysics*, **36**, 167–179.

— 1982. Upper Palaeozoic basins in the British Isles—Caledonide inheritance versus Hercynian plate margin processes. *Journal of the Geological Society of London*, **139**, 479–491.

LEES, G. M. and COX, P. T. 1937. The geological basis of the present search for oil in Great Britain. *Quarterly Journal of the Geological Society of London*, **93**, 156–194.

— and TAITT, A. H. 1946. The geological results of the search for oilfields in Great Britain. *Quarterly Journal of the Geological Society of London*, **101**, 255–317.

LEGGETT, J. K. 1980. British Lower Palaeozoic black shales and their palaeo-oceanographic significance. *Journal of the Geological Society of London*, **137**, 139–156.

— MCKERROW, W. S. and EALES, M. H. 1979. The Southern Uplands of Scotland: a Lower Palaeozoic accretionary prism. *Journal of the Geological Society of London*, **136**, 755–770.

LEWIS, A. D. and BLOXAM, T. W. 1977. Petrotectonic environments of the Girvan–Ballantrae lavas from rare-earth element distributions. *Scottish Journal of Geology*, **13**, 211–222.

LISTER, C. R. B. 1972. On the thermal balance of a mid-ocean ridge. *Geophysical Journal of the Royal Astronomical Society*, **26**, 515–535.

LLEWELLYN, P. G. and STABBINS, R. 1968. Lower Carboniferous evaporites and mineralization in the eastern and central Midlands of Britain. *Transactions of the Institution of Mining and Metallurgy* (Section B), **77**, B170–173.

— — 1970. The Hathern Anhydrite Series, Lower Carboniferous, Leicestershire, England. *Transactions of the Institution of Mining and Metallurgy* (Section B), **79**, B1–15.

MACDONALD, R., GASS, K. N., THORPE, R. S. and GASS, I. G. 1984. Geochemistry and petrogenesis of the Derbyshire Carboniferous basalts. *Journal of the Geological Society of London*, **141**, 147–159.

MAGUIRE, P. K. H. 1987. CHARM II—A deep reflection profile within the central England microcraton. *Journal of the Geological Society of London*, **144**, 661–670.

— HALL, I. A. and KHAN, M. A. 1983. Charm I: A deep seismic reflection profile across Charnwood Forest and the Caledonian Mountsorrel granodiorite intrusions of central England. *First Break* **1**, 38.

McCABE, P. J. 1978. The Kinderscoutian Delta (Carboniferous) of northern England; a slope influenced by density currents. In *Sedimentation in submarine canyons, fans and trenches* STANLEY, D. J. and KELLING, G. eds (Stroudsburg, Pennsylvania: Dowden, Hutchinson and Ross), 116–126.

McKENZIE, D. 1978. Some remarks on the development of sedimentary basins. *Earth and Planetary Science Letters*, **40**, 25–32.

McKERROW, W. S. 1988. The development of Iapetus from Arenig to Wenlock. In *Caledonian–Appalachian orogen* HARRIS, A. L. and FETTES, D. J. *Special Publication of the Geological Society of London no. 38*, 405–412.

MAROOF, S. I. 1973. Geophysical investigations of the Carboniferous and pre-Carboniferous formation of the East Midlands of England. Unpublished PhD thesis, University of Leicester.

— 1976. The structure of the concealed pre-Carboniferous basement of the Derbyshire Dome from gravity data. *Proceedings of the Yorkshire Geological Society*, **41**, 59–69.

MASSON-SMITH, D., HOWELL, P. M., ABERNETHY-CLARKE, A. B. D. E. and PROCTOR, D. W. 1974. The National Gravity Reference Net 1973 (NGRN73). *Ordnance Survey Professional Papers, New Series*, **26**.

MENEISY, M. Y. and MILLER, J. A. 1963. A geochronological study of the crystalline rocks of Charnwood Forest, England. *Geological Magazine*, **100**, 507–523.

MILLER, J. and GRAYSON, R. F. 1982. The regional context of Waulsortian facies in northern England. In *Symposium on the paleoenvironmental setting and distribution of the Waulsortian facies*, BOLTON, K., LANE, H. R. and LeMOME, D. V. eds (El Paso: El Paso Geological Society and the University of Texas), 17–33.

MITCHELL, G. H. and STUBBLEFIELD, C. J. 1941. The Carboniferous Limestone of Breedon Cloud, Leicestershire, and associated inliers. *Geological Magazine*, **78**, 201–219.

— Stephens, J. V., Bromehead, C. E. N. and Wray, D. A. 1947. Geology of the country around Barnsley. *Memoir of the Geological Survey of Great Britain*. Sheet 87, 182p.

Mitchell, M. and Reynolds, M. J. 1981. Early Tournaisian rocks at Lilleshall, Shropshire. *Geological Magazine*, 118, 699–702.

Mitchell, R. H. and Krouse, H. R. 1971. Isotopic composition of sulphur and lead in galena from the Greenhow-Skyreholme area, Yorkshire, England. *Economic Geology*, 66, 243–251.

Moorbath, S. 1962. Lead isotope abundance studies on mineral occurrences in the British Isles and their geological significance. *Philosophical Transactions of the Royal Society*, 254A, 295–360.

Moseley, F. 1972. A tectonic history of northwest England. *Journal of the Geological Society of London*, 128, 561–598.

Mostaghel, M. A. 1984. Trace elements in sulphide minerals and their genesis in the South Pennine Orefield. Unpublished PhD thesis, University of Leicester.

— 1985. Classification of the South Pennine Orefield. *Mercian Geologist*, 10, 27–38.

— and Ford, T. D. 1986. A sedimentary basin evolution model for ore genesis in the South Pennine Orefield. *Mercian Geologist*, 10, 209–224.

Mottl, M. J. 1983. Metabasalts and hot springs and the structure of hydrothermal systems at mid-ocean ridges. *Bulletin of the Geological Society of America*, 94, 161–180.

Mueller, G. 1954. The distribution of coloured varieties of fluorites within the thermal zones of the Derbyshire mineral deposits. *19th International Geological Congress Algiers*, 15, 523–536.

Murphy, T. 1952. Measurements of gravity in Ireland: Gravity survey of Central Ireland. *Geophysics Memoir, Dublin Institute of Advanced Studies*, 2, (part 3).

Nance, W. B. and Taylor, S. R. 1976. Rare earth element patterns and crustal evolution—I. Australian post-Archaean sedimentary rocks. *Geochimica et Cosmochimica Acta*, 40, 1539–1551.

Noble, E. A. 1963. Formation of ore deposits by water of compaction. *Economic Geology*, 58, 1145–1156.

Nordstrom, D. K. and Jenne, E. A. 1977. Fluorite solubility equilibria in selected geothermal waters. *Geochimica et Cosmochimica Acta*, 41, 175–188.

O'Brien, C., Plant, J. A., Simpson, P. R. and Tarney, J. 1985. The geochemistry, metasomatism and petrogenesis of the granites of the English Lake District. *Journal of the Geological Society of London*, 142, 1139–1157.

Odin, G. S., Curry, D., Gale, N. H. and Kennedy, W. J. 1982. The Phanerozoic Time Scale in 1981. In *Numerical dating in stratigraphy* Odin, G. S. ed. (Chichester: John Wiley & Sons), 957–960.

O'Driscoll, E. S. T. 1986. Observations of the lineament-ore relation. *Philosophical Transactions of the Royal Society*, A317, 195–218.

Ohmoto, H. and Rye, R. O. 1974. Hydrogen and oxygen isotopic compositions of fluid inclusions in the Kuroko deposits, Japan. *Economic Geology*, 69, 947–953.

Okolo, S. A. 1983. Fluvial distributary channels in the Fletcherbank Grit (Namurian R_{2b}), at Ramsbottom, Lancashire, England. *Special Publication of the International Association of Sedimentology* no. 6, 421–433.

Pearce, J. A. 1982. Trace element characteristics of lavas from destructive plate boundaries. In *Andesites* Thorpe, R. S. ed (Chichester: John Wiley & Sons), 525–548.

— and Cann, J. R. 1973. Tectonic setting of basic volcanic rocks investigated using trace element analyses. *Earth and Planetary Science Letters*, 19, 290–300.

— Harris, N. B. W. and Tindle, A. G. 1984. Trace element discrimination diagrams for the tectonic interpretation of granitic rocks. *Journal of Petrology*, 25, 956–983.

Perrin, R. M. S. 1971. *The clay mineralogy of British sediments*. (London: Mineralogical Society, Clay Minerals Group), 247p.

Pharaoh, T. C., Merriman, R. J., Webb, P. C. and Beckinsale, R. D. 1988a. The concealed Caledonides of Eastern England: preliminary results of a multidisciplinary study. *Proceedings of the Yorkshire Geological Society*, 46, 355–369.

— Webb, P. C., Thorpe, R. S. and Beckinsale, R. D. 1988b. Geochemical evidence for the tectonic setting of late Proterozoic volcanic suites in central England. In *Geochemistry and mineralisation of late Proterozoic volcanic suites* Pharaoh, T. C., Beckinsale, R. D. and Rickard, D. T. eds. *Special Publication of the Geological Society of London* no. 32, 541–552.

Phillips, W. E. A., Stillman, C. J. and Murphy, T. 1976. A Caledonian plate tectonic model. *Journal of the Geological Society of London*, 132, 579–609.

Phillips, W. J. 1972. Hydraulic fracturing and mineralisation. *Journal of the Geological Society of London*, 128, 337–359.

— 1986. Hydraulic fracturing effects in the formation of mineral deposits. *Transactions of the Institution of Mining and Metallurgy* (Section B), 95, B17–24.

Pidgeon, R. T. and Aftalion, M. 1978. Cogenetic and inherited zircon U–Pb systems in Palaeozoic granites of Scotland and England. In *Crustal evolution in northwest Britain and adjacent regions* Bowes, D. R. and Leake, B. E. eds. *Geological Journal Special Issue* no. 10, (Liverpool: Seel House Press), 183–220.

Plant, J. A. 1986. Models for granites and their mineralising systems in the British and Irish Caledonides. In *Geology and genesis of mineral deposits in Ireland* Andrew, C. J. *et al.* eds (Dublin: Irish Association for Economic Geology), 121–156

— Forrest, M. D., Hodgson, J. F., Smith, R. T. S. and Stevenson, A. G. 1986. Regional geochemistry in the detection and modelling of mineral deposits. In *Applied geochemistry in the 1980s* Thornton, I. and Howarth, R. J. eds (London: Graham and Trotman), 103–139.

— Watson, J. V. and Green, P. M. 1984. Moine-Dalradian relationships and their palaeotectonic significance. *Proceedings of the Royal Society*, A395, 185–202.

Poole, E. G., Williams, B. J. and Hains, B. A. 1968. Geology of the country around Market Harborough. *Memoir of the Institute of Geological Sciences*. Sheet 170, 92p.

POTTER, P. E., MAYNARD, J. B. and PRYOR, W. A. 1980. *Sedimentology of shale* (New York: Springer-Verlag), 306p.

RAMSBOTTOM, W. H. C., CALVER, M. A. EAGAR, R. M. C., HODSON, F., HOLLIDAY, D. W., STUBBLEFIELD, C. J. and WILSON, R. B. 1978. A correlation of Silesian rocks in the British Isles. *Special Report of the Geological Society of London* no. 10, 81p.

— RHYS, G. H. and SMITH, E. G. 1962. Boreholes in the Carboniferous rocks of the Ashover district, Derbyshire. *Bulletin of the Geological Survey of Great Britain* no. 19, 75–168.

— SABINE, P. A., DANGERFIELD, J. and SABINE, P. W. 1981. Mudrocks in the Carboniferous of Britain. *Quarterly Journal of Engineering Geology*, London, **14**, 257–262.

READING, H. G., WATERSON, J. and WHITE, S. H. eds. 1986. Major crustal lineaments and their influence on the geological history of the continental lithosphere. *Philosophical Transactions of the Royal Society*, London, Series A, **317**, 1–290.

RICHARDSON, S. W. and OXBURGH, E. R. 1978. Heat flow, radiogenic heat production and crustal temperatures in England and Wales. *Journal of the Geological Society of London*, **135**, 323–337.

ROBERTS, D. E. and HUDSON, G. R. T. 1983. The Olympic Dam copper–uranium–gold deposit, Roxby Downs, South Australia. *Economic Geology*, **78**, 799–822.

ROBINSON, B. W. and INESON, P. R. 1979. Sulphur, oxygen and carbon isotope investigations of lead–zinc–barite–fluorite–calcite mineralization, Derbyshire, England. *Transactions of the Institution of Mining and Metallurgy* (Section B), **88**, B107–117.

ROEDDER, E. 1977. Fluid inclusions as tools in mineral exploration. *Economic Geology*, **72**, 503–525.

ROGERS, D. E. 1983. Seismic studies on the Derbyshire Dome. Unpublished PhD thesis, University of Leeds.

ROGERS, P. J. 1977. Fluid inclusion studies in fluorite from the Derbyshire orefield. *Transactions of the Institution of Mining and Metallurgy* (Section B), **86**, B128–132.

ROLLIN, K. E. 1978. Interpretation of the main features of the Humber-Trent 1:250 000 Bouguer gravity anomaly map. *Institute of Geological Sciences, Applied Geophysics Unit Report* no. 74.

— 1980. Interpretation of the Widmerpool Gulf from gravity data. *Institute of Geological Sciences, Applied Geophysics Unit Report* no. 93.

— 1982. A review of data relating to hot dry rock and selection of targets for detailed study. *Investigations of the geothermal potential of the United Kingdom* (London: Institute of Geological Sciences).

RONOV, A. B. and MIGDISOV, A. A. 1971. Geochemical history of the crystalline basement and the sedimentary cover of the Russian and North American Platforms. *Sedimentology*, **16**, 137–185.

RUSSELL, M. J. 1978. Downward-excavating hydrothermal cells and Irish-type ore deposits: importance of an underlying thick Caledonian prism. *Transactions of the Institution of Mining and Metallurgy* (Section B), **87**, B168–171.

— and SMITH, F. W. 1979. Plate separation, alkali magmatisms and fluorite mineralization in northern and central England (Abstract). *Transactions of the Institution of Mining and Metallurgy* (Section B), **88**, B30.

— and SMYTHE, D. K. 1978. Evidence for an early Permian oceanic rift in the northern North Atlantic. In *Petrology and geochemistry of continental rifts* NEUMANN, E. R. and RAMBERG, I. B. eds (Dordrecht: Reidel), 173–179.

— — 1983. Origin of the Oslo Graben in relation to the Hercynian-Alleghenian orogeny and lithospheric rifting in the North Atlantic. *Tectonophysics*, **94**, 451–472.

SAKAI, H. and MATSUBAYA, O. 1974. Isotopic geochemistry of the thermal waters of Japan and its bearing on the Kuroko ore solutions. *Economic Geology*, **69**, 974–991.

SAMPSON, R. J. 1978. Surface II graphics system. Kansas Geological Survey.

SAMSON, I. M. and RUSSELL, M. J. 1983. Fluid inclusion data from Silvermines base-metal-baryte deposits, Ireland. *Transactions of the Institution of Mining and Metallurgy* (Section B), **92**, B67–71.

SANDERSON, D. J. and MARCHINI, W. R. D. 1984. Transpression. *Journal of Structural Geology*, **6**, 449–458.

SANGSTER, D. F. 1983. Mississippi Valley-type deposits: a geological mélange. In *International conference on Mississippi Valley-type lead–zinc deposits. Proceedings volume:* KISVARSANYI, G., GRANT, S. K., PRATT, W. P. and KOENIG, J. W. eds (Rolla, Missouri: University of Missouri-Rolla Press), 7–19.

SANGSTER, D. F. 1986. Age of mineralization in Mississippi Valley-type (MVT) deposits: a critical requirement for genetic modelling. In *Geology and genesis of mineral deposits in Ireland* ANDREW, C. J. *et al.* eds (Dublin: Irish Association for Economic Geology), 625–634.

SCHOFIELD, K. 1982. Sedimentology of the Woo Dale limestone formation of Derbyshire. Unpublished PhD thesis, University of Manchester.

SHEPHERD, T. J., DARBYSHIRE, D. P. F., MOORE, G. R. and GREENWOOD, D. A. 1982. Rare earth element and isotope geochemistry of the North Pennine ore deposits. In *Gîtes filoniens Pb Zn F Ba de basse température du domain varisque d'Europe et d'Afrique du Nord. Symposium Orléans. Bulletin BRGM, section II*, (2-3-4), 371–377.

SHEPPARD, S. M. F. and LANGLEY, K. M. 1984. Origin of saline formation waters in northeast England: application of stable isotopes. *Transactions of the Institution of Mining and Metallurgy* (Section B), **93**, B195–201.

— and TAYLOR, H. P. JR. 1974. Hydrogen and oxygen isotope evidence for the origins of water in the Boulder Batholith and the Butte ore deposits, Montana. *Economic Geology*, **69**, 926–946.

SHERIDAN, D. J. R. 1972. The stratigraphy of the Trim No. 1 well, Co. Meath and its relationship to Lower Carboniferous outcrop in east-central Ireland. *Bulletin of the Geological Survey of Ireland*, **1**, 311–334.

SHOTTON, F. W. 1935. The stratigraphy and tectonics of the Cross Fell Inlier. *Quarterly Journal of the Geological Society of London*, **91**, 639–704.

SIMPSON, P. R., BROWN, G. C., PLANT, J. A. and OSTLE, D. 1979. Uranium mineralisation and granite magmatism in the British Isles. *Philosophical Trans-*

actions of the Royal Society, **A291**, 385–412.

SMALL, A. T. 1978. Zonation of Pb–Zn–Cu–F–Ba mineralization in part of the North Yorkshire Pennines. *Transactions of the Institution of Mining and Metallurgy (Section B)*, **87**, B10–13.

SMIRNOV, K. I. 1976. *Geology of ore deposits* (Moscow: MIR Publications).

SMITH, E. G., RHYS, G. H. and EDEN, R. A. 1967. Geology of the country around Chesterfield, Matlock and Mansfield. *Memoir of the Geological Survey of Great Britain*. Sheet 112, 430p.

—— and GOOSSENS, R. F. 1973. Geology of the country around East Retford, Worksop and Gainsborough. *Memoir of the Geological Survey of Great Britain*. Sheet 101, 348p.

SMITH, K., SMITH, N. J. P. and HOLLIDAY, D. W. 1985. The deep structure of Derbyshire. *Geological Journal*, **20**, 215–225.

SOLOMON, M., RAFTER, T. A. and DUNHAM, K. C. 1971. Sulphur and oxygen isotope studies in the northern Pennines in relation to ore genesis. *Transactions of the Institution of Mining and Metallurgy (Section B)*, **80**, B259–275.

SOPER, N. J. 1986. The Newer Granite problem: a geotectonic view. *Geological Magazine*, **123**, 227–236.

— and HUTTON, D. H. W. 1984. Late-Caledonian sinistral displacement in Britain: implications for a three plate collision model. *Tectonics*, **3**, 781–794.

—— In press. Timing and geometry of collision, terrain accretion and sinistral strike-slip events the British Caledonides. In *The evolution of the Caledonian–Appalachian orogen* HARRIS, A. L. and FETTES, D. J. eds. *Special Publication of the Geological Society of London*.

SPEARS, D. A. and AMIN, M. A. 1981a. Geochemistry and mineralogy of marine and non-marine Namurian black shales from the Tansley Borehole, Derbyshire. *Sedimentology*, **28**, 407–417.

—— 1981b. A mineralogical and geochemical study of turbidite sandstones and interbedded shales, Mam Tor, Derbyshire, United Kingdom. *Clay Minerals*, **16**, 333–345.

SPOONER, E. T. C. and FYFE, W. S. 1973. Sub-sea-floor metamorphism, heat and mass transfer. *Contributions to Mineralogy and Petrology*, **42**, 287–304.

STEPHENS, W. E., WATSON, S. W., PHILIP, P. R. and WEIR, J. A. 1975. Element associations and distributions through a Lower Palaeozoic graptolite shale sequence in the Southern Uplands of Scotland. *Chemical Geology*, **16**, 269–294.

STEVENSON, I. P. and GAUNT, G. D. 1971. Geology of the country around Chapel en le Frith. *Memoir of the Institute of Geological Sciences*. Sheet 99, 444p.

— and MITCHELL, G. H. 1955. Geology of the country between Burton upon Trent, Rugeley and Uttoxeter. *Memoir of the Geological Survey of Great Britain*. Sheet 140, 178p.

STILLMAN, C. J. 1981. Caledonian igneous activity. In *Geology of Ireland* HOLLAND, C. H. ed. (Edinburgh: Scottish Academic Press), 83–106.

STONE, P., FLOYD, J. D., BARNES, R. P. and LINTERN, B. C. 1987. A sequential back-arc and foreland basin thrust duplex model for the Southern Uplands of Scotland. *Journal of the Geological Society of London*, **144**, 753–764.

STRANK, A. R. E. 1985. The Dinantian stratigraphy of a deep borehole near Eyam, Derbyshire. *Geological Journal*, **20**, 227–237.

STUBBLEFIELD, J. 1967. Some results of a recent Geological Survey borehole in Huntingdonshire. *Proceedings of the Geological Society of London*, no. 1637, 35–40.

SUGGATE, P. 1976. Coal ranks and geological history of the Nottinghamshire-Yorkshire Coalfield. *Mercian Geologist*, **6**, 1–24.

SUGGATE, R. R. 1981. Coal ranks on the Alston Block, north-east England: a discussion. *Proceedings of the Yorkshire Geological Society*, **43**, 451–453.

SYLVESTER-BRADLEY, P. C. and FORD, T. D. eds. 1968. *The geology of the East Midlands* (Leicester: Leicester University Press).

TARNEY, J. and SAUNDERS, A. D. 1979. Trace element constraints on the origin of Cordillera batholiths. In *Origin of granite batholiths: geochemical evidence* ATHERTON, M. P. and TARNEY, J. eds (Orpington, Kent: Shiva Publications), 34–44.

TAYLOR, H. P. JR. 1974. The application of oxygen and hydrogen isotope studies to problems of hydrothermal alteration and ore deposition. *Economic Geology*, **69**, 843–883.

— 1977. Water-rock interactions and the origin of H_2O in granite batholiths. *Journal of the Geological Society of London*, **133**, 509–558.

TAYLOR, K. and RUSHTON, A. W. A. 1971. The pre-Westphalian geology of the Warwickshire Coalfield. *Bulletin of the Geological Survey of Great Britain* no. 35, 150p.

TAYLOR, S. and ANDREW, C. J. 1978. Silvermines orebodies, County Tipperary, Ireland. *Transactions of the Institution of Mining and Metallurgy (Section B)*, **87**, B111–124.

TAYLOR, S. R. 1964. Abundance of chemical elements in the continental crust: a new table. *Geochimica et Cosmochimica Acta*, **28**, 1273–1285.

— and MCLENNAN, S. H. 1985. *The continental crust: its composition and evolution* (Oxford: Blackwells), 312p.

TELFORD, W. M., GELDART, L. P., SHERIFF, R. E. and KEYS, D. A. 1976. *Applied geophysics* (Cambridge: University Press), 860p.

THORPE, R. S. 1972. The geochemistry and correlation of the Warren House, the Uriconian and the Charnian volcanic rocks from the English Precambrian. *Proceedings of the Geologists' Association*, **83**, 269–286.

— 1974. Aspects of magmatism and plate tectonics in the Precambrian of England and Wales. *Geological Journal*, **9**, 115–136.

— 1982. Precambrian igneous rocks. In *Igneous rocks of the British Isles* SUTHERLAND, D. E. ed. (Chichester: John Wiley & Sons), 19–35.

— BECKINSALE, R. D., PATCHETT, P. J., PIPER, J. D. A., DAVIES, G. R. and EVANS, J. A. 1984. Crustal growth and late Precambrian-early Palaeozoic plate tectonic evolution of England and Wales. *Journal of the Geological Society of London*, **141**, 521–536.

TINDLE, A. G. and PEARCE, J. A. 1981. Petrogenetic modelling of in situ fractional crystallisation in the zoned Loch Doon pluton, Scotland. *Contributions to Mineralogy and Petrology*, **78**, 196–207.

TOURTELOT, E. B. 1970. Selected annotated bibliography of minor-element content of marine black shales and related sedimentary rocks, 1930–65. *Bulletin of the United States Geological Survey* 1293, 118p.

TRACE, R. D. 1973. Illinois-Kentucky fluorspar district. In *A symposium on the geology of fluorspar* HUTCHESON, D. W. ed. *Kentucky Geological Survey, Special Publication*, **22**, 58–76.

TRAILL, J. G. 1939. The geology and development of Mill Close Mine, Derbyshire. *Economic Geology*, **34**, 851–889.

TREWIN, N. H. 1968. Potassium bentonites in the Namurian of Staffordshire and Derbyshire. *Proceedings of the Yorkshire Geological Society*, **37**, 73–91.

— and HOLDSWORTH, B. K. 1972. Further K-bentonites from the Namurian of Staffordshire. *Proceedings of the Yorkshire Geological Society*, **39**, 87–89.

— — 1973. Sedimentation in the Lower Namurian rocks of the north Staffordshire basin. *Proceedings of the Yorkshire Geological Society*, **39**, 371–408.

TUREKIAN, K. K. and WEDEPOHL, K. H. 1961. Distribution of the elements in some major units of the Earth's crust. *Bulletin of the Geological Society of America*, **72**, 175–192.

TURNER, D. R., DICKSON, A. G. and WHITFIELD, M. 1980. Water-rock partition coefficients and the composition of natural waters—a reassessment. *Marine Chemistry*, **9**, 211–218.

TURNER, J. S. 1949. The deeper structure of central and northern England. *Proceedings of the Yorkshire Geological Society*, **27**, 280–297.

UPTON, B. G. J. 1982. Carboniferous to Permian volcanism in the stable foreland. In *Igneous rocks of the British Isles* SUTHERLAND, D. S. ed. (Chichester: John Wiley & Sons), 255–275.

VAASJOKI, M. and GULSON, B. L. 1986. Carbonate-hosted base metal deposits: lead isotope data bearing on their genesis and exploration. *Economic Geology*, **81**, 156–172.

VAUGHAN, D. J. and IXER, R. A. 1980. Studies of sulphide mineralogy of north Pennine ores and its contribution to genetic models. *Transactions of the Institution of Mining and Metallurgy (Section B)*, **89**, B99–109.

VINE, J. D. and TOURTELOT, E. B. 1970. Geochemistry of black shale deposits—a summary report. *Economic Geology*, **65**, 253–272.

WADGE, A. J., BATESON, J. H. and EVANS, A. D. 1983. Mineral reconnaissance surveys in the Craven Basin. *Mineral Reconnaissance Programme Report, Institute of Geological Sciences* no. 66, 100p.

WALKDEN, G. M. 1977. Volcanic and erosive events on an Upper Visean carbonate platform, north Derbyshire. *Proceedings of the Yorkshire Geological Society*, **41**, 347–367.

WALKER, R. G. 1966. Shale Grit and Grindslow Shales transition from turbidite to shallow water sediments in the Upper Carboniferous of northern England. *Journal of Sedimentary Petrology*, **36**, 90–114.

WALTERS, S. G. 1981. The igneous horizons of the South Pennine Orefield and their interactions with mineralization. Unpublished PhD thesis, University of Sheffield.

— and INESON, P. R. 1981. A review of the distribu-

tion and correlation of igneous rocks in Derbyshire, England. *Mercian Geologist*, **8**, 81–132.

— — 1983. Hydrothermal alteration of dolerite wallrock within the Ible Sill, Derbyshire. *Mercian Geologist*, **9**, 41–48.

WALTON, E. K. 1983. Lower Palaeozoic—stratigraphy; Lower Palaeozoic—structure and palaeogeography. In *Geology of Scotland, 2nd edition* CRAIG, G. Y. ed. (Edinburgh: Scottish Academic Press), 105–138; 139–166.

WATSON, J. 1985. Northern Scotland as an Atlantic-North Sea Divide. *Journal of the Geological Society of London*, **142**, 221–243.

WEAVER, J. D. 1974. Systematic jointing in South Derbyshire. *Mercian Geologist*, **5**, 115–132.

WEBB, P. C., TINDLE, A. G., BARRITT, S. D., BROWN, G. C. and MILLER, J. F. 1985. Radiothermal granites of the United Kingdom: comparison of fractionation patterns and variation of heat production for selected granites. In *High heat production (HHP) granites, hydrothermal fluids and ore genesis* (London: Institution of Mining and Metallurgy), 409–424.

WHEILDON, J. and ROLLIN, K. E. 1986. Heat flow. In *Geothermal energy—the potential in the United Kingdom* DOWNING, R. A. and GRAY, D. A. eds (London: HMSO for British Geological Survey), 8–20.

WHITCOMBE, D. N. and MAGUIRE, P. K. H. 1980. An analysis of the velocity structure of the Precambrian rocks of Charnwood Forest. *Geophysical Journal of the Royal Astronomical Society*, **63**, 405–416.

— — 1981a. Seismic refraction evidence for a basement ridge between the Derbyshire Dome and the west of Charnwood Forest. *Journal of the Geological Society of London*, **138**, 653–659.

— — 1981b. A seismic refraction investigation of the Charnian basement and granitic intrusions flanking Charnwood Forest. *Journal of the Geological Society of London*, **138**, 643–651.

WHITE, D. E. 1958. Liquid of inclusions in sulfides from Tri-State (Missouri–Kansas–Oklahoma) is probably connate in origin. (Abstract.) *Bulletin of the Geological Society of America*, **69**, 1660.

WHITFIELD, M. 1979. The mean oceanic residence time (MORT) concept—a rationalisation. *Marine Chemistry*, **8**, 101–123.

WHITTAKER, A. 1985. *Atlas of onshore sedimentary basins in England and Wales* (Glasgow: Blackie.)

— and CHADWICK, R. A. 1983. Deep seismic reflection profiling onshore United Kingdom. *First Break*, **1**, 9–13.

— HOLLIDAY, D. W. and PENN, I. E. 1985. Geophysical logs in British stratigraphy. *Special Report of the Geological Society of London* no. 18, 74p.

— CHADWICK, R. A. and PENN, I. E. 1986. Deep crustal traverse across southern Britain from seismic reflection profiles. *Bulletin de la Société Géologique de France*, **8**, 55–68.

WILCOX, R. E., HARDING, T. P. and SEELY, D. R. 1973. Basic wrench tectonics. *Bulletin of the American Association of Petroleum Geologists*, **57**, 74–96.

WILLIAMS, B. 1982. Review of current geophysical exploration techniques for base metals in Ireland. In *Mineral exploration in Ireland: progress and developments,*

1971–1981 Brown, A. G. ed. (Dublin: Irish Association for Economic Geology), 135–47.

Wills, L. J. 1956. *Concealed coalfields* (London: Blackie).

— 1973. A palaeogeological map of the Palaeozoic floor beneath the Permian and Mesozoic Formations in England and Wales. *Memoir of the Geological Society of London*, no. 7.

— 1978. A palaeogeological map of the Lower Palaeozoic floor below the cover of Upper Devonian, Carboniferous and later formations. *Memoir of the Geological Society of London*, no. 8.

Wilson, A. A. and Cornwell, J. D. 1982. The Institute of Geological Sciences borehole at Beckermonds Scar, North Yorkshire. *Proceedings of the Yorkshire Geological Society*, **44**, 59–88.

Wilson, G. V. 1926. The concealed coalfield of Yorkshire and Nottinghamshire. *Memoir of the Geological Survey of Great Britain*, 283p.

Wood, D. A. 1979. A variably veined sub-oceanic upper mantle-genetic significance for mid-ocean ridge basalts from geochemical evidence. *Geology*, **7**, 499–503.

Woodall, R. 1984. Success in mineral exploration: confidence in source and ore deposit models. *Geoscience Canada*, **11**, 127–132.

Worley, N. E. 1978. Stratigraphical control of mineralisation in the Peak District of Derbyshire. Unpublished PhD thesis, University of Leicester.

— and Ford, T. D. 1977. Mississippi Valley type orefields in Britain. *Bulletin of Peak District Mines Historical Society*, **6**, 201–208.

Wray, D. A., Stephens, J. V., Edwards, W. N. and Bromehead, C. E. N. 1930. The geology of the country around Huddersfield and Halifax. *Memoir of the Geological Survey of England and Wales*. Sheet 77, 221p.

Wyrobek, S. M. 1959. Well velocity determinations in the English Trias, Permian and Carboniferous. *Geophysical Prospecting*, **7**, 218–230.

APPENDIX 1 PROCESSING OF GEOPHYSICAL DATA AND USE OF THE I²S IMAGE ANALYSING SYSTEM

A. S. D. Walker

GEOPHYSICAL DATA

Gravity

The primary sources of Bouguer gravity data available for the East Midlands region consisted of (1) observations made previously by the BGS Regional Geophysics Research Group (approximately 20 000 stations at an average density of about 0.5 to 1 per square kilometre, randomly distributed) and (2) data obtained by oil companies (BP and Esso) and Leeds University. The results of the latter surveys were digitised on a 2 km grid from the 1:250 000 scale contour maps published by the Geological Survey of Great Britain in 1956.

The digitised data values were recalculated to take account of modifications to the International Gravity Formula (Telford and others, 1976) and to the base gravity values (NGRN73) (Masson-Smith and others, 1974) before they were merged with the other data.

Aeromagnetic

Aeromagnetic data were obtained for the region in 1955, 1956 and 1958 as part of the national coverage of the UK by the Geological Survey of Great Britain (1965). The data were recorded in analogue form and were converted to digital form for the present study by digitising selected values and contour cuts from the flight line work sheets, which were at 1:169 000 and 1:63 360 scales. The data obtained in 1955 required re-calculation to the geomagnetic field model developed in 1956 to define local magnetic field variations.

Gravity data processing

The digital data were first converted to a form compatible with the image analysis system (IAS). This involved interpolation of data points onto a regular grid, the centres of each square representing the grid nodes. This process generated a potential field surface over the survey area, each value representing the value of the potential field at that point. A grid dimension of 1 km was chosen to minimise the smoothing effect of the interpolation process. A 'best fit' surface was calculated by use of a 'nearest neighbour' option of Surface II (Sampson, 1978) whereby the value of each grid node is calculated from the eight nearest data points, each of which is weighted by the inverse square of its distance from the grid node. The grid is represented as an image, the data amplitude values being assigned grey level intensity values in proportion to their magnitude. The grid of amplitude values was converted to a byte grid by use of a Fortran program. This file was loaded, via magnetic tape, onto the I²S and provided the basic gravity data set on which all subsequent operations were performed.

Gravity stripping

Gravity stripping was performed to permit the recog-

nition of anomalies attributable to deep sources, mainly in the pre-Carboniferous basement. The major rock types and their thicknesses in the study area are known from borehole samples and logs and seismic reflection surveys. These data indicate that the Lower Carboniferous strata (predominantly limestone) form a marked density and velocity contrast with the younger rocks, although they are difficult to distinguish from basement. Consequently, 'stripping' was performed only down to the level of the base Namurian.

The Bouguer anomaly field was recalculated by mathematically compensating for the gravity effects of the less dense stratigraphic horizons overlying the Carboniferous limestone sequence (Appendix 1a). The stratigraphic sequence and structural contour maps were prepared from seismic and borehole data and the maps were digitised on a 5 km grid. The average density assigned to each of the major lithological units is given in Tables 4.1 and 4.2. For units composed of more than one lithology the appropriate densities were calculated in proportion to the relative thicknesses of the different lithologies by use of physical properties derived from borehole logs.

The gravitational effect of each unit was computed by first replacing depth contours with a rectangular grid of digitised point values. The gravity effect was then calculated by summing the effect of a series of 5 km × 5 km vertical prisms centered on the grid points and extending upwards from the base of the unit to the zero datum level. The application of this procedure to the units listed in Table 4.1 (except Dinantian) led to the production of a 'stripped' gravity map that shows anomalies due mainly to density variations in the pre-Namurian rocks (Map 2; Appendix 1a equation II).

Second vertical derivative

Second vertical derivative maps were prepared as a means of indentifying changes in the Bouguer anomaly gradient that might indicate density boundaries associated with faults or other near-surface structures. The method proved particularly valuable in indicating listric faulting affecting Carboniferous and later cover rocks in the study region. Calculation of the second derivative is a filtering operation that enhances steeper gradients and near-surface short-wavelength anomalies, giving rise to a pseudo-residual field (Elkins, 1951). This process was performed on the I²S by use of a convolution method with a 5 km × 5 km filter kernel (a 9 × 9 pixel filter) applied to a 1 km gravity grid (Appendix 1a, equation III). Tests with a larger kernel resulted in enhancement of broad regional structure and a reduced kernel size produced data increasingly dominated by random 'noise'.

Aeromagnetic data processing

The aeromagnetic data, which were recorded along flight-lines, are not randomly distributed. The flight-line separation was, on average, approximately 20 times greater than the distance between data points along any one flight-line. Hence conventional octant, quadrant and nearest-neighbour gridding algorithms proved unsuitable. The best approximation to the magnetic potential surface was therefore achieved, first by using a kriging technique (Davis and McCullagh, 1975) and, secondly, by use of a binning method followed by an interpolation procedure (Plate 1). Kriging is a technique that provides a best linear unbiased estimator of a regional variable across a map area and can thus be used to generate a

155

regular grid. Kriging, being an exact interpolation method, can also be useful in providing an estimation of variance at any point and an indication of the integrity of the grid value generated. As in the case of the gravity data, a 1 km grid was generated on tape before transfer to the I²S for further processing. The 'binning' method involved identification of the largest square cell to satisfy the condition that no cell contain more than a single data point. For this study a cell of side 0.2 km was found to be the most satisfactory with observed data values occupying a single element in a two-dimensional data array. Elements not represented by a data value were labelled as absent data cells. The binned array was written to the I²S in byte format by use of the same procedure as for the other data sets. The kriged grid gave rise to a full grey tone image directly on the I²S (Plate 1). In the case of the binned array, a two-dimensional interpolation filter (a 7 × 7 kernel, each coefficient of which was set to unity) was applied to fill absent data pixels. Comparisons between images prepared using the kriged and interpolated-bin methods indicated that the kriged version was the better surface approximation of the potential field; the bin-interpolation technique applied on the I²S generated an over-smoothed image. The kriged magnetic potential surface was therefore used for all subsequent operations.

Reduction to pole

The geomagnetic field at the latitude of the British Isles is inclined downwards at 68 degrees, 8 degrees W of N. The earth's magnetic (dipole) field therefore results in a di-polar anomaly over any magnetic body. This effect makes it difficult to compare gravity and magnetic anomalies. The effect can be removed by the use of Fast Fourier transforms (Cribb, 1976) whereby the magnetic values are recalculated to the approximate values that would be obtained if magnetism were related only to induction in a vertical magnetic field (reduction to the pole). The assumptions of the procedure are that the anomaly is due purely to induced magnetisation with negligible remanent magnetisation and adjacent magnetic anomalies do not overlap.

Application of the method over a large area, in which many magnetic sources interact, should be treated with caution, however, since the method is designed principally for processing single anomalies and multiple (possibly overlapping) anomalies could give rise to spurious magnetic features. The flight-line data for the study area were processed on the assumption that the anomalies are due to induction alone and the results contoured to produce Map 3.

IMAGING PROCESSING

Introduction

Methods of processing that tend to be of particular value in the analysis of geophysical maps include re-scaling the histograms of individual images by stretching selected parts of their distribution to fit the range 1 – 255. Another process involves equalisation of a histogram distribution whereby the data are re-calculated to fit a linear cumulative frequency plot (Plate 2). Structural trends in geophysical data can also be enhanced by generating shadowgrams to give a pseudo-topographical appearance to the data surface (Broome and Teskey, 1985). For example, in the East Midlands the predominant trend is NW – SE and this was enhanced by convolution with a

filter of dimension 5 × 5 coefficients of the type shown below. The effect is that of an apparent illumination from 45° above the horizontal from a NE direction (Plate 3b).

– 0.1	– 0.5	0.0	0.5	1.0
– 0.1	– 0.5	0.0	0.5	0.5
– 0.1	– 0.5	0.0	0.0	0.0
– 0.1	– 0.5	– 0.5	– 0.5	– 0.5
– 0.1	– 0.1	– 0.1	– 0.1	– 0.1

As a result of the convolution process, the positive coefficients in the filter mathematically increase the value of the pixels (and hence their whiteness) on NE-facing gradients, the negative ones having the reverse effect on SW-facing gradients.

Two methods of colour enhancement were employed. The first, known as 'level slicing', continuously assigns combinations of red, green and blue to the grey-tone image in an additive sense such that ratios of 1,1,1 produce white and 1,1,0 yellow. This method of colouring particular levels of the image allows significant contours to be highlighted or interesting structures to be enhanced. Colour ratios can be calculated to provide either a continuous spectrum or a discontinuous suite of colours. For this study, positive data values were indicated generally by red tones and negative values by blue tones (e.g. Plates 3c, 4a).

The second method, pseudo-colouring, allows a continuous spectrum of colour (colours controlled by movement of a trackball/cursor) to be applied across grey-level images. This method was mainly used for the study because of its speed of application—particularly to colour images that contained either all positive or all negative data values; it is also the simplest method of colouring (e.g. Plates 1c, 5a and 9b). Multiplication of grey shadowgrams by 'flat' coloured images was used to generate coloured pseudo-topographic images (Plate 3). This process is particularly useful for combining different images. For example, a coloured gravity image was multiplied by a magnetic shadowgram, providing a direct means of comparing the two fields (Plate 7c) and helping to identify coincident anomalies and structures.

Primary and secondary data

The main regional databases prepared for this study comprised those of gravity (Plate 4a), aeromagnetic (Plate 6a), heat flow (Plate 9b), structural, stratigraphic and metallogenic information. These data, together with secondary derived databases, including 'stripped' gravity (Plate 7c), second vertical derivative (Plate 4b) and reduced-to-the-pole (Plate 6b) grids, were prepared for the I²S by use of a GEC 4090 computer. Second derivative simulation was also achieved directly on the I²S by a convolution method, filter coefficients being derived from the Elkins algorithm (Appendix 1a, equation III). A stripped (pre-Namurian) gravity map was generated by multiplication of structural contour images (continuous surfaces) by a series of factors that represented average density contrasts between adjacent layers (Appendix 1a, equations I and II). Subtraction of the different layers sequentially from the 'raw' gravity data allowed examination of changes in the gravity field at successively deeper stratigraphic levels.

Lineaments

Regional geophysical data frequently provide evidence of major discontinuities or 'lineaments' that may be fault zones or other structural features, such as fold axes. They may be difficult to observe by conventional geological mapping with the use of surface outcrop, although they can provide important controls on the location of ore deposits (Bhan and Hedge, 1985).

In this study, lineaments were identified by the examination of different types of regional geophysical data images — for example, the second vertical derivative map was used to identify probable near-surface faults. Lineaments identified in the different data sets were drawn directly onto the colour monitor of the I^2S and colour-coded according to their source (i.e. gravity or magnetic). Major faults that affected the cover sequence and had been identified independently on the basis of seismic and borehole evidence were also colour-coded. The similarity of the trends and, in some cases, the location of near-surface faults and deep lineaments provided strong evidence that geophysical lineaments are geologically significant structures in the study region (Plates 8a, 8b). The application of the I^2S to the analysis of geophysical data is a powerful and rapid means of identifying lineaments. Moreover, the facility of scanning a range of data sets ensures that all the evidence is assessed.

Certain characteristics of the lineaments are apparent from an examination of the data sets — for example, the interrelationship between the different structural units defined by the different geophysical datasets. In places, lineaments defined by the gravity and magnetic data indicate slightly different positions, suggesting either displacements due to the dip of fault plane or responses at different depths in the crust (Plate 8c).

Mapping and classification

Following analysis of single images and simple combinations of images and their evaluation for the preparation of mineral deposit models, sophisticated combined images were prepared. For example, anomalies defined on the basis of the frequency distribution of aeromagnetic 'highs' were combined with comparable gravity 'highs' to enable mapping areas likely to comprise basic intrusions at depth of be mapped (Plate 7a). This approach was used to prepare prospectivity maps for Irish- and Pennine-style mineralisation. In the case of Pennine-style deposits, for example, gravity lows (possible high heat production granites), areas of Brigantian facies Carboniferous limestone and postulated pre-Dinantian basins were combined in single images (Plate 12). A further type of image integration, which allows a qualitative assessment of interrelationships between datasets, is the addition of three images prepared in different primary colours, the frequency response distributions of which have been equalised. Red, green and blue colour guns are applied to the respective images. The resultant multi-coloured composite image can be interpreted directly by looking for similar colour combinations, which represent lithological units with the same geophysical signature (Plate 10).

Appendix 1a   Processing of geophysical data as an aid to interpretation

The regional Bouguer gravity anomaly data were processed by the mathematical procedures described below to produce additional maps and/or images.

Gravity stripping

This procedure is intended to remove the gravity effect of lower-density sedimentary 'cover' rocks so that the 'stripped' gravity map reflects only density variations at a deeper level in the crust.

The cover rocks are considered to occur as a series of horizontal layers 1, 2, ... n and with average densities $\varrho_1, \varrho_2, \ldots \varrho_n$, extending to depths of $z_1, z_2, \ldots, z_n$

The Bouguer gravity anomaly observed can be defined as the sum of the gravity effects of these layers such that

$$BA = K \sum_{i=1}^{\infty} \varrho_i \,(z_i - z_{i-1}) \qquad (I)$$

where $K = 2\pi G$ (for a layer of infinite horizontal extent) and G is the Universal gravitational constant

This can be re-written as

$$BA = K \sum_{i=1}^{\infty-1} z_i(\varrho_i - \varrho_{i+1}) \qquad (II)$$

It follows from this last equation that if z_i is represented over the study region as a two-dimensional array (grid) of depths to a particular interface, and if both ϱ_i and ϱ_{i+1} are known, the gravitational effect of any unit due to density contrasts at its interfaces with other units can be calculated. This gravity effect can then be subtracted (stripped) from BA. The calculation is carried out by assuming that each grid square represents the top of a vertical prism, extending down to an appropriate depth, and summing at the centre of each square the gravity effects of these prisms, using the formula for a thick prism (Telford and others, 1976).

This procedure allows the gravity effect of a layer of variable thickness to be computed — rather than that of a simple horizontal layer of infinite lateral extent as indicated by equation (I).

The application of this procedure in the East Midlands enabled the removal of the effects of cover rocks overlying the Carboniferous Limestone to isolate those anomalies due mainly to density variations in the pre-Carboniferous basement.

Further, let us consider those terms (omitting K) (of the type derived from equation II) associated with a layer n, i.e. the Namurian.

$$z_n \,(\varrho_n - \varrho_{n+1}) \qquad\qquad A$$
$$z_{n+1} \,(\varrho_{n+1} - \varrho_{n+2}) \qquad\qquad B$$
$$z_{n+2} \,(\varrho_{n+2} - \varrho_{n+3}) \qquad\qquad C$$

where z_n denotes the depth to the base Namurian and ϱ_n, ϱ_{n+1} are densities for the Namurian and Carboniferous limestone units respectively.

Term A can be subtracted (stripped) from BA (since all variables are known) to leave B, C and further terms that may be defined.

C and further terms represent the regional field (which can be estimated and removed), leaving B defining the residual anomalies to be interpreted.

Given the nature of term B, an inverse stripping operation can be applied to map the depth to the base of the Carboniferous limestone by adopting a reasonable value for ϱ_{n+2}.

Second vertical derivative

The following formula, derived by Elkins (1951), provided the basis for calculation of the second vertical derivative:

$$\frac{\partial^2 g}{\partial z^2} = 1/62s^2 \left(44H(0) + 4H'(s) - 3H'(s\sqrt{2}) - 6H'(s\sqrt{5})\right)$$

$$(\text{III})$$

where s is grid size for the calculation, $H(0)$ is grid value at the centre point, $H'(s)$ is the sum of the four grid values at distance s from the centre, $H'(s\sqrt{2})$ is the sum of the four grid values at distance $s\sqrt{2}$ from the centre and $H'(s\sqrt{5})$ is the sum of the eight grid values at distance $s\sqrt{5}$ from the centre.

Equation (III) is used to define the values for filter coefficients of a convolution matrix that is applied to Bouguer gravity anomaly data to generate a grid of the derivative values. This process can be performed on the I²S by utilisation of a convolve 'sub-routine' that requires input of the filter coefficients.

A value of $s = 2$ km was selected for processing the 1 km gravity grid of the East Midlands gravity data and the following weighting values were used for a 9 × 9 (pixel) filter.

0.0	0.0	-0.02419	0.0	0.0	0.0	-0.02419	0.0	0.0
0.0	0.0	0.0	0.0	0.0	0.0	0.0	0.0	0.0
-0.02419	0.0	-0.012097	0.0	0.01613	0.0	-0.012097	0.0	-0.02419
0.0	0.0	0.0	0.0	0.0	0.0	0.0	0.0	0.0
0.0	0.0	0.01613	0.0	0.1774	0.0	0.01613	0.0	0.0
0.0	0.0	0.0	0.0	0.0	0.0	0.0	0.0	0.0
-0.02419	0.0	-0.012097	0.0	0.01613	0.0	-0.012097	0.0	-0.02419
0.0	0.0	0.0	0.0	0.0	0.0	0.0	0.0	0.0
0.0	0.0	-0.02419	0.0	0.0	0.0	-0.02419	0.0	0.0

The second vertical derivative is utilised to emphasise short-wavelength anomalies due to near-surface structures, at the same time suppressing those due to broad regional structure.

Appendix 2 Physical properties of Pre-Carboniferous rocks based on outcrop sample data and boreholes from BGS and other acknowledged sources.

J. D. Cornwell

	Area	Densities (Mg/m³) Saturated	Grain	Porosity %	N*	Susceptibility ×10⁻³SI	n*	Velocity km/s	n*
Sedimentary rocks									
DEVONIAN (ORS)									
	Nuneaton	2.47±0.10	2.73±0.10	15.0±6.5	8	0.1±0.1	10		
SILURIAN[3]									
Welsh Borders	Wales	2.70[3]							
Radnor (geosynclinal type)		2.85[3]							
CAMBRIAN									
Monks Park Shales	Nuneaton	2.47±0.09	2.77±0.04	16.6±4.7	10				
Outwoods Shales	Nuneaton	2.65±0.09	2.81±0.03	8.9±3.8	7				
Stockingford Shales	Leicester	2.75±0.00	2.84±0.00		7[5]				
PRECAMBRIAN									
Brand Series (Swithland Slates)	Charnwood Forest	2.78±0.00	2.83±0.00		7[5]				
Maplewell Series									
Woodhouse and Bradgate Beds		2.70±0.06	2.73±0.06		10[5]	0.1		5.65[6]	
Woodhouse and Bradgate Beds		2.69±0.06	2.73±0.06	2.54±0.4	3			5.98	1
Beacon Hill Beds (hornstone)		2.72	2.73	0.6	2			6.13	2
Blackbrook Formation (pyroclastics)		2.64±0.05	2.67±0.05	2.4±1.7	13[5]	0.3±0.1	5	5.40[6]	
Slate and agglomerate		2.78±0.01							
Igneous rocks									
S. Leicestershire Diorites									
Croft		2.65±0.04	2.68±0.02	2.1±1.5	4	8.8±6.5		5.30±0.35	
Barrow Hill		2.72±0.02	2.73±0.02	0.7±0.3	10				
Enderby		2.74	2.75	0.7	1	0.4[4]		5.60	
Sapcote						0.7±0.3[4]			
Nuneaton diorite		2.71	2.75		1			5.60	
Mountsorrel granodiorite		2.66±0.02	2.67±0.02	1.2±1.3	60[5]	26.4±37.7	42	4.90[6]	
Southern Diorites (Charnwood Forest)									
Bradgate		2.77	2.82	2.8	1	6.8	1	5.28	1
Cliff Hill (W)		2.80±0.04	2.83±0.04		39[5]	78.7±103.2	42[5]		
Groby		2.73±0.02	2.75±0.02	1.0±0.4	34[5]	4.5±13.0	34[5]		
Northern Diorites (Charnwood Forest)									
Newhurst		2.88	2.90	1.0	2	1.1	2	6.30	2
'Porphyroids'									
Whitwick		2.70±0.05	2.74±0.03		4[5]				
Bardon Hill Pyroclastics		2.84±0.04	2.85±0.04	0.1	29[5]	0.3±0.8	26[5]		

* Number of samples or values: from borehole logs (L). C = chipping samples only.

Borehole	Drilled in	by	Rock type	Age*	Saturated	Density, Mg/m³ Grain	Porosity %	n	Velocity (km/s)	n	Suscept-ibility SI×10⁻³	n
Bardney	1966	BP	Quartzite	C?	2.77	2.77	0.1	2			0.6±0.5	10(C)
Beckermonds Scar	1976	BGS	Siltstone/sst	O	2.77±0.05	2.78±0.05	0.3±0.2	25	5.85±0.26	22	31.0±22.0	22
Burmah 47/29-A1	1968	Burmah	Shale	O	2.75±0.04	2.76±0.04		3			0.4±0.2	3
Byfield	1980	NCB	Mdst./sst.	C	2.61±0.08	2.68±0.05	4.4±2.0	5			5.4±5.1	9
Caldon Low	1977	BGS	Sandstone	LC/D	2.64±0.05	2.76±0.09	6.6±5.4	6				
Cox's Walk	1975	NCB	Acid volcanic	PC?	2.69	2.72	1.4	1			0.9	1
Eakring 146	1944	D'Arcy	Conglom./sst.	LC	2.58±0.09			60[5]				
Eyam	1972	BGS	Mudstone	O	2.73±0.02	2.78±0.13	2.8±0.68	5			1.2±0.9	5
Ellingham	1965	Superior	Mudstone/slate	D	2.63±0.03			7[2]			0.1±0.0	8
Galley Hill	1976	NCB	Sandstone	C?	2.65±0.08	2.67±0.05	1.0±0.4	5			0.6±0.2	3
Glinton	1961	BP	Tuff	PO?	2.73±0.05	2.73±0.04	0.1±0.0	3	5.49[7]		1.4±2.8	16
G. Osgrove Wood	1977	NCB	Siliceous seds	PC	2.53±0.03	2.59±0.00	3.3±1.7	6			0.5	2
Great Paxton	1966	BGS	Mudstone	O	2.55[1]				3.41[1]			
Grove 3	1981	BP	Metasediments	PC?	2.72±0.02	2.73±0.02	0.5±0.2	4	5.86	L		
Home Farm	1978	NCB	Mudstone	C	2.54±0.02	2.74±0.01	11.1±1.3	5			0.2±0.0	12
Huntingdon	1966	BGS	Mudstone	C	2.68±0.01	2.69±0.01	0.8±0.4	3				
Hunstanton	1969	PLACE	Volcanic	PC?					5.64	L	0.4±0.5	25
Kirby Lane	1975	NCB	Granite	CAL	2.58	2.63	3.3	1				
Leicester Forest East	1978	BGS	Mudstone	C	2.69±0.04	2.82±0.02	6.7±1.6	3			0.2±0.1	26
Lexham	1971	NRC	Igneous?	PC?	2.75			L				
Nocton	1943	D'Arcy	Quartzite	C?		2.54		C			0.4±0.5	38
North Creake	1945	D'Arcy	Agglomerate	PC?					5.60[7]			
Raydale	1974	BGS	Granite	CAL	2.59±0.01	2.60±0.01	0.9±0.4	6	5.59±0.11	6	0.1±0.1	6
Lakenheath	1965	Superior	Siltstone	D	2.62±0.04			6[2]				
Rotherwood	1977	BGS	Mudstone	C	2.76±0.02	2.82±0.03	3.0±1.8	6	4.67±0.65	3	0.3±0.2	3
Saxthorpe	1970	DUT	Pelite	S	2.71	2.71	0.2	2	5.35	L		
Sibsey	1970	BAC	Phyllites	?	2.72			L				
Soham	1955	BGS	Mudstone	D	2.65±0.03	2.75±0.04	5.4±3.6	4				
South Creake	1969	BP	Shale/quartzite		2.66±0.01	2.67±0.03	0.6±0.8	5	4.7 – 5.5	L		
Spalding	1971	Texaco	Quartzite	C?					5.37	L		
Sproxton	1945	D'Arcy	Phyllitic shales	PC?		2.84±0.08		8(C)[5]				
Stixwould	1944	D'Arcy	Sandstones	PC?	2.65±0.04	2.73±0.03	4.2±3.0	5	5.40	2	0.1±0.1	12
Thorpe	1972	BGS	Mudstone	O/S?	2.74±0.03	2.76±0.03	5.7±0.2	26	5.68±0.24	24	0.2±0.1	17
Twycross	1978	BGS	Mudstone	C	2.51±0.08	2.75±0.05	13.8±3.2	2			0.2±0.1	4
Upwood	1965	BGS	Agglom./tuffs	PC?	2.66±0.04	2.76±0.03	4.2±0.6	9	4.24±0.55	9	0.4±0.1	12
Warboys	1965	BGS	Diorite	CAL	2.84±0.03	2.87±0.02	2.1±0.8	28	6.10±0.23	26	20.0±11.0	26
Wiggenhall	1971	Texaco	Pyroclastic	PC?		2.65±0.08		5(C)	5.60 – 5.90	L		
Wisbech	1971	Texaco	Quartzite	C?					5.86?	L		
Wittering	1966	B.Gas	Tuff	PC?	2.79±0.10	2.79±0.01	0.2±0.0	3			3.4±1.6	8
Woodale	1949	ICI	Volcanics	PC?	2.64±0.10	2.73±0.10	4.6±1.5	5[5]				
Wyboston	1955	BGS	Mdst./sst.	C	2.77±0.11	2.80±0.01	1.7±0.6	3				
Welton	1981	BP	Phyllite	?	2.73			L	6.35	L		

Drilled by: BAC Ball & Collins, DUT Duntex Petroleum, NRC Norris Oil Co.

*Age: LC/D, Lower Carboniferous/Devonian; D, Devonian; S, Silurian; O, Ordovician; C, Cambrian; CAL, Caledonian; PC, Precambrian.

References
1 Bullerwell (in Stubblefield, 1967).
2 Chroston and Sola (1982).
3 Cook and Thirlaway (1955).
4 Duff (private communication).
5 Maroof (1973).
6 Whitcombe and Maguire (1980).
7 Kent (1962).

Appendix 3 Geochemical sampling and analysis

D. G. Jones and P. C. Webb

In the case of both the igneous rocks and shales surface exposures and borehole cores were sampled in order to provide as systematic coverage as possible. Borehole core was used, when available, in preference to outcrop material. Surface samples of about 2 kg of igneous rocks and 1 kg of shale were obtained. Care was taken to ensure that the samples were unweathered and as homogeneous as possible. Borehole samples were generally of smaller size, owing to the limited amount of material available. In addition to fresh rock, samples of hydrothermally altered igneous lithologies, particularly granites, were collected in order to assess the chemical effects associated with alteration. Sample coverage is variable, the basement lithologies being least well represented because of the limited extent of surface exposures and the scattered occurrence of deep boreholes. For example, the Weardale and Wensleydale Granites are sampled only by single boreholes at Rookhope and Raydale, which penetrate to 400 and 100 m, respectively, below the weathered surface of the plutons. In contrast Carboniferous shales and igneous rocks are generally well represented by cored intervals from a large number of oil, coal, mineral exploration and BGS stratigraphic boreholes.

Approximately half of the samples were provided in powder form from BGS and Lancaster University. The remainder were cleaned, jaw-crushed and ground to a fine powder in an agate mill. More than 300 igneous rocks and 1000 argillaceous rocks were investigated. All samples were analysed by X-ray fluorescence (XRF) spectrometry on pressed powder pellets (PVP/methyl cellulose bound) using the Philips PW1400 instruments at Nottingham University and Midland Earth Science Associates for the following: SiO_2, Al_2O_3, TiO_2, Fe_2O_3, MgO, CaO, Na_2O K_2O, P_2O_5, MnO, As, Ba, Ce, Cl, Co, Cr, Cu, Mo, Ni, Nb, Pb, Rb, S, Sc, Sr, Th, V, Y, Zn and Zr. The major-element data on the pellets were checked against approximately 20 fusion bead analyses.

Data for other elements were obtained on selected samples by a number of methods. The light elements Li, Be and B were determined by DC-arc emission spectrometry at the BGS Geochemical Directorate laboratories. Instrumental neutron activation analysis for La, Ce, Nd, Sm, Eu, Gd, Tb, Ho, Tm, Yb, Lu, Sc, Cs, Hf, Ta, Th and U was performed at the HERALD Reactor Centre, AWRE Aldermaston, at the London University Reactor Centre and at the Open University. Uranium data were obtained by the delayed neutron method at AWRE and at ICI, Billingham. The rare earth elements (REE) (La, Ce, Pr, Nd, Sm, Eu, Gd, Dy, Er, Yb and Lu) were determined on a number of shales in the laboratories of the BGS Geochemistry Directorate by inductively coupled plasma atomic emission spectrometry, following cation-exchange separation.